STUDENT STUDY GUIDE & SELECTED SOLUTIONS MANUAL

VOLUME 2

JOSEPH BOYLE
MIAMI DADE COLLEGE

PHYSICS
PRINCIPLES WITH APPLICATIONS

7TH EDITION

GIANCOLI

PEARSON
San Francisco Boston New York
Cape Town Hong Kong London Madrid Mexico City
Montreal Munich Paris Singapore Sydney Tokyo Toronto

President, Science, Business & Technology: Paul Corey
Publisher: Jim Smith
Executive Development Editor: Karen Karlin
Senior Managing Editor: Corinne Benson
Project Manager: Elisa Mandelbaum
Production Management and Compositor: PreMediaGlobal
Manufacturing Buyer: Jeffrey Sargent
Senior Marketing Manager: Will Moore
Cover Design: Derek Bacchus and Seventeenth Street Design
Cover Photo Credits: Douglas C. Giancoli
Cover and Text Printer: LSC Communications

ISBN-10: 0-321-76808-6
ISBN-13: 978-0-321-76808-7

www.pearsonhighered.com

40 2021

DEDICATED TO MY WIFE, MOTHER, FATHER AND SISTER

CONTENTS

PREFACE

This study guide was written to accompany PHYSICS: PRINCIPLES AND APPLICATIONS, seventh edition by Douglas C. Giancoli. The study guide is intended to provide additional help in understanding the basic principles covered in the textbook and add to the student's problem solving skills.

Each chapter in the study guide begins with a list of course objectives based on the information covered in the chapter. This section is followed by a list of key terms and phrases and a list of basic mathematical equations used in the textbook. The concept summary section of the study guide summarizes the main topics covered in the corresponding chapter of the textbook. The concept summary section also includes the answers to three or more end-of-chapter questions from the textbook as well as the solution to example problems similar to the type of problems found in the textbook. Each chapter of the study guide concludes with the step-by-step process to the solution of six or more representative end-of-chapter problems found in the textbook.

Because beginning physics course emphasize problem solving, hints on problem solving skills are placed just before the representative problems found in the textbook. A suggestion for the proper use of the mathematical problem section is included in chapter 1. The section of the textbook to which the representative problem corresponds is included as part of the solution. The student should be aware that the study guide is meant to complement, not replace, the textbook as a learning tool. Because of this, it is suggested that the student carefully read the chapter in the textbook before using the study guide.

I wish to acknowledge the help given by the professors who reviewed the solution to each mathematical problem and question included in the study guide. The help provided by Karen Karlin of Pearson Education was invaluable. Also, I could not have completed this study guide without the assistance given by my wife in typing and editing each page.

Every effort has been made to avoid errors; however, I alone have responsibility for any errors which remain and corrections and comments are most welcome.

Joseph J. Boyle
Professor Emeritus
Miami Dade College

Email - jboyle @ mdc.edu

16

Electric Charge and Electric Field

OBJECTIVES

After studying the material of this chapter, the student should be able to:

- state from memory the magnitude and sign of the charge on an electron and a proton and also state the mass of both particles.
- apply Coulomb's law to determine the magnitude of the electric force between point charges separated by a distance *r* and state whether the force will be one of attraction or repulsion.
- state from memory the law of conservation of electric charge.
- distinguish among insulators, conductors, and semiconductors and give examples of each.
- explain the concept of electric field and determine the resultant electric field at a point some distance from two or more point charges.
- determine the magnitude and direction of the electric force on a charged particle placed in an electric field.
- sketch the electric field lines in the region between charged objects.
- use Gauss's law to determine the magnitude of the electric field in problems in which static electric charge is distributed on a symmetrical surface.

KEY TERMS AND PHRASES

electrostatics is the study of interaction between electric charges that are not moving.

electric charge is a fundamental property of matter. Electric charge appears as two kinds, arbitrarily designated **positive** and **negative**. Negative charges are carried by particles called **electrons**; the positive charge carriers are known as **protons**.

law of conservation of electric charge states that the net amount of electric charge produced in any process is zero. Another way of saying this is that in any process, electric charge cannot be created or destroyed; however, it can be transferred from one object to another.

coulomb (C) is the SI unit of charge. The charge carried by the electron is represented by the symbol $-e$, and the charge carried by the proton is $+e$; $e = 1.60\times10^{-19}$ C.

insulators are materials in which the electrons are tightly held by the nucleus and are not free to move through the material. There is no such thing as a perfect insulator, but examples of good insulators are substances such as glass, rubber, plastic, and dry wood.

conductors are materials through which electrons are free to move. There is no such thing as a perfect conductor, but examples of good conductors are metals, such as silver, copper, gold, and mercury.

semiconductors are materials in which there are a few free electrons and the material is a poor conductor of electricity. As the temperature rises, electrons break free and move through the material. Examples of elements that are semiconductors are silicon, germanium, and carbon.

charging by conduction occurs when electric charge is transferred from a charged object to an uncharged object.

charging by induction occurs when an electrically charged object is brought near but does not touch an uncharged second object. The negative and positive charges on the second object separate. Overall, the second conductor is still electrically neutral; if the first object is removed, then a redistribution of the negative charge occurs.

Coulomb's law states that two **point charges** exert a force (F) on one another that is directly proportional to the product of the magnitudes of the charges (Q) and inversely proportional to the square of the distance (r) between their centers.

electric fields exist in the space surrounding a charged particle or object. Electric fields are represented by **electric field lines**, or **lines of force**, that start on a positive charge and end on a negative charge.

SUMMARY OF MATHEMATICAL FORMULAS

electric charge	$q = ne$, where $e = 1.60\times10^{-19}$ C	Relates the total charge on an object (q) to the fundamental unit of charge (e) and the total number of charges on the object (n)
Coulomb's law	$F = \frac{kQ_1Q_2}{r^2}$, where $k = 9\times10^9\ \text{N}\cdot\text{m}^2/\text{C}^2$	Coulomb's law states that two point charges exert a force (F) on one another that is directly proportional to the product of the magnitudes of the charges (Q) and inversely proportional to the square of the distance (r) between their centers.
electric field (E)	$\vec{\mathbf{E}} = \frac{\vec{\mathbf{F}}}{q}$ or $\vec{\mathbf{F}} = q\vec{\mathbf{E}}$	The magnitude of the electric field (E) at any point in space can be determined by the ratio of the force (F) exerted on a test charge placed at the point to the magnitude of the charge on the test particle (q).
electric field due to a point charge	$E = \frac{kQ}{r^2}$	The magnitude of the electric field (E) a distance (r) from a single point charge is related to the magnitude of the charge (q) and is inversely proportional to the square of the distance from the charge.

CONCEPT SUMMARY

Electric Charge

There are two types of **electric charge**, arbitrarily called **positive** and **negative**. Rubbing certain electrically neutral objects together, such as a glass rod and a silk cloth, tends to cause the electric charges to separate. The glass rod loses negative charge and therefore becomes positively charged, while the silk cloth gains negative charge and therefore becomes negatively charged. After separation, the negative charges and positive charges attract one another.

If a glass rod is suspended from a string and a second positively charged glass rod is brought near, then a force of electrical repulsion results. Negatively charged objects also exert a repulsive force on one another. These results can be summarized as follows: **unlike charges attract; like charges repel.**

TEXTBOOK QUESTION 2. (Section 16-1) Why does a shirt or blouse taken from a clothes dryer sometimes cling to your body?

ANSWER: The air in a clothes dryer is hot and dry. As the clothes tumble, they rub against one another and acquire a net charge. When you remove the clothes from the dryer, they tend to cling to one another (that is, exhibit static cling) and also to you.

Conservation of Electric Charge

In the process of rubbing two solid objects together, electrical charges are *not* created. Instead, the two objects contain both positive and negative charges. During the rubbing process, the negative charge is transferred from one object to the other, leaving one object with an excess of positive charge and the other with an excess of negative charge. The quantity of excess charge on each object is exactly the same. This concept is summarized by the **law of conservation of electric charge**: the net amount of electric charge produced in any process is zero. Another way of saying this is that in any process, electric charge *cannot* be created or destroyed, but it can be transferred from one object to another.

During the past century, the negative charges have been shown to be carried by particles that are now called electrons, whereas the positive charge carriers are known as protons. The neutron, a third type of particle, carries no electric charge.

These particles, as well as many others, are found in the atoms that make up a substance. Note: In a later chapter, we will discuss the wave properties of electrons, protons, and other particles. However, for ease of visualization and discussion, we will use the term "particle" for now.

The SI unit of charge is the **coulomb** (C). The amount of charge transferred when objects such as glass and silk are rubbed together is on the order of microcoulombs (μC):

$$1\ \mu\text{C} = 10^{-6}\ \text{C}$$

The charge carried by the electron is represented by the symbol $-e$, and the charge carried by the proton is $+e$:

$$e = 1.6\times10^{-19}\ \text{C}$$

Experiments performed early in the twentieth century led to the conclusion that protons and neutrons are confined to the nucleus of the atom while the electrons exist outside the nucleus. When solids are rubbed together, it is the electrons that are transferred from one object to the other. The positive charges, which are located in the nucleus, do not move.

Insulators, Semiconductors, and Conductors

An **insulator** is a material in which the electrons are tightly held by the nucleus and are not free to move through the material. There is no such thing as a perfect insulator, but examples of good insulators are substances such as glass, rubber, plastic, and dry wood.

A **conductor** is a material through which electrons are free to move. Just as in the case of the insulators, there is no such thing as a perfect conductor. Examples of good conductors are metals, such as silver, copper, gold, and mercury.

A few materials, such as silicon, germanium, and carbon, are called **semiconductors**. At ordinary temperatures, they contain a few free electrons and the material is a poor conductor of electricity. As the temperature rises, electrons break free and move through the material. As a result, the ability of a semiconducting material to conduct improves with temperature.

Charging by Induction and Charging by Conduction

As shown in the diagram, when a negatively charged rod is brought near an uncharged electrical conductor, the negative charges in the conductor travel to the far end of the conductor.

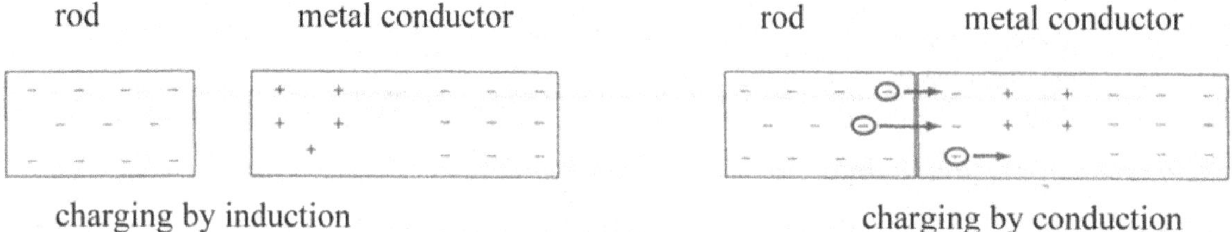

The positive charges are not free to move through a solid object, and a charge is temporarily *induced* at the two ends of the conductor; the object has been **charged by induction**. Overall, the conductor is still electrically neutral. If the rod were removed, then a redistribution of the negative charge would occur.

If the metal conductor is touched by a person's finger or a wire connected to the ground directly or through a conducting pipe leading to the ground, then it is said to be **grounded**. The negative charges would flow from the conductor to ground. If the ground is removed and then the rod is removed, a permanent positive charge would be left on the conductor. The electrons would move until the excess positive charge was uniformly distributed over the conductor.

If the rod touches the metal conductor, then some of the negative charges on the rod transfers to the metal. This charge distributes uniformly over the metal. The metal has been **charged by conduction**, and a permanent charge remains when the rod is removed.

TEXTBOOK QUESTION 5. (Section 16-1) A positively charged rod is brought close to a neutral piece of paper, which it attracts. Draw a diagram showing the separation of charge in the paper, and explain why attraction occurs.

ANSWER: Paper is a poor conductor, and only a slight displacement of the electric charge occurs when the positively charged rod is brought near the neutral piece of paper. The negative charges in the paper are attracted toward the rod, while the positive charges are located in the atomic nucleus and cannot move. In the diagram, the piece of paper remains neutral overall, but a temporary separation of electric charge is induced in the paper. Because the negative charges in the paper are closer to the rod than the positive charges, the force of attraction is greater than the force of repulsion between the rod and the positive charges in the paper. Thus, the paper is attracted to the rod.

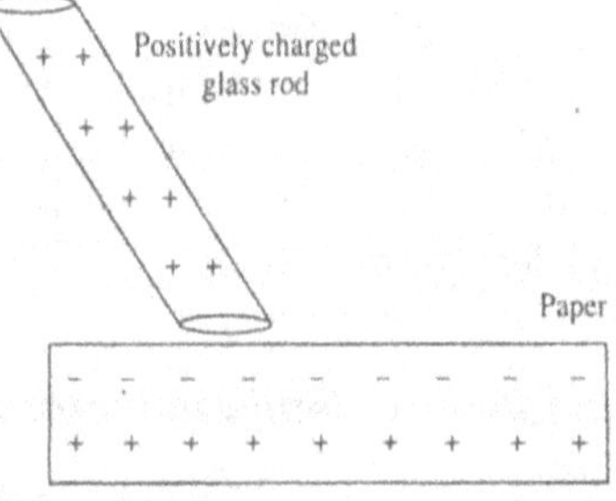

TEXTBOOK QUESTION 4. (Section 16-4) Why does a plastic ruler that has been rubbed with a cloth have the ability to pick up small pieces of paper? Why is this difficult to do on a humid day?

ANSWER: Based on the convention used by Benjamin Franklin, the rubbed plastic ruler carries a negative charge. Paper is a poor conductor, and only a slight displacement of the electric charge occurs when the negatively charged ruler is brought near the neutral piece of paper. The negative charges in the paper are displaced away from the ruler, while the positive charges are located in the atomic nucleus and cannot move. The piece of paper remains neutral overall, but a temporary separation of electric charge is induced in the paper. Because the positive charges in the paper are closer to the ruler than the negative charges, the force of attraction is greater than the force of repulsion between the ruler and the positive charges in the paper. Thus, the paper is attracted to the ruler.

As shown in Fig. 16-4 of the textbook, water molecules in air are **polar**. On a humid day, water molecules are attracted to the ruler and remove some of the excess negative charge on the ruler. The ruler has a reduced negative charge, and the displacement of negative charge in the paper is reduced. If the humidity is high enough, then the force of attraction between the ruler and the paper may not be large enough to overcome the weight of the paper.

TEXTBOOK QUESTION 8. (Section 16-4) When an electroscope is charged, its two leaves repel each other and remain at an angle. What balances the electric force of repulsion so that the leaves don't separate further?

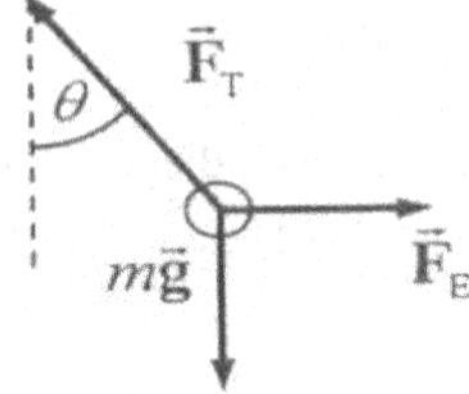

ANSWER: As shown in Fig. 16-12a of the textbook, the greater the charge on the leaves, the greater the repulsive force and the greater the deflection. As shown in this diagram, three forces act on the deflected leaves: the tension in the leaf (F_T), electrostatic force (F_E), and the weight of the leaf (mg). When the vector sum of the three forces equals zero, the leaf will be in static equilibrium.

TEXTBOOK QUESTION 11. (Section 16-1) When a charged ruler attracts small pieces of paper, sometimes a piece jumps quickly away after touching the ruler. Explain.

ANSWER: Let us assume that the rod in Textbook Question 5 (above) represents a plastic ruler. Plastic is a very poor conductor of electricity and paper is a poor conductor. However, if the paper touches the plastic, then some of the negative charges in the plastic could to transfer to the paper. Then both the plastic ruler and the paper would contain an excess of negative charge. The paper would be repelled from the ruler.

Coulomb's Law

Coulomb's law states that two **point charges** exert a force (F) on one another that is directly proportional to the product of the magnitudes of the charges (Q) and inversely proportional to the square of the distance (r) between their centers. The formula relating the force to the charges and the distance is

$$F = \frac{kQ_1Q_2}{r^2}, \text{ where } k = 9\times10^9\,\text{N}\cdot\text{m}^2/\text{C}^2$$

For objects in which the electric charge is distributed in such a way that they can be considered to act like point charges, such as small metal spheres with uniform charge distribution, the distance r is measured from the center of one object to the center of the second object.

The constant k may also be expressed in terms of the permittivity of free space (ε_0), where

$$\varepsilon_0 = \frac{1}{4\pi k} = 8.85\times10^{-12}\ \mathrm{C^2/N\cdot m^2}.$$

EXAMPLE PROBLEM 1. (Sections 16-5 and 16-6) Charges $Q_1 = +8.0\ \mu\mathrm{C}$ and $Q_2 = -2.0\ \mu\mathrm{C}$ are located on the x axis at $x = 0$ and $x = 6.0\,\mathrm{m}$, respectively. Determine the point on the x axis where a charge $Q_3 = +3.0\ \mu\mathrm{C}$ can be placed and experience no net force.

Part a. Step 1.

Locate the point on the x axis where the net force equals zero.

At the point in question, the vector sum of the two forces equals zero. The force exerted by Q_1 on Q_3 must be equal in magnitude but opposite in direction from the force exerted by Q_2 on Q_3. Since the magnitude of the charge on Q_1 is much greater than the charge on Q_2, the only location that would satisfy this condition would be a location on the x axis to the right of charge Q_2.

Part a. Step 2.

Draw a diagram locating each charge.

Let r represent the distance from Q_2 to Q_3 and $6.0\ \mathrm{m} + r$ represent the distance from Q_1 to Q_3.

Q_1 ⊕ ¦← 6.0 m →¦ Q_2 ⊖ ¦← r →¦ Q_3: $F_{2\text{ on }3}$ ⇐ ⊕ ⇒ $F_{1\text{ on }3}$

Part a. Step 3.

Use Coulomb's law to write an equation for the force that Q_1 exerts on Q_3 and Q_2 exerts on Q_3.

$$F_{1\text{on}3} = \frac{kQ_1Q_3}{(6.0\ \mathrm{m}+r)^2} \quad \text{and} \quad F_{2\text{on}3} = \frac{kQ_2Q_3}{r^2}$$

$$F_{1\text{on}3} = -F_{2\text{on}3}$$

Note: The magnitudes of the forces are equal; however, $F_{2\text{on}3}$ has a negative value because $Q_2 = -2.0\,\mu\mathrm{C}$.

$$\frac{kQ_1Q_3}{(6.0\ \mathrm{m}+r)^2} = \frac{-kQ_2Q_3}{r^2}$$

Both k and Q_3 cancel algebraically. Rearranging gives

$$\frac{(6.0\ \mathrm{m}+r)^2}{r^2} = \frac{Q_1}{-Q_2}$$

$$\text{but } \frac{Q_1}{Q_2} = \frac{(8\,\mu\text{C})}{-(-2\,\mu\text{C})} = 4$$

$$\frac{(6.0\text{ m}+r)^2}{r^2} = 4 \text{ and } (6.0\text{ m}+r)^2 = 4r^2$$

$$r^2 + (12\text{ m})r + 36\text{ m} = 4r^2$$

$$-3r^2 + (12\text{ m})r + 36\text{ m} = 0$$

Using the quadratic formula to determine the value of r gives either $r = 6.0\,\text{m}$ or $r = -2.0\,\text{m}$. Based on the diagram, r must be positive. Therefore, Q_3 is located on the x axis 6.0 m to the right of charge Q_2, so Q_3 is located at $x = 12$ m.

EXAMPLE PROBLEM 2. (Sections 16-5 and 16-6) Two small, identical metal spheres contain excess charges of $-7.0\,\mu$C and $+3.0\,\mu$C, respectively. The spheres are mounted on insulated stands and placed 0.50 m apart. (*a*) Determine the magnitude and direction of the force between the spheres. (*b*) The spheres are touched together and then returned to their original 0.50-m separation. Determine the magnitude and direction of the force between the spheres.

Part a. Step 1.

Use Coulomb's law to determine the force between the charges.

$$Q_1Q_2 = (-7.0\times10^{-6}\text{ C})(+3.0\times10^{-6}\text{ C}) = -2.1\times10^{-11}\text{ C}^2$$

$$F = \frac{kQ_1Q_2}{r^2} = \frac{(9.0\times10^9\text{ N}\cdot\text{m}^2/\text{C}^2)(-2.1\times10^{-11}\text{ C}^2)}{(0.50\text{ m})^2}$$

$F = -0.76$ N The negative value indicates that the force between the charges is one of attraction.

Part b. Step 1.

Determine the magnitude of the charge on each sphere.

When the spheres touch, electrons travel from the negatively charged sphere to the positively charged sphere. The magnitude of the negative charge is greater than that of the positive charge; therefore, the negative charge completely neutralizes the positive charge. Because the two metal spheres are identical, the excess charge will distribute equally between them. The magnitude of the charge on each sphere after separation can be determined as follows. The overall excess charge on the spheres is

$$-7.0\,\mu\text{C} + 3.0\,\mu\text{C} = -4.0\,\mu\text{C}$$

$$Q_1 = Q_2 = \frac{-4\,\mu\text{C}}{2} = -2\,\mu\text{C}$$

Part b. Step 2.

Apply Coulomb's law to determine the force between the charges.

$$Q_1 = Q_2 = (-2.0\times10^{-6}\ \text{C})(-2.0\times10^{-6}\ \text{C})$$

$$Q_1Q_2 = +4.0\times10^{-12}\ \text{C}^2$$

$$F = \frac{kQ_1Q_2}{r^2} = \frac{(9.0\times10^{9}\ \text{N}\cdot\text{m}^2/\text{C}^2)(4.0\times10^{-12}\ \text{C}^2)}{(0.50\ \text{m})^2}$$

$F = +0.14$ N Note: The positive value indicates that the force between the charges is one of repulsion.

Electric Field

If an electric charge experiences an **electric force** at a particular point in space, then it is in the presence of an **electric field**. The magnitude of the electric field (E) at any point in space can be determined by the ratio of the force (F) exerted on a **test charge** placed at the point to the magnitude of the charge on the test particle (q):

$$\vec{\mathbf{E}} = \frac{\vec{\mathbf{F}}}{q}$$

where the electric field (E) is measured in units of newtons/coulomb (N/C).

Electric field is a vector quantity. Its direction is the same as the direction of the force vector if the test charge is positive, and it is directed opposite the force vector if the test charge is negative. For a single point charge (Q), the electric field a distance (r) from the charge is $E = \dfrac{kQ}{r^2}$.

In order to visualize the path taken by a charged particle placed in an electric field, **electric field lines,** also known as **lines of force**, are drawn. The diagrams shown here represent the electric field lines for certain arrangements of charge.

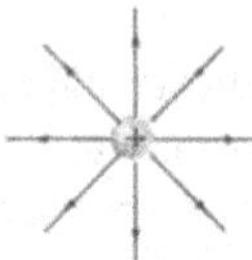

a) positive point charge

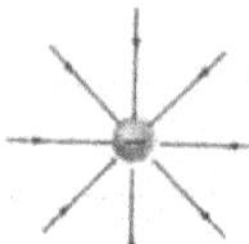

b) negative point charge

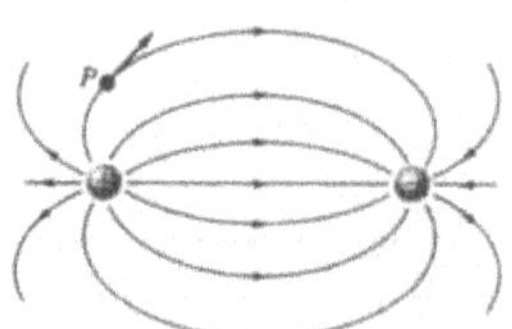

c) positive and negative point charges

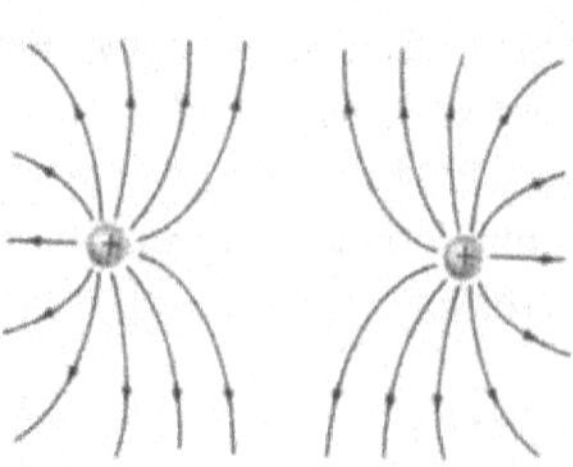

d) two positive point charges

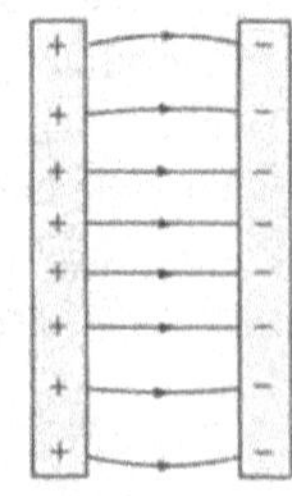

e) oppositely charged parallel plates

Since a **positive test charge** is arbitrarily chosen to analyze the field, electric field lines are drawn away from an object with an excess of positive charge and toward an object with an excess of negative charge. The electric field lines in diagrams (c), (d), and (e) start on the positive charge and terminate on the negative charge.

The electric field is strong in regions where the lines are close together and weak where the lines are farther apart. Thus, in diagrams (a) and (b), the field is strongest closest to the point charges. In diagram (c), the field weakens as the lines diverge as they leave the positive charge and strengthens as the lines converge on the negative charge. In diagram (e), two parallel plates of opposite charge produce an electric field where the lines are parallel near the center of the plates. In this region, the electric field is uniform and E is constant in magnitude and direction.

Coulomb's law can be used to predict that the electric field inside a closed conductor is zero. An example of a closed conductor is a hollow metal sphere that contains an excess of static electric charge. The charge on the conductor tends to reside on its outer surface. Inside the conductor, the electric field is zero. Outside the conductor, the electric field is not zero, and the electric field lines are drawn perpendicular to the surface.

Electric charges tend to distribute throughout the volume of a nonconductor that contains an excess of static charge. It can be shown that a charged nonconductor has an electric field inside as well as outside the nonconductor.

TEXTBOOK QUESTION 18. (Sections 16-7 and 16-8) Why can electric field lines never cross?

ANSWER: An electric field line indicates the direction of the force on a positive test charge and the path the test charge would follow. At a point where two electric field lines cross, the force on the test charge would be in two different directions at the same time. Since a test charge cannot move in two directions at the same time, electric field lines cannot cross.

EXAMPLE PROBLEM 3. (Section 16-7) Two point charges, $Q_1 = +5.0\ \mu\text{C}$ and $Q_2 = -5.0\ \mu\text{C}$, are located on the y axis at $y = +3.0$ m and $y = -3.0$ m, respectively. Determine the magnitude and direction of the electric field on the x axis at $x = 4.0$ m.

Part a. Step 1.

Draw a diagram locating the position of each charge and the direction of the electric field at point P.

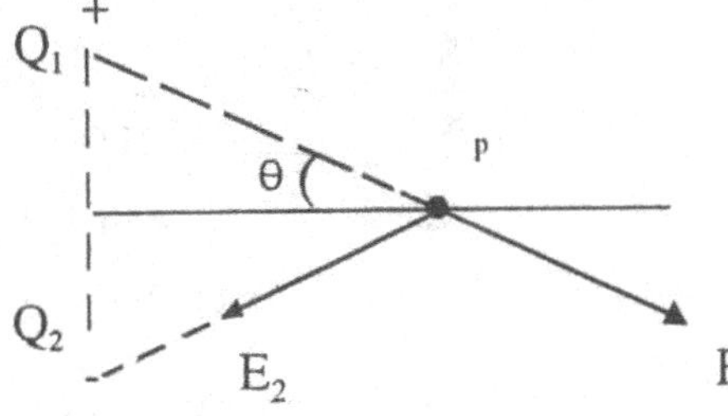

Part a. Step 2.

Determine the distance r and the angle θ shown in the diagram.

$$r = \sqrt{(3.0\ \text{m})^2 + (4.0\ \text{m})^2} = 5.0\ \text{m}$$

$$\theta = \tan^{-1}\frac{3.0\ \text{m}}{4.0\ \text{m}} = 37°$$

Part a. Step 3.

Determine the magnitude and direction of the electric field produced by each point charge at point P.

$$E_1 = \frac{k\,Q_1}{r^2}$$

$$= \frac{(9.0\times10^9\ \text{N}\cdot\text{m}^2/\text{C}^2)(5.0\times10^{-6}\ \text{C})}{(5.0\ \text{m})^2}$$

$\vec{\mathbf{E}}_1 = 1.8\times10^3$ N/C, 37° below the horizontal and away from Q_1

$$E_2 = \frac{k\,Q_2}{r_2}$$

$$= \frac{(9.0\times10^9\ \text{N}\cdot\text{m}^2/\text{C}^2)(-5.0\times10^{-6}\ \text{C})}{(5.0\ \text{m})^2}$$

$\vec{\mathbf{E}}_2 = -1.8\times10^3$ N/C, 37° below the horizontal and toward Q_2

Part a. Step 4.

Use the tail-to-tip or parallelogram method to determine the magnitude and direction of the resultant vectors.

Note: Use the sign convention adopted in Chapter 3 in designating the direction of each component.

$\mathbf{E}_{1x}$ $\mathbf{E}_{1y}$ $\mathbf{E}_{2x}$ $\mathbf{E}_{2y}$

$$E_{1x} = E_1 \cos\ 37° = (1.8\times10^3\ \text{N/C})\ \cos 37°$$
$$E_{1x} = +1.4\times10^3\ \text{N/C}$$
$$E_{1y} = E_1 \sin 37° = (1.8\times10^3\ \text{N/C})\ \sin 37° = -1.1\times10^3\ \text{N/C}$$
$$E_{2x} = E_2 \cos 37° = (1.8\times10^3\ \text{N/C})\ \cos 37° = -1.4\times10^3\ \text{N/C}$$
$$E_{2y} = E_2 \sin 37° = (1.8\times10^3\ \text{N/C})\ \sin 37° = -1.1\times10^3\ \text{N/C}$$

$$\sum X = E_{1x} + E_{2x}$$
$$\sum X = +1.4\times10^3\ \text{N/C} + -1.4\times10^3\ \text{N/C} = 0$$
$$\sum Y = E_{1y} + E_{2y} = -1.1\times10^3\ \text{N/C} + -1.1\times10^3\ \text{N/C}$$
$$\sum Y = -2.2\times10^3\ \text{N/C} = -2200\ \text{N/C}$$

The resultant electric field has a magnitude of 2200 N/C and is directed in the negative y direction at point P.

Gauss's Law

The **electric flux** (Φ_E) passing through a flat area A due to a uniform electric field (E) is given by $\Phi_E = EA\cos\theta$, where θ is the angle between the electric field vector and a line drawn perpendicular to the area.

Gauss's law states that the total electric flux passing through any closed surface is equal to the net charge (Q) enclosed by the surface divided by ε_0. The total flux through the closed surface also equals the sum of the flux through the incremental areas ΔA that make up the surface. Gauss's law is written as follows:

$$\sum_{\substack{\text{closed}\\\text{surface}}} E_{\perp}\Delta A = \frac{Q_{\text{encl}}}{\varepsilon_0}$$

where $E_{\perp} = E\cos\theta$ and is perpendicular to A, $\varepsilon_0 = 8.85\times10^{-12}\ \text{N}\cdot\text{m}^2/\text{C}^2$, Q_{encl} is the net charge enclosed by the surface, and $\sum$ means "sum of."

EXAMPLE PROBLEM 4. (Section 16-10) A very long, thin cylindrical shell of length L and radius R carries a charge $+Q$ that is uniformly distributed over the outer part of the shell. Use Gauss's law to determine the electric field at (*a*) points inside the shell $(r < R)$ and (*b*) points outside the shell $(r > R)$. Assume that the points are far from the ends but not too far from the shell.

Part a. Step 1.

Draw a diagram showing the direction of the lines around the shell.

It is necessary to make assumptions based on symmetry. As shown in the diagram, we expect that field to be radially outward and away from the shell.

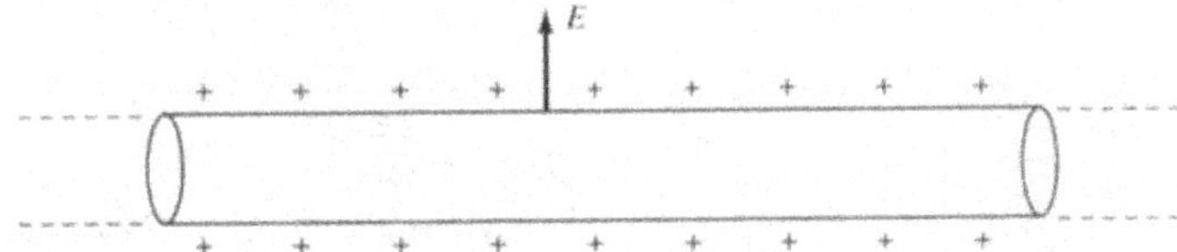

Part a. Step 2.

Draw a gaussian surface for $r < R$ and determine the charge enclosed by the gaussian surface.

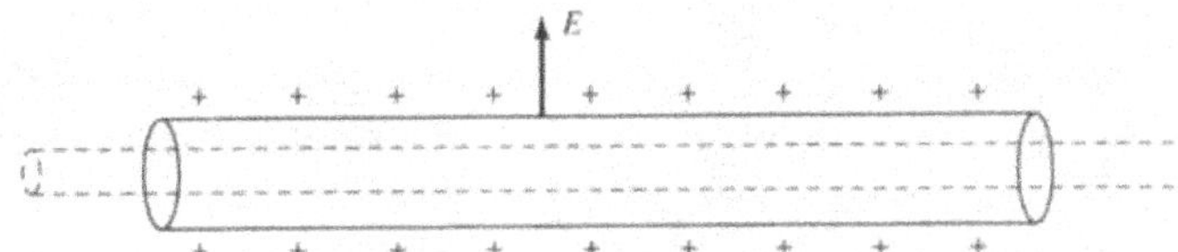

There is no charge enclosed by the gaussian surface.

Part a. Step 3.

Use Gauss's law to determine the electric field strength.

$$\sum E\,\Delta A = \frac{Q}{\varepsilon_0}$$

$E = \dfrac{Q}{\varepsilon_0 A}$ but $Q = 0$; therefore,

$E = 0$ inside the cylindrical shell

Part b. Step 1.

Draw a gaussian surface for $r > R$ and determine the charge enclosed by the surface.

The charge enclosed by the gaussian surface equals the total charge.

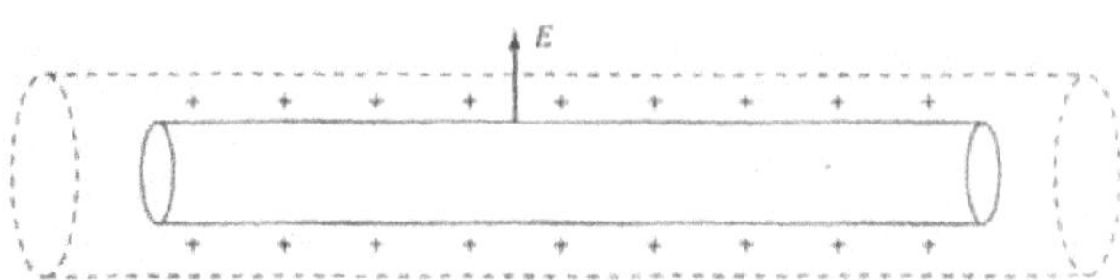

Part b. Step 2.

Use Gauss's law to determine the magnitude of the electric field at $r > R$.

The cylinder is very long, and the question asks for the electric field at points far from the ends. Therefore, only the lateral surface area of the gaussian surface need be considered $(A = 2\pi rL)$:

$$\sum E\,\Delta A = EA\cos 0° = \frac{Q}{\varepsilon_0}$$

$$E = \frac{Q}{\varepsilon_0 A} \quad \text{but} \quad A = 2\pi rL$$

$$E = \frac{Q}{2\pi\varepsilon_0 rL}$$

Alternate solution: The charge on the cylinder can be expressed in terms of the surface charge density (σ) and the lateral surface area of the cylinder $A = 2\pi RL$:

$$E = \frac{Q}{2\pi\varepsilon_0 rL} = \frac{A}{2\pi\varepsilon_0 rL} = \frac{\sigma 2\pi RL}{2\pi\varepsilon_0 rL}$$

$$E = \frac{\sigma R}{\varepsilon_0 r}$$

Just above the cylinder's surface, *r* is approximately equal to *R*. At this point, $E = \frac{\sigma}{\varepsilon_0}$, which agrees with the solution to Example 16-13 in the textbook.

PROBLEM SOLVING SKILLS

For problems involving Coulomb's law:

1. Complete a data table listing the charge on each object and the distance between the objects. If more than two charged objects are given, draw a diagram showing the position of each object.
2. If the objects touch, then charge transfer occurs and the law of conservation of charge must be applied to determine the charge on each object.

3. If more than two charges are given, then it may be necessary to use the methods of vector algebra discussed in Chapter 3 to solve the problem.
4. Apply Coulomb's law and solve the problem.

For problems involving the motion of a charged particle in an electric field:

1. Draw an accurate diagram showing the motion of the particle in the field.
2. Complete a data table based on information both given and implied in the problem.
3. Determine the magnitude and direction of the electric force acting on the particle. Use Newton's second law to determine the rate of acceleration.
4. Use the kinematic equations of Chapter 2 to solve the problem.

For problems involving the magnitude and direction of the resultant electric field due to two or more point charges:

1. Draw an accurate diagram locating the position of each charge.
2. Determine the magnitude and direction of the electric field at the point in question due to each charge.
3. Use the vector algebra methods of Chapter 3 to solve for the resultant electric field.

For problems involving Gauss's law:

1. Draw a diagram showing the electric field lines around the object in question.
2. Draw a gaussian surface to take advantage of the object's symmetry.
3. If the electric flux is perpendicular to the surface, then $\theta = 0°$. If the electric flux is parallel to the surface, then $\theta = 90°$.
4. Determine the surface area of both the object and the gaussian surface. The lateral surface area of a cylinder is $2\pi L$, and the surface area of each end cap is πR^2. The surface area of a sphere is $4\pi r^2$.
5. If necessary, express the charge enclosed by the gaussian surface in terms of the charge density on the object. For example, a charge uniformly distributed along a line can be written as $Q = \lambda L$, where λ is the charge per unit length (coulombs/meter). A charge uniformly distributed across a surface of area A can be written as $Q = \sigma A$, where σ is the charge per unit area (coulombs/meter^2).
6. Use Gauss's law to solve the problem.

SOLUTIONS TO SELECTED TEXTBOOK PROBLEMS

TEXTBOOK PROBLEM 9. (Section 16-5) What is the total charge of all the electrons in a 12-kg bar of gold? What is the net charge of the bar? (Gold has 79 electrons per atom and an atomic mass of 197 u.)

Part a. Step 1.

Determine the number of moles in 12 kg of gold.	1 mole of gold = 197 g = 0.197 kg (12 kg gold)(1 mol/0.197 kg) = 61 mol

Part a. Step 2.

Determine the number of electrons in 61 moles of gold.

1 mole of gold $= 6.02\times10^{23}$ gold atoms

1 gold atom contains 79 electrons:

$$61\text{ mol}\left(\frac{6.02\times10^{23}\text{ gold atoms}}{1.0\text{ mol}}\right)\left(\frac{79\text{ eletrons}}{1\text{ atom}}\right)$$

$$= 2.9\times10^{27}\text{ electrons}$$

Part a. Step 3.

Determine the electric charge in coulombs.

$$Q = 2.9\times10^{27}\text{ electrons}\left(\frac{1\text{ C}}{6.25\times10^{18}\text{ electrons}}\right)$$

$$Q = -4.6\times10^{8}\text{ C}$$

Part b. Step 1.

Determine the net charge on the bar.

If the are no excess electrons on the gold bar, then the net charge on the gold bar is 0.

TEXTBOOK PROBLEM 16. (Sections 16-5 and 16-6) A large electroscope is made with "leaves" that are 78-cm-long wires with tiny 21-g spheres at the ends. When charged, nearly all the charge resides on the spheres. If the wires each make a $26°$ angle with the vertical (Fig. 16-55), what total charge Q must have been applied to the electroscope? Ignore the mass of the wires.

Part a. Step 1.

Draw an accurate diagram locating the forces acting on one of the charges.

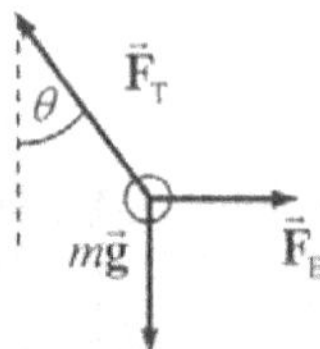

Part a. Step 2.

From Fig. 16-55, determine the distance (d) between the spheres.

$$\sin 26° = \frac{d/2}{78\text{ cm}}$$

$$0.44 = \frac{d/2}{78\text{ cm}}$$

$$d = 2(0.44)(78\text{ cm}) = 68.4\text{ cm}$$

Part a. Step 3.

Solve for the tension (F_T) in the wire.

Apply the first condition of equilibrium and solve for the tension (F_T) in the wire:

$$\sum F = 0, F_T \cos 26° - mg = 0$$

$$F_T(0.90) - (0.021 \text{ kg})(9.8 \text{ m/s}^2) = 0$$

$$F_T = 0.23 \text{ N}$$

Part a. Step 4.

Solve for the electric force between the charged spheres.

Apply the first condition of equilibrium and solve for the electric force between the charged spheres:

$$\sum F_x = 0, \quad F_T = \sin 26° - F_E = 0$$

$$(0.23 \text{ N})(0.44) - F_E = 0$$

$$F_E = 0.10 \text{ N}$$

Part a. Step 5.

Use Coulomb's law to determine the total charge on the spheres.

$$F_E = \frac{kQ_1Q_2}{d^2} \text{ but } Q_1 = Q_2 = \frac{Q}{2}$$

$$0.10 \text{ N} = \left(\frac{(9\times10^9 \text{ N}\cdot\text{m}^2/\text{C}^2)(Q/2)^2}{(0.684 \text{ m})^2}\right)$$

$$\left(\frac{Q}{2}\right)^2 = \frac{(0.10 \text{ N})(0.684 \text{ m})^2}{9\times10^9 \text{ N}\cdot\text{m}^2/\text{C}^2} = 5.2\times10^{-12} \text{ C}^2$$

$$\frac{Q}{2} = 2.3\times10^{-6} \text{C}^2$$

$$Q = 4.6\times10^{-6} \text{C}$$

TEXTBOOK PROBLEM 32. (Sections 16-7 and 16-8) Two point charges, $Q_1 = -32\ \mu\text{C}$ and $Q_2 = +45\ \mu\text{C}$, are separated by a distance of 12 cm. The electric field at the point P (see Fig. 16-57) is zero. How far from Q_1 is P?

Part a. Step 1.

Based on Fig. 16-57, determine the value of x.

$$E_{\text{total}} = E_1 - E_2$$

$$0 = E_1 - E_2$$

$$E_2 = -E_1$$

$$\frac{kQ_2}{(x+12\text{ cm})^2} = -\frac{kQ_1}{x^2}$$ Note: The constant k cancels.

$$\frac{(+45\ \mu\text{C})}{(x+12\text{ cm})^2} = -\frac{(-32\ \mu\text{C})}{x^2}$$

$$1.4x^2 = (x+12\text{ cm})^2 = x^2 + (24\text{ cm})x + 144\text{ cm}^2$$

$$0 = -0.4\,x^2 + (24\text{ cm})x + 144\text{ cm}^2$$ and, using the quadratic formula,

$$x = 65\text{ cm}$$

TEXTBOOK PROBLEM 49. (Section 16-7) A water droplet of radius 0.018 mm remains stationary in the air. If the downward-directed electric field of the Earth is 150 N/C, how many excess electron charges must the water droplet have?

Part a. Step 1.

Draw a diagram locating the forces acting on the water droplet.

+ + + + + +

⇑ F_{electric}

⊖

⇓ F_{gravity}

\- - - - - - -

Part a. Step 2.

Determine the mass of the droplet.

$m = \rho V$, where $\rho_{\text{water}} = 1000\text{ kg/m}^3$ and $V = \frac{4}{3}\pi r^3$

$$m = (1000\text{ kg/m}^3)\tfrac{4}{3}\pi[(0.018\text{ mm})(1\text{ m}/1000\text{ mm})]^3$$

$$m = 2.44\times10^{-11}\text{ kg}$$

Part a. Step 3.

The electric force balances gravity. Use $F = qE$ to determine the magnitude of the net electric charge.

The droplet is suspended at rest; therefore, net $F = 0$:

$$F_{\text{electric}} = -F_G = 0$$

$$qE - mg = 0$$

$$q = \frac{mg}{E} = \frac{(2.44\times10^{-11}\ \text{kg})(9.8\ \text{m/s}^2)}{150\ \text{N/C}}$$

$$q = 1.59\times10^{-12}\ \text{C}$$

Part a. Step 4.

Determine the number of excess electrons on the droplet.

$q = ne$, where $n =$ number of excess electrons, $e = 1.6\times10^{-19}\ \text{C}$:

$$1.59\times10^{-12}\ \text{C} = n(1.6\times10^{-19}\ \text{C})$$

$$n = 9.96\times10^{6}\ \text{electrons} \approx 1.0\times10^{7}\ \text{electrons}$$

TEXTBOOK PROBLEM 57. (Section 16-7) A point charge $(m = 1.0\ \text{gram})$ at the end of an insulating cord of length 55 cm is observed to be in equilibrium in a uniform horizontal electric field of 9500 N/C, when the pendulum's position is as shown in Fig. 16-66, with the charge 12 cm above the lowest (vertical) position. If the field points to the right in Fig. 16-66, determine the magnitude and sign of the point charge.

Part a. Step 1.

Complete a data t able.

$m = 1.0\ \text{g} = 0.0010\ \text{kg}$ $\quad \ell = 55\ \text{cm} = 0.55\ \text{m}$ $\quad y = 0.43\ \text{m}$

$E = 12{,}000\ \text{N/C}$ $\quad Q = ?$ $\quad k = 9\times10^{9}\ \text{N}\cdot\text{m}^2/\text{C}^2$

Part a. Step 2.

Draw an accurate diagram locating the forces acting on the point charge.

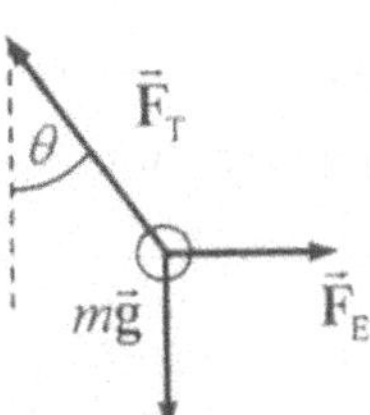

Part a. Step 3.

Determine the angle θ shown in the diagram.

$$\cos\theta = \frac{0.43\ \text{m}}{0.55\ \text{m}}$$

$$\cos\theta = 0.782$$

$$\theta = 38.6°$$

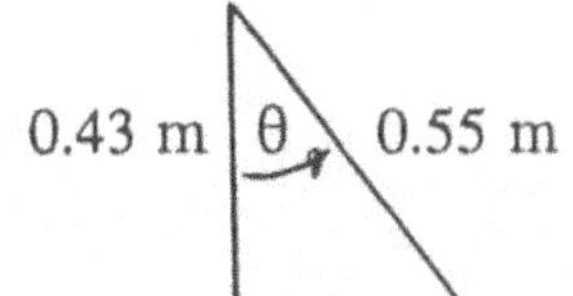

Part a. Step 4.

Apply the first condition of static equilibrium and solve for Q.

$\Sigma F_y = 0 \qquad F_{Ty} - mg = 0$

$F_T \cos 38.6° - (0.0010 \text{ kg})(9.8 \text{ m/s}^2) = 0$

$F_T = \dfrac{0.0098 \text{ N}}{\cos 38.6°} = 0.0125 \text{ N}$

$\sum F_x = 0, \quad \sum F_{Tx} - QE = 0$

$(0.0125 \text{ N})(\sin 38.6°) - Q(9500 \text{ N/C}) = 0$

$Q = (7.80 \times 10^{-3} \text{ N})/(9500 \text{ N/C}) = 8.21 \times 10^{-7} \text{ C}$

Part a. Step 5.

Determine the sign on the point charge.

Based on the diagram, the direction of the force on the point charge is the same as that of the electric field. Since the electric field is toward the right, the charge must be positive.

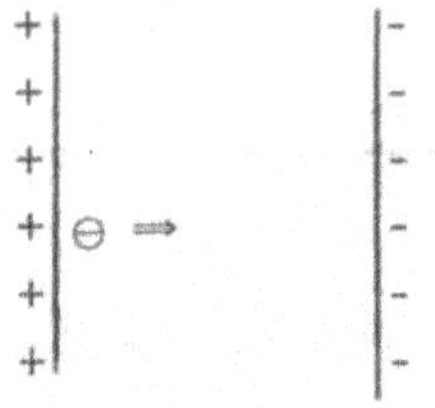

TEXTBOOK PROBLEM 62. (Section 16-7) An electron with speed $v_0 = 5.32 \times 10^6$ m/s is traveling parallel to an electric field of magnitude $E = 9.45 \times 10^3$ N/C. (*a*) How far will the electron travel before it stops? (*b*) How much time will elapse before it returns to its starting point?

Part a. Step 1.

Use the work-energy theorem to solve for the distance traveled.

$W = \Delta\overline{\text{KE}}$

$Fd \cos 180° = \frac{1}{2}mv^2 - \frac{1}{2}mv_0^2 \quad \text{but} \quad v_0 = 5.32 \times 10^6 \text{ m/s}$

$$(1.6 \times 10^{-19} \text{ C})(9.45 \times 10^3 \text{ N/C})d \cos 180° = 0 - \tfrac{1}{2}(9.1 \times 10^{-31} \text{ kg})(5.32 \times 10^6 \text{ m/s})^2$$

$d = 8.52 \times 10^{-3} \text{ m}$

Part b. Step 1.

Use Newton's second law to solve for the deceleration.

$F = ma$ but $F = qE$

$$a = \frac{qE}{m} = \frac{(-1.6\times10^{-19}\text{ C})(9.45\times10^{3}\text{ N/C})}{9.1\times10^{-31}\text{ kg}}$$

$$a = -1.66\times10^{15}\text{ m/s}^2$$

Part b. Step 2.

Use the kinematic equations to solve for the time to return to the starting point.

$$x = \tfrac{1}{2}at^2 + v_0t + x_0$$

The electron returns to the starting point. Therefore, $x = x_0 = 0$:

$$0 = \tfrac{1}{2}(-1.66\times10^{15}\text{ m/s}^2)t^2 + (5.32\times10^{6}\text{ m/s})t + 0$$

Factoring gives

$0 = [(-8.31\times10^{14}\text{ m/s}^2)t + 5.32\times10^{6}\text{ m/s}]t$, so either

$t = 6.40\times10^{-9}\text{ s}$ or $t = 0$ (starting time)

It cannot take zero time to return to the starting point. Therefore, the correct answer is $t = 6.40\times10^{-9}$ s.

ELECTRIC POTENTIAL

OBJECTIVES

After studying the material of this chapter, the student should be able to:

- write from memory the definitions of electric potential and electric potential difference.
- distinguish among electric potential, electric potential energy, and electric potential difference.
- draw the electric field pattern and equipotential line pattern that exist between charged objects.
- determine the magnitude of the potential at a point a known distance from a point charge or an arrangement of point charges.
- state the relationship between electric potential and electric field and determine the potential difference between two points a fixed distance apart in a region where the electric field is uniform.
- determine the kinetic energy in both joules and electric volts of a charged particle that is accelerated through a given potential difference.
- explain what is meant by an electric dipole and determine the magnitude of the electric dipole moment between two point charges.
- given the dimensions, the distance between plates, and the dielectric constant of the material between the plates, determine the magnitude of the capacitance of a parallel-plate capacitor.
- given the capacitance, the dielectric constant, and either the potential difference or the charge stored on the plates of a parallel-plate capacitor, determine the energy and the energy density stored in the capacitor.

KEY TERMS AND PHRASES

electric potential at point a (V_a) equals the electric potential energy (PE_a) per unit charge (q) placed at that point.

electric potential difference between two points (V_{ab}) is measured by the work required to move a unit of electric charge from point b to point a.

equipotential lines are lines along which each point is at the same potential. On an **equipotential surface**, each point on the surface is at the same potential.

voltage is a common term used for potential difference. The SI unit of electric potential and potential difference is the volt (V), where $1\text{ V} = 1\text{ J/C}$.

electron volt (eV) is the unit used for the energy gained by a charged particle which is accelerated through a potential difference; $1\text{ eV} = 1.60 \times 10^{-19}\text{ J}$.

electric dipole is two equal point charges (q), of opposite sign, separated by a distance ℓ. The SI unit of the dipole moment (p) is the **debye,** where $1 \text{ debye} = 3.33 \times 10^{-33} \text{ C} \cdot \text{m}$.

capacitor stores electric charge and consists of two conductors separated by an insulator known as a **dielectric.**

capacitance (*C*) is the ability of a capacitor to store electric charge.

SUMMARY OF MATHEMATICAL FORMULAS

electric potential	$V_a = \frac{\text{PE}_a}{q}$	The electric potential at point a (V_a) equals the electric potential energy (PE_a) per unit charge (q) placed at that point.
electric potential due to a point charge	$V = \frac{kQ}{r}$	The electric potential (V) due to a point charge is related to the magnitude of the charge and the distance from the charge (r) to the point in question. If more than one point charge is present, then the potential at a particular point is equal to the arithmetic sum of the potential due to each charge at the point in question.
potential difference	$V_{ab} = V_a - V_b = \frac{W_{ab}}{q}$	The electric potential difference between two points (V_{ab}) is measured by the work required to move a unit of electric charge from point b to point a.
	$\Delta\text{PE}_{ab} = \text{PE}_a - \text{PE}_b = qV_{ab}$	The potential difference can also be discussed in terms of the change in potential energy of a charge q when it is moved between points a and b.
	$V_{ab} = \frac{\Delta\text{PE}}{q}$ $V_{ab} = Ed \cos\theta$	If the charged particle is in an electric field which is uniform (constant in magnitude and direction), then the potential difference (V_{ab}) equals the product of the electric field strength (E), the distance (d) between points a and b, and the cosine of the angle ($\cos\theta$) between the electric field vector and the displacement vector.
electric potential due to an electric dipole	$V = kQ\ell \cos\frac{\theta}{r^2}$ or $V = \frac{kp\cos\theta}{r^2}$	The potential due to two equal point charges (Q), of opposite sign, separated by a distance ℓ, depends on the angle θ and the distance r. Here $Q\ell$ is the dipole moment (p).

capacitance	$C = \frac{Q}{V}$ $C = \frac{K\varepsilon_0 A}{d}$	Capacitance (C) is the ratio of the charge stored (Q) to the potential difference (V) between the conducting surfaces. For a parallel-plate capacitor, the capacitance (C) depends on the dielectric constant (K) of the insulating material between the plates, the permittivity of free space (ε_0), the surface area (A) of one side of one plate that is opposed by an equal area of the other plate, and the distance between the plates (d).
potential energy stored in a charged capacitor	$\text{PE} = \frac{1}{2}QV$ $\text{PE} = \frac{1}{2}CV^2$ $\text{PE} = \frac{1}{2}\frac{Q^2}{C}$ $\text{PE} = \frac{1}{2}\varepsilon_0 E^2 Ad$	The electric energy stored in charging a capacitor depends on the charge stored (Q) and the potential difference (V) across the conducting surfaces. The potential energy stored by a parallel-plate capacitor is related to the electric field existing between the plates. The product of the area (A) and the distance between the plates (d) equals the volume between the conducting surfaces.
energy density	$\text{energy density} = \frac{1}{2}\varepsilon_0 E^2$	The energy density is the energy stored per unit volume.

CONCEPT SUMMARY

Electric Potential and Potential Difference

The **electric potential** at point a (V_a) equals the electric potential energy (PE_a) per unit charge (q) placed at that point:

$$V_a = \frac{\text{PE}_a}{q}$$

The **electric potential difference** between two points (V_{ab}) is measured by the work required to move a unit of electric charge from point b to point a:

$$V_{ab} = V_a - V_b = \frac{W_{ab}}{q}$$

The potential difference can also be discussed in terms of the change in potential energy of a charge (q) when it is moved between points a and b:

$$\Delta\text{PE} = \text{PE}_a - \text{PE}_b = qV_{ab}; \quad \text{therefore,} \quad V_{ab} = \frac{\Delta\text{PE}}{q}$$

Potential difference is often referred to as **voltage**. Both potential and potential difference are scalar quantities that have dimensions of joules/coulomb. The SI unit of electric potential and potential difference is the **volt** (V), where $1\text{ V} = 1\text{ J/C}$.

If the charged particle is in an electric field that is uniform—that is, constant in magnitude and direction—then the potential difference is related to the electric field as follows:

$$V_{\text{ab}} = Ed \cos \theta$$

where E is the electric field strength in N/C, d is the distance between points a and b in meters, and θ is the angle between the electric field vector and the displacement vector.

EXAMPLE PROBLEM 1. (Section 17-5) (*a*) Determine the potential at point p due to a point charge of magnitude $0.60\ \mu\text{C}$ if point p is 0.60 m from the charge. (*b*) Determine the work that must be done in order to move a $0.10\ \mu\text{C}$ charge from infinity to point p.

Part a. Step 1.

Use the formula for the electric potential a distance r from a single point charge.	Assume that the potential at infinity is zero. $V = \frac{kq}{r} = \frac{(9.0 \times 10^9\ \text{N}\cdot\text{m/C})(0.60 \times 10^{-6}\ \text{C})}{0.60\ \text{m}}$ $V = 9.0 \times 10^3\ \text{V}$

Part b. Step 1.

Determine the work required to move a point charge of $0.10\ \mu\text{C}$ from infinity to point p.	At infinity, the potential due to the $0.60\ \mu\text{C}$ charge is arbitrarily defined as being equal to zero. The work done in moving the $0.10\ \mu\text{C}$ charge from infinity to point p can be determined by the product of the magnitude of $0.10\ \mu\text{C}$ and the potential difference between the potential due to the $0.60\ \mu\text{C}$ charge at point p as compared to the $0.60\ \mu\text{C}$ charge's potential at infinity: $W = qV = (0.10 \times 10^{-6}\ \text{C})(9 \times 10^3\ \text{V})$ $W = 9.0 \times 10^{-4}\ \text{J}$

EXAMPLE PROBLEM 2. (Section 17-1) The two horizontal parallel plates of a capacitor are 0.050 m apart, and the electric field between the plates is 1000 N/C. A helium nucleus is initially at rest near the positive plate. Determine the (*a*) magnitude of the electric force acting on the helium nucleus, (*b*) work done on the helium nucleus as it passes between the plates, and (*c*) kinetic energy and speed of the helium nucleus just before it strikes the negative plate. Note: The mass of the nucleus is 6.68×10^{-27} kg, and the charge is 3.2×10^{-19} C.

Part a. Step 1.

The magnitude of the electric field is known. Determine the magnitude of the electric force.

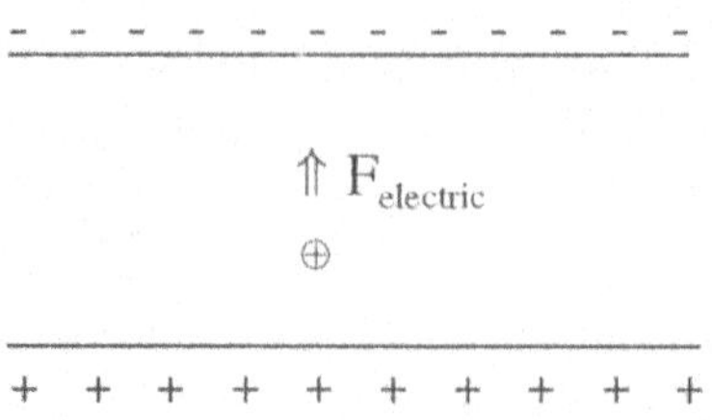

$$F = qE = (3.2 \times 10^{-19}\ \text{C})(1000\ \text{N/C})$$

$$F = 3.2 \times 10^{-16}\ \text{N}$$

Part b. Step 1.

Determine the work done on the nucleus as it passes between the plates.

$$W = Fd \cos\theta = (3.2 \times 10^{-16}\ \text{N})(0.050\ \text{m})\cos 0°$$

$$W = 1.6 \times 10^{-17}\ \text{J}$$

Alternate method:

$$W = qV, \quad \text{where} \quad V = Ed$$

$$W = qEd = (3.2 \times 10^{-19}\ \text{C})(1000\ \text{N/C})(0.050\ \text{m}) = 1.6 \times 10^{-17}\ \text{J}$$

Part c. Step 1.

Use the work-energy theorem to determine the final kinetic energy. Assume that the nucleus was initially at rest and determine the velocity just before it struck the negative plate.

$$\text{Work} = \Delta\text{KE} = \text{KE}_\text{f} - \text{KE}_\text{i}$$

$$1.6 \times 10^{-17}\ \text{J} = \text{KE}_\text{f} - 0$$

$$\text{KE}_\text{f} = 1.6 \times 10^{-17}\ \text{J}$$

$$\text{KE}_\text{f} = \tfrac{1}{2}mv^2$$

$$1.6 \times 10^{-17}\ \text{J} = \tfrac{1}{2}(6.68 \times 10^{-27}\ \text{kg})v_\text{f}^2$$

$$v_\text{f} = 6.9 \times 10^{4}\ \text{m/s}$$

EXAMPLE PROBLEM 3. (Sections 17-1 and 17-2) The two horizontal parallel plates of a capacitor are 0.0400 m apart, and the potential difference between the plates is 100 V. A positively charged particle of mass 8.16×10^{-17} kg remains motionless between the plates. Calculate the (*a*) magnitude and direction of the electric field between the plates and (*b*) charge on the particle.

Part a. Step 1.

Determine the magnitude and direction of the electric field between the plates.

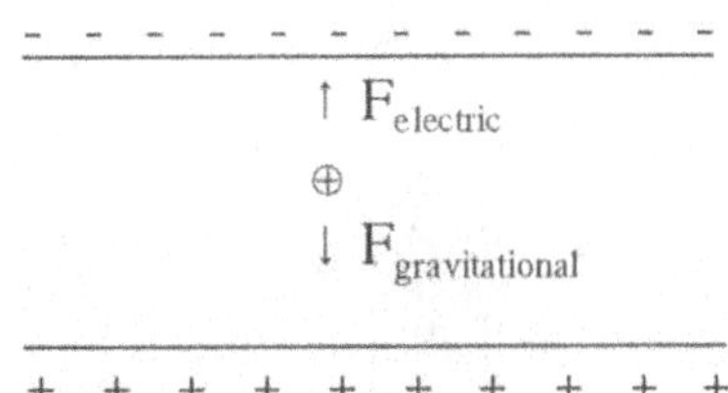

The electric field is uniform in the region between the two parallel plates and is directed from the positive plate toward the negative plate.

The magnitude of the electric field is given by the following:

$$V = Ed \cos \theta$$

$$100 \text{ V} = E(0.0400 \text{ m})(\cos 0°)$$

$$E = \frac{100 \text{ V}}{0.0400 \text{ m}} = 2500 \text{ V/m} = 2500 \text{ N/C}$$

Part b. Step 1.

Determine the charge on the particle.

Since the particle is in static equilibrium, the electric force must be equal to but oppositely directed from the gravitational force:

$$F_{\text{electric}} = F_{\text{G}}$$

$$qE = mg$$

$$q(2500 \text{ N/C}) = (8.16 \times 10^{-17} \text{ kg})(9.80 \text{ m/s}^2)$$

$$q = \frac{(8.16 \times 10^{-17} \text{ kg})(9.80 \text{ m/s}^2)}{2500 \text{ N/C}}$$

$$q = 3.2 \times 10^{-19} \text{ C}$$

Equipotential Lines

Equipotential lines are lines along which each point is at the same potential. On an **equipotential surface**, each point on the surface is at the same potential. The equipotential line or surface is perpendicular to the direction of the electric field lines at every point. Thus, if the electric field pattern is known, then it is possible to determine the pattern of equipotential lines or surfaces and vice versa.

In the following diagrams, the dashed lines represent equipotential lines, and the solid lines represent the electric field lines.

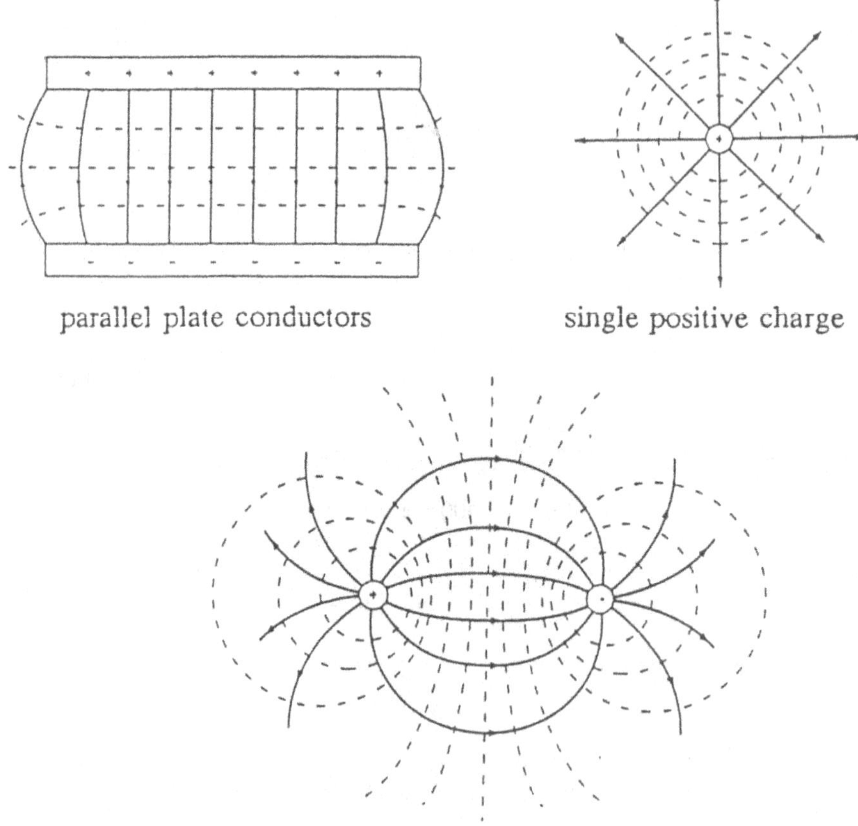

TEXTBOOK QUESTION 2. (Section 17-1) If a negative charge is initially at rest in an electric field, will it move toward a region of higher potential or lower potential? What about a positive charge? How does the potential energy of the charge change in each instance? Explain.

ANSWER: The answer can be best explained by using Fig. 17-1. As that figure shows, the positively charged plate of the parallel-plate capacitor is at the higher potential, and the negatively charged plate is at the lower potential. A negative charge placed in the electric field between the plates will move toward the positive plate and therefore moves toward the region of higher potential. On the other hand, a positive charge placed in the electric field between the plates will move toward the negative plate and therefore moves toward the region of lower potential. In each case, the charged particle will lose potential energy while gaining kinetic energy as the particle travels between the plates.

TEXTBOOK QUESTION 8. (Section 17-3) Can two equipotential lines cross? Explain.

ANSWER: All points on equipotential lines (or surfaces) are at the same potential. If two different equipotential lines were to cross, then the point at which they crossed would be at two different potentials. Because it is not possible to have two different potentials at the same point, two equipotential lines cannot cross.

Also, equipotential lines are always perpendicular to electric field lines. Electric field lines never cross. Therefore, it is not possible for equipotential lines to cross.

Electron Volt

The energy gained by a charged particle that is accelerated through a potential difference can be expressed in **electron volts** (eV) as well as joules. Higher amounts of energy can be measured in KeV or MeV, where $1\text{ eV} = 1.60 \times 10^{-19}\text{ J}$, $1\text{ KeV} = 10^3\text{ eV}$, and $1\text{ MeV} = 10^6\text{ eV}$.

Electric Potential Due to a Point Charge

The electric potential due to a point charge (Q) at a distance r from the charge is given by

$$V = \frac{kQ}{r}$$

Notice that the zero of potential is arbitrarily taken to be at infinity ($r = \infty$). For a negative charge, the potential at distance r from the charge is less than zero. As r increases, the potential increases toward zero, reaching zero at $r = \infty$. If more than one point charge is present, then the potential at a particular point is equal to the arithmetic sum of the potentials due to each charge at the point in question.

TEXTBOOK QUESTION 5. (Section 17-5) Is there a point along the line joining two equal positive charges where the electric field is zero? Where the electric potential is zero? Explain.

ANSWER: Electric field is a vector quantity. As shown in the diagram, at the midpoint between the two charges, the two electric field vectors, E_1 and E_2, are equal in magnitude but oppositely directed.

$$Q_1 + \qquad E_2 \Leftarrow \bullet \Rightarrow E_1 \qquad + Q_2$$

The magnitude of the resultant electric field is zero.

Electric potential is a scalar quantity. Electric potential has magnitude but no direction. At the midpoint between the charges, the electric potential due to each charge is positive and equal in magnitude. The resultant electric potential is the arithmetic sum of the two potentials and is greater than zero. Both charges are positive; therefore, the arithmetic sum of the potential of each charge would be greater than zero at every point on the line between the two charges.

TEXTBOOK QUESTION 7. (Section 17-2) If $V = 0$ at a point in space, must $\vec{E} = 0$? If $\vec{E} = 0$ at some point, must $V = 0$ at that point? Explain. Give examples for each.

ANSWER: At a point where the potential equals zero, the electric field need NOT equal zero. As shown in the diagram, the two point charges are equal in magnitude, but one is positive and the other negative.

$$Q_1 \; + \qquad \Rightarrow \quad \Rightarrow \qquad - \; Q_1$$

At the midpoint between the two charges, the electric field due to the positive charge is directed away from the positive charge and toward the negative. The electric field due to the negative charge is directed toward the negative charge. The resultant electric field equals the sum of the two vectors and is directed toward the negative charge.

Electric potential is a scalar quantity. At the midpoint between the two charges, the magnitude of the potential due to the positive charge equals the potential due to the negative charge. However, one potential is positive and the other is negative, and their arithmetic sum is zero.

In the answer to Textbook Question 5, it was shown that the electric potential at the midpoint between two positive point charges is not zero, while the electric field at the midpoint equals zero. The reason is that electric field is a vector quantity, but electric potential is a scalar quantity.

EXAMPLE PROBLEM 4. (Section 17-5) Four point charges are placed at the four corners of a square 0.100 m on a side. Determine the potential at the center of the square if (*a*) each charge has a magnitude of 2.00 μC and (*b*) two of the charges have a magnitude of 2.00 μC and the other two have a magnitude of −2.00 μC.

Part a. Step 1.

Draw a diagram locating the four charges. Determine the distance *r* from each charge to the center of the square.

$$r = \sqrt{(0.0500\text{ m})^2 + (0.0500\text{ m})^2}$$

$$r = 0.0710\text{ m}$$

Q_1 Q_2 r p Q_3 Q_4

Part a. Step 2.

Determine the magnitude of the potential due to one of the charges.

Each charge has the same magnitude, and the distance from point p is the same. The potential due to one of the charges is given by

$$V = \frac{kQ}{r}$$

$$V_1 = \frac{(9.0 \times 10^9 \text{ N}\cdot\text{m}^2/\text{C}^2)(2.00 \times 10^{-6} \text{ C})}{0.0707 \text{ m}}$$

$$V_1 = 2.54 \times 10^5 \text{ V}$$

Part a. Step 3.

Determine the total potential of the four charges.

Potential is a scalar quantity. The total potential equals the arithmetic sum of the potential due to the individual charges.

$$V_{\text{total}} = V_1 + V_2 + V_3 + V_4$$

$$V_{\text{total}} = 4\frac{kQ}{r} = 4V_1$$

$$V_{\text{total}} = 4(2.54 \times 10^5 \text{ V}) = 1.0^2 \times 10^6 \text{ V}$$

Part b. Step 1.

Assume that charges Q_3 and Q_4 are negative. Determine V_{total}.

$$V_1 = V_2 = 2.54 \times 10^5 \text{ V}$$

$$V_3 = V_4 = -2.54 \times 10^5 \text{ V}$$

$$V_{\text{total}} = V_1 + V_2 + V_3 + V_4 = 0$$

Note: Although the potential at point p is zero, the electric field strength does not equal zero. It is left to the student to show that if charges Q_1 and Q_4 are positive and Q_2 and Q_3 are negative, then both the potential and the electric field strength equal zero.

Electric Dipole

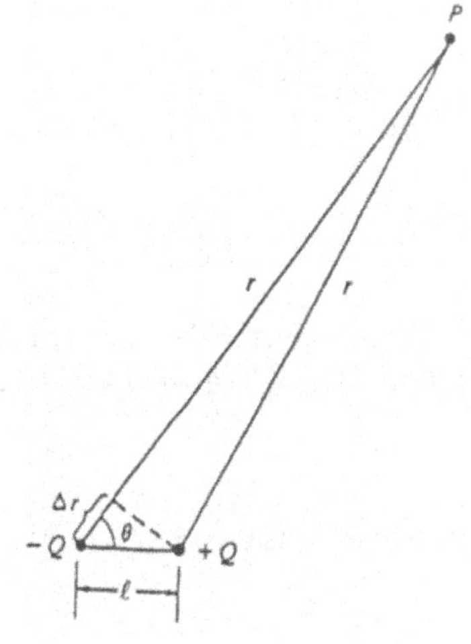

Two equal point charges (Q) of opposite sign and separated by a distance ℓ are called an **electric dipole**. If $r >> \ell$, then the potential at point P due to the dipole is given by

$$V = kQ\ell \cos\frac{\theta}{r^2}$$

The product $Q\ell$ is the **dipole moment** (p), and the potential can be written as

$$V = kp \cos\frac{\theta}{r^2}$$

The SI unit of the dipole moment is the **debye**, where $1\text{ debye} = 3.33\times10^{-33}\text{ C}\cdot\text{m}$. In polar molecules, such as water, the molecule is electrically neutral, but there is a separation of charge in the molecule. Such molecules have a net dipole moment.

Capacitance and Dielectrics

A **capacitor** stores electric charge and consists of two conductors separated by an insulator known as a **dielectric**. The ability of a capacitor to store electric charge is referred to as **capacitance** (*C*) and is found by the following equation:

$$C = \frac{Q}{V}$$

Here *Q* is the charge stored in coulombs and *V* is the potential difference between the conducting surfaces in volts.

The SI unit of capacitance is the **farad** (F), where 1 farad = 1 coulomb/volt (1 F = 1 C/V). Typical capacitors have values that range from 1 picofarad (1 pF) to 1 microfarad (1 μF), where

$$1\text{ pF} = 1\times10^{-12}\text{ F} \qquad \text{and} \qquad 1\ \mu\text{F} = 1\times10^{-6}\text{ F}$$

The capacitance of a capacitor depends on the physical characteristics of the capacitor as well as the insulating material that separates the conducting surfaces that store the electric charge. For a parallel-plate capacitor, the capacitance is given by

$$C = \frac{K\varepsilon_0 A}{d} \text{ or } C = \frac{\varepsilon A}{d}, \quad \text{where} \quad \varepsilon = K\varepsilon_0 \quad \text{and} \quad \varepsilon_0 = 8.85\times10^{-12}\text{ C}^2/\text{N}\cdot\text{m}^2$$

Here *K* is the **dielectric constant** of the insulating material between the plates. The constant is dimensionless and depends on the material. For dry air at 20°C, $K = 1.0006$; for a vacuum, $K = 1$; for distilled water, $K = 80$; and for wax paper, $K = 2.25$. The constant ε_0 is the *permittivity of free space*. *A* is the surface area of one side of one plate that is opposed by an equal area of the other plate, and *d* is the distance between the plates. ε is the permittivity of the material between the plates.

Energy Stored in a Capacitor

A charged capacitor stores potential energy. The electric energy stored in charging a capacitor from an uncharged condition to a charge of *Q* and a potential difference *V* is given by

$$\text{PE} = \tfrac{1}{2}QV = \tfrac{1}{2}CV^2 = \tfrac{1}{2}\frac{Q^2}{C}$$

The energy stored by a parallel-plate capacitor is related to the electric field existing between the plates:

$$C = \varepsilon_0\frac{A}{d} \quad \text{and} \quad V = Ed; \quad \text{then} \quad \text{PE} = \tfrac{1}{2}CV^2 = \tfrac{1}{2}\varepsilon_0 E^2 Ad$$

The product of the area (*A*) and the distance between the plates (*d*) equals the volume between the conducting surfaces. The **energy density** is the energy stored per unit volume:

$$\text{energy density} = \tfrac{1}{2}\varepsilon_0 E^2$$

If a dielectric is present, then ε_0 is replaced with ε.

EXAMPLE PROBLEM 5. (Sections 17-7 to 17-9) A parallel-plate capacitor has circular plates that are 0.100 m in diameter and separated by a piece of mica 0.00200 m thick. The dielectric constant of mica is 7.00. (*a*) Determine the capacitance of the capacitor.

Suppose that now the capacitor is connected to a 100-volt battery. Determine the (*b*) charge stored on each plate and (*c*) energy stored in the capacitor.

Part a. Step 1.

Determine the area of one plate.

$$A = \pi r^2$$

$$A = \pi(0.0500\text{ m})^2 = 7.85 \times 10^{-3}\text{ m}^2$$

Part a. Step 2.

Determine the capacitance.

$$C = \frac{K\varepsilon_0 A}{d}$$

$$= \frac{(7.00)(8.85 \times 10^{-12}\text{ C}^2/\text{N}\cdot\text{m}^2)(7.85 \times 10^{-3}\text{ m}^2)}{0.00200\text{ m}}$$

$$C = 2.43 \times 10^{-10}\text{ F, or } 243\text{ pF}$$

Part b. Step 1.

Determine the charge stored on each plate.

$$Q = CV$$

$$Q = (2.43 \times 10^{-10}\text{ F})(100\text{ V}) = 2.43 \times 10^{-8}\text{ C}$$

Part c. Step 1.

Determine the energy stored in the capacitor.

$$\text{PE} = \tfrac{1}{2}QV$$

$$= \tfrac{1}{2}(2.43 \times 10^{-8}\text{ C})(100\text{ V})$$

$$\text{PE} = 1.22 \times 10^{-6}\text{ J}$$

EXAMPLE PROBLEM 6. (Sections 17-7 to 17-9) A parallel-plate capacitor consists of two metal plates separated by 0.0060 m of dry air. The capacitor is connected to a 100-volt source. The area of each plate is 0.040 m^2. Determine the (*a*) capacitance, (*b*) charge on each plate, (*c*) energy stored, and (*d*) electric field between the plates. (*e*) A sheet of mica 0.0060 m thick is inserted between the plates. The dielectric constant of mica is 7.0. Determine the capacitance with the mica inserted, the total energy stored, and the increase in the charge stored on the plates.

Part a. Step 1.

Determine the initial capacitance (C_i).

The dielectric constant of dry air to three significant figures is 1.00.

$$C = \frac{K\varepsilon_0 A}{d}$$

$$= \frac{(1.00)(8.85 \times 10^{-12}\ \text{C}^2/\text{N}\cdot\text{m}^2)(0.040\ \text{m}^2)}{0.0060\ \text{m}}$$

$$C_i = 5.9 \times 10^{-11}\ \text{F}, \text{ or } 59\ \text{pF}$$

Part b. Step 1.

Determine the charge stored.

$$Q_i = CV = (5.9 \times 10^{-11}\ \text{F})(100\ \text{V})$$

$$Q_i = 5.9 \times 10^{-9}\ \text{C}$$

Part c. Step 1.

Determine the energy stored.

$$\text{PE} = \tfrac{1}{2}QV = \tfrac{1}{2}(5.9 \times 10^{-9}\ \text{C})(100\ \text{V})$$

$$\text{PE} = 3.0 \times 10^{-7}\ \text{J}$$

Part d. Step 1.

Determine the initial electric field (E_i).

$V = Ed$ and $E = V/d = (100\ \text{V})/(0.0060\ \text{m})$

$$E_i = 1.7 \times 10^4\ \text{V/m} = 17{,}000\ \text{N/C}$$

Part e. Step 1.

Determine the capacitance with the dielectric inserted (C_f).

Since the battery remains connected to the capacitor, the amount of charge stored on the plates will increase. The electric field will decrease immediately after the dielectric is inserted but will return to 17,000 N/C when the charge stops forming on the plates. The final voltage will be 100 volts. Thus:

$C_f = K\varepsilon_0 A/d$ but $\varepsilon_0 A/d = 59\ \text{pF}$

$$C_f = (7.0)(59\ \text{pF}) = 410\ \text{pF}$$

Part e. Step 2.	
Determine the energy stored.	$\text{PE} = \frac{1}{2}CV^2$ $\text{PE} = \frac{1}{2}(410 \times 10^{-12}\text{ F})(100\text{ V})^2$ $\text{PE} = 2.1 \times 10^{-6}\text{ J}$
Part e. Step 3.	
Determine the final charge stored on the plates and the increase in the charge stored.	$Q_\text{f} = CV = (410 \times 10^{-12}\text{ F})(100\text{ V})$ $Q_\text{f} = 4.1 \times 10^{-8}\text{ C}$ $\Delta Q = Q_\text{f} - Q_\text{i} = 4.1 \times 10^{-8}\text{ C} - 5.9 \times 10^{-9}\text{ C} = 3.5 \times 10^{-8}\text{ C}$

PROBLEM SOLVING SKILLS

For problems involving a charged particle accelerated through a potential difference:

1. Draw an accurate diagram showing the motion of the particle.
2. Complete a data table.
3. If necessary, review the work-energy theorem, the concept of kinetic energy, Newton's laws of motion, and the kinematic equations.
4. Use the concepts of electric field, electric force, and electric potential along with the concepts listed in skill 3 to solve the problem.

For problems involving the electric potential due to a point charge (or point charges):

1. Remember that potential is a scalar quantity.
2. Solve for the potential due to each charge. The potential due to a negative charge is negative, and the potential due to a positive charge is positive.
3. The total potential equals the arithmetic sum of the potentials due to the individual point charges.

For problems involving a parallel-plate capacitor based on the physical characteristics of the capacitor and the dielectric constant of the material between the plates:

1. Note the value of the dielectric constant.
2. Determine the area of one plate and the distance between the plates.
3. Apply the formula for the capacitance of a parallel-plate capacitor.

For problems involving capacitance when a dielectric is inserted after the initial conditions are described:

1. If the battery is disconnected before the dielectric is inserted, then the charge stored cannot increase. However, the magnitude of both the electric field and potential difference decreases after the dielectric is inserted.

2. If the battery remains connected, then the charge stored increases. The magnitude of both the final electric field and the potential difference between the plates equals the value before the dielectric was inserted.
3. In each case, the final capacitance is greater than the original.

SOLUTIONS TO SELECTED TEXTBOOK PROBLEMS

TEXTBOOK PROBLEM 14. (Sections 17-1 to 17-4) An alpha particle (which is a helium nucleus, $Q = +2e$, $m = 6.64 \times 10^{-27}$ kg) is emitted in a radioactive decay with KE = 5.53 MeV. What is its speed?

Part a. Step 1.

Convert 5.53 MeV to joules.

$$(5.53\ \text{MeV})\left(\frac{1 \times 10^{6}\ \text{eV}}{1\ \text{MeV}}\right)\left(\frac{1.6 \times 10^{-19}\ \text{J}}{1\ \text{eV}}\right) = 8.85 \times 10^{-13}\ \text{J}$$

Part a. Step 2.

Use the formula for the kinetic energy to determine the alpha particle's speed.

$$\text{KE}_\text{f} = \tfrac{1}{2}mv^2$$

$$8.85 \times 10^{-13}\ \text{J} = \tfrac{1}{2}(6.64 \times 10^{-27}\ \text{kg})v_\text{f}^2$$

$$v_\text{f}^2 = 2.67 \times 10^{14}\ \text{m}^2/\text{s}^2$$

$$v_\text{f} = 1.63 \times 10^{7}\ \text{m/s}$$

TEXTBOOK PROBLEM 20. (Section 17-5) A $+35\ \mu\text{C}$ point charge is placed 46 cm from an identical $+35\ \mu\text{C}$ charge. How much work would be required to move a $+0.50\ \mu\text{C}$ test charge from a point midway between them to a point 12 cm closer to either of the charges?

Part a. Step 1.

Determine the initial potential energy

$$\text{PE}_\text{initial} = \frac{kQ_1q}{r_1} + \frac{kQ_2q}{r_2} \quad \text{Note: } Q_1 = Q_2 \text{ and } r_1 = r_2 = 0.23\ \text{m}.$$

$$= \frac{2(9.0 \times 10^{9}\text{N}\cdot\text{m}^2/\text{C}^2)(+35 \times 10^{-6}\ \text{C})(0.50 \times 10^{-6}\ \text{C})}{0.23\ \text{m}}$$

$$\text{PE}_\text{initial} = 1.37\ \text{J}$$

Part a. Step 2.

Determine the final potential energy. Note: $r_1 = 11.0\text{ cm}$ and $r_2 = 35.0\text{ cm}$.

$$\text{PE}_{\text{final}} = \frac{kQ_1q}{r_1} + \frac{kQ_2q}{r_2} \quad \text{but} \quad Q_1 = Q_2$$

$$= \frac{(9.0 \times 10^9\ \text{N}\cdot\text{m}^2/\text{C}^2)\ (+35 \times 10^{-6}\ \text{C})(0.50 \times 10^{-6}\ \text{C})}{0.11\text{m}}$$

$$+ \frac{(9.0 \times 10^9\ \text{N}\cdot\text{m}^2/\text{C}^2)\ (+35 \times 10^{-6}\ \text{C})(0.50 \times 10^{-6}\ \text{C})}{0.35\ \text{m}}$$

$$= 1.43\ \text{J} + 0.45\ \text{J}$$

$$\text{PE}_{\text{final}} = 1.88\ \text{J}$$

Part a. Step 3.

Determine the change in potential energy.

$$\Delta\text{PE} = 1.88\ \text{J} - 1.37\ \text{J}$$

$$\Delta\text{PE} = 0.51\ \text{J}$$

Part a. Step 4.

Determine the work required to move the test charge.

$$W = \Delta\text{PE}$$

$$W = 0.51\ \text{J}$$

TEXTBOOK PROBLEM 44. (Sections 17-7 and 17-8) How strong is the electric field between the plates of a $0.80\ \mu\text{F}$ air-gap capacitor if they are 2.0 mm apart and each has a charge of $62\ \mu\text{C}$?

Part a. Step 1.

Determine the potential difference between the plates.

$$C = \frac{Q}{V}$$

$$0.80\ \mu\text{F} = \frac{62\ \mu\text{C}}{V}$$

$$V = \frac{62\ \mu\text{C}}{0.80\ \mu\text{F}} = 78\ \text{V}$$

Part a. Step 2.

Determine the electric field between the plates.

$$V = Ed$$

$$78\ \text{V} = E(2.0 \times 10^{-3}\ \text{m})$$

$$E = 3.9 \times 10^4\ \text{V/m} = 3.9 \times 10^4\ \text{N/C}$$

TEXTBOOK PROBLEM 50. (Sections 17-7 and 17-8) A 3500-pF air-gap capacitor is connected to a 32-V battery. If a piece of mica is placed between the plates, how much charge will flow from the battery?

Part a. Step 1.

Determine the charge stored before the mica is inserted.

$$Q_i = CV = (3500 \times 10^{-12}\ \text{F})(32\ \text{V})$$

$$Q_i = 1.1 \times 10^{-7}\ \text{C}$$

Part a. Step 2.

Determine the capacitance with the dielectric inserted. Note: $K_{\text{mica}} = 7.0$.

$$C_f = \frac{K\varepsilon_0 A}{d} \quad \text{but} \quad \frac{\varepsilon_0 A}{d} = 3500\ \text{pF}$$

$$C_f = (7.0)(3500\ \text{pF}) = 2.45 \times 10^4\ \text{pF}$$

Part a. Step 3.

Determine the charge stored on the plates after the dielectric is inserted and also the increase (ΔQ) in the charge stored.

Since the battery remains connected to the capacitor, the potential difference between the plates stays at 32 V. However, because of the dielectric, the amount of charge stored on the plates will increase:

$$Q_f = CV = (24{,}500 \times 10^{-12}\ \text{F})(32\ \text{V})$$

$$Q_f = 7.8 \times 10^{-7}\ \text{C}$$

$$\Delta Q = Q_f - Q_i = 7.8 \times 10^{-7}\ \text{C} - 1.1 \times 10^{-7}\ \text{C}$$

$$\Delta Q = 6.7 \times 10^{-7}\ \text{C}$$

TEXTBOOK PROBLEM 53. (Section 17-9) A cardiac defibrillator is used to shock a heart that is beating erratically. A capacitor in this device is charged to 5.0 kV and stores 1200 J of energy. What is its capacitance?

Part a. Step 1.

Use Eq. 17-10 of the textbook to solve for the capacitance.

$$\text{PE} = \tfrac{1}{2}CV^2$$

$$1200\ \text{J} = \tfrac{1}{2}(5000\ \text{V})^2$$

$$C = 9.6 \times 10^{-5}\ \text{F} = 96\ \mu\text{F}$$

TEXTBOOK PROBLEM 55. (Sections 17-1, 17-8, and 17-9) A homemade capacitor is assembled by placing two 9-in. pie pans 4 cm apart and connecting them to the opposite terminals of a 9-V battery. Estimate (*a*) the capacitance, (*b*) the charge on each plate, (*c*) the electric field halfway between the plates, and (*d*) the work done by the battery to charge them. (*e*) Which of the above values change if a dielectric is inserted?

Part a. Step 1.

Determine the diameter and the radius of one of the pie pans in meters.

$$\text{Diameter} = (9\text{ in.})\left(\frac{2.54\text{ cm}}{1\text{ in.}}\right)\left(\frac{1.00\text{ m}}{100\text{ cm}}\right) = 0.23\text{ m}$$

$$\text{Radius} = \tfrac{1}{2}(\text{diameter}) = 0.11\text{ m}$$

Part a. Step 2.

Determine the surface area of one of the pans.

$$A = \pi r^2$$

$$= \pi(0.11\text{ m})^2$$

$$A = 4.1 \times 10^{-2}\text{ m}^2$$

Part a. Step 3.

Use Eq. 17-10 to determine the capacitance.

The dielectric constant of dry air to three significant figures is 1.00:

$$C = \frac{K\varepsilon_0 A}{d}$$

$$= \frac{(1.00)(8.85 \times 10^{-12}\text{ C/N}\cdot\text{m}^2)(0.041\text{ m}^2)}{0.04\text{ m}}$$

$$C = 9.1 \times 10^{-12}\text{ F} \approx 9\text{ pF}$$

Part b. Step 1.

Use Eq. 17-3 to determine the charge stored.

$$Q = CV = (9.1 \times 10^{-12}\text{ F})(9\text{ V})$$

$$Q = 8.2 \times 10^{-11}\text{ C} \approx 8 \times 10^{-11}\text{ C}$$

Part c. Step 1.

Use Eq. 17-4b to determine the magnitude of the electric field.

Assume that the electric field near the center of the plates is uniform:

$$V = Ed \quad \text{and} \quad E = \frac{V}{d} = \frac{9\text{ V}}{0.040\text{ m}}$$

$$E = 225\text{ V/m} \approx 200\text{ N/C}$$

Part d. Step 1.

Use Eq. 17-10 to determine the energy stored.

The work done by the battery equals the energy stored in the capacitor:

$$\text{PE} = \tfrac{1}{2}QV = \tfrac{1}{2}(8.2 \times 10^{-11}\ \text{C})(9\ \text{V})$$

$$\text{PE} = 3.7 \times 10^{-10}\ \text{J} \approx 4 \times 10^{-10}\ \text{J}$$

Part e. Step 1.

Which of the above values change if a dielectric is inserted?

Since the battery remains connected to the capacitor, the amount of charge stored on the plates will increase. The electric field will decrease immediately after the dielectric is inserted but will return to 200 N/C when the charge stops forming on the plates. Because a dielectric is inserted, the capacitance will increase. The final voltage will be 9 V. Since the charge on the plates increases and the capacitance changes, the energy stored will change.

TEXTBOOK PROBLEM 84. (Sections 17-5 and 16-7) A $+3.5\ \mu\text{C}$ charge is 23 cm to the right of a $-7.2\ \mu\text{C}$ charge. At the midpoint between the two charges, (*a*) determine the potential and (*b*) the electric field.

Part a. Step 1.

Draw a diagram locating each charge as well as the point between the charges where $V = ?$.

Let p_1 represent the midpoint between the charges. Let $Q_1 = -7.2\ \mu\text{C}$ and $Q_2 = +3.5\ \mu\text{C}$. The distance from each charge to point $\text{p}_1 = \tfrac{1}{2}(0.23\ \text{m}) = 0.115\ \text{m}$:

Q_1 ⊖ $\qquad$ p_1 • $\qquad$ Q_2 ⊕

|← 0.115 m →|← 0.115 m →|

Part a. Step 2.

Mathematically determine the value of the total potential.

Electric potential is a scalar quantity. At the point in question, the total potential equals the arithmetic sum of the two potentials:

$$V_T = V_1 + V_2$$

$$= \frac{kQ_1}{0.115\text{ m}} + \frac{kQ_2}{0.115\text{ m}}$$

$$= \frac{k(Q_1 + Q_2)}{0.115\text{ m}}$$

$$= \frac{(9.0 \times 10^9\text{ N}\cdot\text{m}^2/\text{C}^2)\,(-7.2 \times 10^{-6}\text{ C} + 3.5 \times 10^{-6}\text{ C})}{0.115\text{ m}}$$

$$V_T = -2.9 \times 10^5\text{ V}$$

Part b. Step 1.

Determine the electric field at the midpoint between the two charges.

Electric field is a vector quantity. The electric field E_1 points toward the negative charge (Q_1), and E_2 points away from the positive charge (Q_2). Therefore, the resultant electric field (E_T) points toward Q_1:

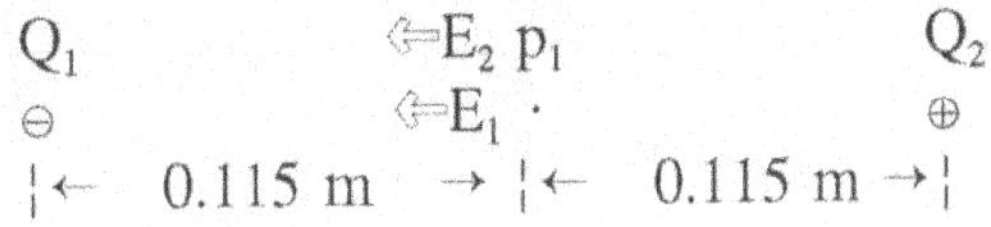

$$E_T = |E_1| + |E_2|$$

$$= \left|\frac{kQ_1}{(0.115\text{ m})^2}\right| + \left|\frac{kQ_2}{(0.115\text{ m})^2}\right|$$

$$= \frac{k(|Q_1| + |Q_2|)}{(0.115\text{ m})^2}$$

$$= \frac{(9.0 \times 10^9\text{ N}\cdot\text{m}^2/\text{C}^2)(|-7.2 \times 10^{-6}\text{ C}| + |+3.5 \times 10^{-6}\text{ C}|)}{(0.115\text{ m})^2}$$

$$E_T = 7.3 \times 10^6\text{ V/m toward } Q_1$$

TEXTBOOK SEARCH AND LEARN PROBLEM 3. (Sections 17-7 to 17-9) In lightning storms, the potential difference between the Earth and the bottom of thunderclouds may be 35,000,000 V. The bottoms of the thunderclouds are typically 1500 m above the Earth and can have an area of $110\ \text{km}^2$. Modeling the Earth–cloud system as a huge capacitor, calculate (*a*) the capacitance of the Earth–cloud system, (*b*) the charge stored in the "capacitor," and (*c*) the energy stored in the "capacitor."

Part a. Step 1.

Convert $110\ \text{km}^2$ to m^2.

$$110\ \text{km}^2 = \left(\frac{1000\ \text{m}}{1\ \text{km}}\right)^2 = 1.1 \times 10^8\ \text{m}^2$$

Part a. Step 2.

Use Eq. 17-8 to determine the capacitance.

$$C = \frac{K\varepsilon_0 A}{d}$$

$$= \frac{(1.00)(8.85 \times 10^{-12}\ \text{C}^2/\text{N}\cdot\text{m}^2)(1.1 \times 10^8\ \text{m}^2)}{1500\ \text{m}}$$

$$C = 6.5 \times 10^{-7}\ \text{F} = 0.65\ \mu\text{F}$$

Part b. Step 1.

Use Eq. 17-7 to determine the charge stored.

$$Q = CV$$

$$Q = (6.5 \times 10^{-7}\ \text{F})(35{,}000{,}000\ \text{V}) = 22.7\ \text{C} \approx 23\ \text{C}$$

Part c. Step 1.

Use Eq. 17-10 to determine the energy stored.

$$\text{PE} = \frac{1}{2}\frac{Q^2}{C}$$

$$= \frac{1}{2}\frac{(23\ \text{C})^2}{6.5 \times 10^{-7}\ \text{F}}$$

$$\text{PE} = 4.0 \times 10^8\ \text{J}$$

18

ELECTRIC CURRENTS

OBJECTIVES

After studying the material of this chapter, the student should be able to:

- explain how a simple battery can produce an electric current.
- define current, ampere, voltage, resistance, resistivity, and temperature coefficient of resistance.
- write the symbols used for electric current, resistance, resistivity, temperature coefficient of resistivity, and power and state the unit associated with each quantity.
- distinguish between conventional current and electron flow and between direct current and alternating current.
- given the length of a wire and its cross-sectional area, resistivity, and temperature coefficient of resistivity, determine the wire's resistance at room temperature and at some higher or lower temperature.
- use Ohm's law to solve simple dc circuit problems.
- use the equations for electric power to determine the power and energy dissipated in a resistor and to calculate the cost of this energy to the consumer.
- distinguish between the rms and peak values for current and voltage and apply these concepts in solving problems involving a simple ac circuit.

KEY TERMS AND PHRASES

electric current is the rate of flow of electric charge. The magnitude of the current is measured in **amperes** (I), where 1 ampere = 1 coulomb/second.

conventional current is the direction of positive charge flow. Both positive and negative ions move through gases and liquids. Only negative charges—electrons—move through solids. For historical reasons, conventional current is used in referring to the direction of electron flow.

electrical resistance refers to the opposition offered by a substance to the flow of electric current. The resistance of a metal conductor is a property that depends on its dimensions, material, and temperature. The unit of resistance is the **ohm** (**Ω**).

Ohm's law states that magnitude of the electric current that flows through a closed circuit is directly proportional to the voltage between the ends of a wire and inversely proportional to the electrical resistance.

electric power is the rate at which work is done to maintain an electric current in a circuit. The SI unit of power is the watt (W), where $1\text{ W} = 1\text{ J/s}$. The kilowatt is a commonly used unit, where $1\text{ kilowatt} = 1000\text{ watts}$.

direct current (dc) refers to a current that flows in one direction only.

alternating current (ac) refers to a current for which the direction of current flow through a circuit changes, usually at a particular frequency (f). The frequency used in the United States is 60 cycles per second, or 60 Hz.

rms (root-mean-square) current (I_{rms}) is the square root of the mean of the square of the current flowing through a circuit.

SUMMARY OF MATHEMATICAL FORMULAS

electric current	$I = \dfrac{\Delta Q}{\Delta t}$	Electric current is measured in amperes (I), where 1 ampere = 1 coulomb/second.
resistance	$R = \rho \dfrac{\ell}{A}$	At a specific temperature, the resistance (R) of a metal wire is related to the length (ℓ), the cross-sectional area (A), and a constant of proportionality called the **resistivity** (ρ).
	$\rho_{\text{T}} = \rho_0 (1 + \alpha\ \Delta T)$ $R_T = R_0 (1 + \alpha\ \Delta T)$	As the temperature of a conductor changes, both the resistivity and resistance of a conductor change. ρ_T and R_T are the values of the resistivity and resistance at temperature T, respectively, while ρ_0 and R_0 are the values of the resistivity and resistance at 20°C. α is the temperature coefficient of resistivity, which remains constant over a certain range of temperature. The value of α depends on the material, and its units are $(\text{C}°)^{-1}$.
Ohm's law	$I = \dfrac{V}{R}$ or $V = IR$	The magnitude of the electric current (I) that flows through a closed circuit is directly proportional to the voltage (V) between the battery terminals and inversely proportional to the circuit resistance (R). The current (I) is measured in amperes, the voltage (V) in volts, and the resistance (R) in ohms.
electric power in a dc circuit	$P = IV$	Electric power equals the product of the current I and the potential difference V.
	$P = I^2 R$ $P = \dfrac{V^2}{R}$	Since $P = IV$ and $I = V/R$, alternate formulas for power can be written in terms of the current and the resistance and in terms of the voltage and the resistance.

alternating current electricity	$V = V_0 \sin 2\pi ft$	The voltage (V) produced by an ac electric generator is sinusoidal.
	$I = I_0 \sin 2\pi ft$	The current produced in a closed circuit connected to an ac generator is sinusoidal. V_0 and I_0 represent the peak or maximum voltage and current, respectively, and t represents the time in seconds.
	$P = I_0^2 R \sin^2 2\pi ft$	The power delivered to a resistance R at any instant (t).
	$\overline{P} = \frac{1}{2} I_0^2 R$ $\overline{P} = I_{\text{rms}}^2 R = I_{\text{rms}} V_{\text{rms}}$	The average power ($\overline{P}$) delivered to the resistance, where I_{rms} is the root-mean-square current: $I_{\text{rms}} = 0.707 I_0$ and $V_{\text{rms}} = 0.707 V_0$.

CONCEPT SUMMARY

The Electric Battery

A **battery** is a source of electric energy. A simple battery contains two dissimilar metals, called **electrodes**, and a solution called the **electrolyte**, in which the electrodes are partially immersed. An example of a simple battery would be one in which zinc and carbon are used as the electrodes, while a dilute acid, such as sulfuric acid (dilute), acts as the electrolyte. The acid dissolves the zinc and causes zinc ions to leave the electrode. Each zinc ion that enters the electrolyte leaves two electrons on the zinc plate. The carbon electrode also dissolves but at a slower rate. The result is a difference in potential between the two electrodes.

Electric Current

An electric **current** exists whenever electric charge flows through a region, for example, a simple lightbulb circuit. The magnitude of the current is measured in **amperes** (I), where 1 ampere = 1 coulomb/second and $I = \dfrac{\Delta Q}{\Delta t}$. The direction of **conventional current** is in the direction in which positive charge flows. In gases and liquids, both positive and negative ions move. Only negative charges, that is, electrons, move through solids; this is referred to as electron current. For historical reasons, conventional current is used in referring to the direction of electric charge flow.

TEXTBOOK QUESTION 3. (Section 18-20) What quantity is measured by a battery rating in ampere-hours $(\text{A} \cdot \text{h})$? Explain.

ANSWER: An ampere-hour is a way of expressing the total charge that the battery can supply at its rated voltage. For example, if the battery is rated at 55 $\text{A} \cdot \text{h}$, then $Q = It = (55 \text{ C/s})(3600 \text{ s})$ and $Q = 198{,}000 \text{ C}$.

Resistivity

When electric charge flows through a circuit it encounters electrical **resistance**. The resistance of a metal conductor is a property that depends on its dimensions, material, and temperature. A resistor is represented by a jagged line, for example:

At a specific temperature, the resistance (R) of a metal wire of length ℓ and cross-sectional area A is given by

$$R = \rho \frac{\ell}{A}$$

where ρ is a constant of proportionality called the **resistivity**. The unit of resistance is the ohm (Ω), and the unit of resistivity is the ohm-meter ($\Omega \cdot \text{m}$). Within a certain range of temperature, the resistivity of a conductor changes according to the following equation:

$$\rho_T = \rho_0 (1 + \alpha\, \Delta T)$$

while the resistance changes according to the equation

$$R_T = R_0 (1 + \alpha\, \Delta T)$$

ρ_0 and R_0 are the values of the resistivity and resistance at 20°C, while ρ_T and R_T are the values of the resistivity and resistance at temperature T, where T is measured in °C. α is the **temperature coefficient of resistivity**. The value of α depends on the material; its units are $(\text{C}°)^{-1}$.

EXAMPLE PROBLEM 1. (Section 18-4) According to the electrical code, the maximum allowable current for 14-gauge copper wire is 15.0 A. The resistivity of copper at 20.0°C is $1.68 \times 10^{-8}\ \Omega \cdot \text{m}$, and the diameter of the wire is 1.628 mm. Calculate the (*a*) electrical resistance of 20.0 m of 14-gauge wire and (*b*) potential difference between the ends of the wire when 15.0 A flows through it.

Part a. Step 1.

Complete a data table.

$$\text{Radius} = (\tfrac{1}{2})\text{diameter} = \tfrac{1}{2}(1.628\text{ mm}) = 0.814\text{ mm}$$

$$A = \pi r^2 = \pi \left(0.814\text{ mm}^2\right)\left(\frac{1.0\text{ m}}{1000\text{ mm}}\right)^2 = 2.08 \times 10^{-6}\text{ m}^2$$

$$\ell = 20.0\text{ m} \qquad \rho = 1.68 \times 10^{-8}\ \Omega \cdot \text{m}$$

Part a. Step 2.

Determine the resistance of 20.0 m of wire.

$$R = \rho \frac{\ell}{A}$$

$$= \frac{(1.68 \times 10^{-8}\ \Omega \cdot \text{m})(20.0\text{ m})}{2.08 \times 10^{-6}\text{ m}^2}$$

$$R = 0.161\ \Omega$$

Part b. Step 1.

Determine the potential difference in volts between the ends of the wire.

$V = IR$

$= (15.0\ \text{A})(0.161\ \Omega)$

$V = 2.42\ \text{V}$

TEXTBOOK QUESTION 16. (Section 18-3) The heating element in a toaster is made of Nichrome wire. Immediately after the toaster is turned on, is the current magnitude (I_{rms}) in the wire increasing, decreasing, or staying constant? Explain.

ANSWER: The resistance of a Nichrome wire is given by Eq. 18-3, $R = \rho\dfrac{\ell}{A}$, where ρ refers to the resistivity of the wire. From Table 18-1, the temperature coefficient of Nichrome is +0.0004/C°. Based on Eq. 18-4, as the temperature increases, the resistivity (ρ) of Nichrome increases and the resistance of the wire increases. Assuming that the voltage is constant, from Ohm's law, $I = V/R$ and the current through the heating element decreases.

EXAMPLE PROBLEM 2. (Sections 18-4 and 18-5) A certain type of copper wire is rated at 160 Ω per 300 m at 0.00°C. If the wire is cylindrical, determine the (*a*) radius of the wire and (*b*) resistance of 75.0 m of wire. The resistivity of copper is $1.68 \times 10^{-8}\ \Omega \cdot \text{m}$.

Part a. Step 1.

Complete a data table.

$A = \pi r^2$ $\quad T_0 = 0.00°\text{C}$

$\ell = 300\ \text{m}$ $\quad R_0 = 160\ \Omega$ $\quad \rho = 1.68 \times 10^{-8}\ \Omega \cdot \text{m}$

Part a. Step 2.

Determine the cross-sectional area of the wire.

$$R = \rho\frac{\ell}{A}$$

$$160\ \Omega = \frac{(1.68 \times 10^{-8}\ \Omega \cdot \text{m})(300\ \text{m})}{A}$$

$$A = 3.15 \times 10^{-8}\ \text{m}^2$$

Part a. Step 3.

Determine the radius of the wire.

$$A = \pi r^2$$

$$3.15 \times 10^{-8}\ \text{m}^2 = \pi r^2$$

$$r = 1.00 \times 10^{-4}\ \text{m}$$

Part b. Step 1.

Determine the resistance of 75.0 m of wire.

$$R = \rho \frac{\ell}{A}$$

$$= \frac{(1.68 \times 10^{-8}\ \Omega \cdot \text{m})(75.0\ \text{m})}{3.15 \times 10^{-8}\ \text{m}^2}$$

$$R = 40.0\ \Omega$$

EXAMPLE PROBLEM 3. (Section 18-4) (*a*) The resistance of the tungsten-filament lightbulb is $80.0\ \Omega$ at $20.0°\text{C}$. Determine the resistance of the same wire if the temperature is raised to $50.0°\text{C}$. (*b*) The resistance of a carbon-filament lightbulb is $80.0\ \Omega$ at $20.0°\text{C}$. Determine the resistance of the wire at $50.0°\text{C}$. Explain why the resistance of each substance changes the way it does with increasing temperature. Note: The temperature coefficient of resistivity is $0.0045/\text{C°}$ for tungsten and $-0.00050/\text{C°}$ for carbon.

Part a. Step 1.

Determine the resistance of tungsten at $50.0°\text{C}$.

$$R = R_0(1 + \alpha\,\Delta T)$$

$$R = (80.0\ \Omega)\left[1 + \left(\frac{0.0045}{\text{C°}}\right)(50.0°\text{C} - 20.0°\text{C})\right]$$

$$R = (80.0\ \Omega)(1 + 0.135) = (80.0\ \Omega)(1.14) = 90.8\ \Omega$$

Part a. Step 2.

Explain why the resistance of tungsten increases with the temperature.

Tungsten is a metal, and its resistance increases with temperature. At room temperature the outer electron is free to move throughout the metal. As the temperature increases, the atoms are vibrating more rapidly. The electric field of the individual atoms has a greater probability of interfering with the electrons as they move through metal; therefore, the resistance increases.

Part b. Step 1.

Determine the resistance of carbon at 50.0°C.

$$R = R_0(1+\alpha\,\Delta T)$$

$$R = (80.0\ \Omega)\left[1+\frac{-0.00050}{\text{C}^\circ}(50.0^\circ\text{C}-20.0^\circ\text{C})\right]$$

$$= 80.0\ \Omega(1+-0.015) = (80.0\ \Omega)(0.985)$$

$$R = 78.8\ \Omega$$

Part b. Step 2.

Explain why the resistance of carbon decreases as the temperature increases.

Carbon is a semiconductor, and the increase in temperature causes more electrons to break free and become part of the electron current. Because of the increased number of free electrons, the resistance of carbon wire to the flow of current decreases.

Ohm's Law

The magnitude of the electric current that flows through a closed circuit is directly proportional to the voltage between the battery terminals and inversely proportional to the circuit resistance. The relationship that connects current, voltage, and resistance is known as **Ohm's law** and is written as follows:

$$I = \frac{V}{R} \qquad \text{or} \qquad V = IR$$

The current (I) is measured in amperes, the voltage (V) in volts, and the resistance (R) in ohms.

The following is a schematic for a simple direct current circuit, for example, a flashlight circuit, where ξ represents the battery voltage and R is the electrical resistance of the filament of the lightbulb. I is the current. The arrow represents the direction of conventional current. Conventional current is directed away from the positive side of the battery.

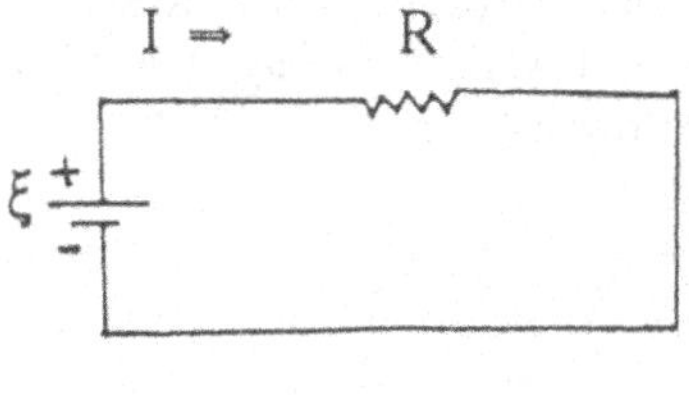

I → R

ξ + −

TEXTBOOK QUESTION 10. (Section 18-6) Explain why lightbulbs almost always burn out just as they are turned on and not after they have been on for some time.

ANSWER: An ordinary incandescent lightbulb has a tungsten filament. Tungsten is a metal, and its length increases or decreases with changes in temperature as the lightbulb is turned on or off. With repeated use, the stress on the filament causes the filament to break. When the filament breaks, it no longer allows electric current to pass through the bulb, and the lightbulb needs to be replaced.

TEXTBOOK QUESTION 13. (Section 18-6) A 15-A fuse blows out repeatedly. Why is it dangerous to replace this fuse with a 25-A fuse?

ANSWER: As shown in Fig. 18-20 of the textbook, a fuse is always placed in series with the circuit element(s). Electric current flowing through the connecting wires causes the wires to heat up. If the circuit is designed to carry 15 A, then the fuse will burn out if the current exceeds 15 A. A 25-A fuse will allow the connecting wires to a carry an excessive current, and if the wires get too hot, then there is the potential for a fire.

Electric Power

Work is required to transfer charge through an electric circuit. The work (W) required depends on the amount of charge transferred through the circuit and the potential difference between the terminals of the battery: $W = QV$. The rate at which work is done to maintain an electric current in a circuit is termed **electric power**. Electric power equals the product of the current I and the potential difference V: $P = IV$. The SI unit of power is the watt (W), where 1 W = 1 J/s. The kilowatt is a commonly used unit; 1 kilowatt = 1000 watts.

Electric energy is dissipated in the circuit in the form of heat. The kilowatt-hour (kWh) is commonly used to represent electric energy production and consumption, where $1\text{ kWh} = 3.6 \times 10^6$ J.

In a circuit of resistance R, the rate at which electrical energy is converted to heat energy is given by $P = IV$. But $V = IR$, so $P = I(IR) = I^2R$.

Since $I = \dfrac{V}{R}$, an alternate formula for power can be written as follows:

$$P = IV = \left(\frac{V}{R}\right)V = \frac{V^2}{R}$$

EXAMPLE PROBLEM 4. (Sections 18-5 and 18-6) A bedroom air conditioner draws 10.0 A of current from a 120-V source. Determine the (*a*) power rating and (*b*) approximate cost of operation of the air conditioner during the month of August (31 days long) if it operates 10 hours each day and electric energy costs 10 cents per kilowatt-hour.

Part a. Step 1.

Determine the power rating.

$$P = IV = (10.0\text{ A})(120\text{ V})$$

$$P = 1200\text{ watts} = 1.20\text{ kW}$$

Part b. Step 1.

Determine the number of kilowatt-hours used.

$$W = Pt = (1.20\text{ kW})\left(\frac{10\text{ hours}}{1\text{ day}}\right)(31\text{ days})$$

$$W = 370\text{ kWh}$$

Part b. Step 2.

Determine the cost of operation.

Cost = (\$0.10/kWh)(370 kWh) ≈ \$37.00

EXAMPLE PROBLEM 5. (Section 18-6) According to recent reports, a typical 8-year-old American child watches approximately 21 hours of television each week. If a particular LCD HDTV draws 1.5 amperes from a 120-volt line, determine the (*a*) power rating of the television in watts, (*b*) number of kilowatt-hours of electrical energy consumed in one year (52 weeks) during the time that the child watches the TV, and (*c*) yearly cost to the parents if electricity costs 8.00 cents per kilowatt-hour.

Part a. Step 1.

Determine the power rating of the TV.

$$P = IV = (1.5\text{ A})(120\text{ V}) = 180\text{ W}$$

Part b. Step 1.

Determine the total number of kWh used in one year.

$$W = Pt$$

$$= (180\text{ W})\left(\frac{1.0\text{ kW}}{1000\text{ W}}\right)\left(\frac{21\text{ h}}{1\text{ week}}\right)\left(\frac{52\text{ weeks}}{1\text{ year}}\right)$$

$$W = 197\text{ kWh}$$

Part c. Step 1.

Calculate the yearly cost.

$$\text{Cost} = (197\text{ kWh})\left(\frac{8.00\text{ cents}}{1.0\text{ kWh}}\right)$$

$$\text{Cost} = 1570\text{ cents, or }\$15.70$$

TEXTBOOK QUESTION 8. (Section 18-5) If the resistance of a small immersion heater (to heat water for tea or soup, Fig. 18-32) was increased, would it speed up or slow down the heating process? Explain.

ANSWER: The rate at which electric energy is converted to heat is given by $P = \frac{V^2}{R}$. Assuming that the voltage does not change during the heating process, then as the resistance increases, the power will decrease. Also, based on Ohm's law, $I = \frac{V}{R}$, as the resistance increases, the current through the heating element decreases and the rate at which electric energy $(P = I^2R)$ is converted to heat decreases, so the heating process would slow down.

Alternating Current

In a **direct current** (dc) circuit, the current flows in one direction only. In an **alternating current** (ac) circuit, the direction of current flow through the circuit changes at a particular frequency (f). The frequency used in the United States is 60 cycles per second, or 60 Hz.

The voltage produced by an ac electric generator is sinusoidal. The current produced in a closed circuit connected to the generator is also sinusoidal. The equations for the voltage and current are as follows:

$$V = V_0 \sin 2\pi ft \quad \text{and} \quad I = I_0 \sin 2\pi ft$$

Here V_0 and I_0 represent the peak or maximum voltage and current, respectively, and t represents the time in seconds.

The power delivered to a resistance R at any instant is

$$P = I^2 R = I_0^2 R \sin^2 2\pi ft$$

The average power delivered to the resistance is

$$\overline{P} = \frac{1}{2} I_0^2 R = I_{\text{rms}}^2 R$$

where I_{rms} is the **root-mean-square** (rms) current. This current is the square root of the mean of the square of the current. It can be shown that $I_{\text{rms}} = 0.707 I_0$ and $V_{\text{rms}} = 0.707 V_0$.

A direct current (dc) whose values of I and V equal the rms values of I and V of an alternating current produces the same amount of power. In ac circuits, it is usually the rms value that is specified. For example, ordinary ac line voltage is 120 volts. The 120 volts is V_{rms}, whereas the peak voltage V_0 is 170 volts.

TEXTBOOK QUESTION 12. Electric power is transferred over large distances at very high voltages. Explain how the high voltage reduces power losses in the transmission lines.

ANSWER: Power is lost in transmission lines due to resistance according to $P = I^2 R$. As a result, it is necessary to have a low current (I) in the line. Since $P = IV$, this can be accomplished by increasing the voltage at the power plant, assuming that the resistance is constant. The higher the voltage, the lower the current must be through the transmission lines, thereby reducing the amount of power lost.

EXAMPLE PROBLEM 6. (Section 18-8) A 12.0-Ω resistor is connected to a 120-volt ac line. Determine the (*a*) average power dissipated in the resistor, (*b*) rms current through the resistor, (*c*) maximum instantaneous current through the resistor, and (*d*) maximum instantaneous power dissipated in the resistor.

Part a. Step 1.

Determine the average power.	$P = IV$, where I and V refer to the rms values of current and voltage, while P is the power dissipated in the resistor:
	Since $I = \dfrac{V}{R}$,

$$\overline{P} = \frac{V^2}{R}$$

$$\overline{P} = \frac{(120\ \text{V})^2}{12.0\ \Omega} = 1200\ \text{W}$$

Part b. Step 1.

Determine I_{rms}.

$$I_{\text{rms}} = \frac{V_{\text{rms}}}{R} = \frac{120\ \text{V}}{12.0\ \Omega}$$

$$I_{\text{rms}} = 10.0\ \text{A}$$

Part c. Step 1.

Determine the maximum instantaneous current that flows through the resistor.

$I_{\text{rms}} = 0.707\ I_0$, where I_0 is the maximum instantaneous (peak) current that flows through the resistor; therefore,

$$I_0 = \frac{I_{\text{rms}}}{0.707}$$

$$I_0 = \frac{10.0\ \text{A}}{0.707} = 14.1\ \text{A}$$

Part d. Step 1.

Determine the maximum instantaneous power dissipated in the resistor.

$$P = I_0^2 R \sin^2 2\pi ft$$

Note: The maximum value of $\sin^2 2\pi ft = 1.0$.

$$P_{\text{max}} = I_0^2 R$$

$$P_{\text{max}} = (14.1\ \text{A})^2 (12.0\ \Omega) = 2400\ \text{W}$$

PROBLEM SOLVING SKILLS

For problems involving resistivity and temperature coefficient of resistivity:

1. Complete a data table listing information both given and implied, for example, the resistivity, length, cross-sectional area, temperature coefficient of resistivity, and change in temperature.
2. Solve for the resistance and the resistance at some higher temperature.

For problems involving electric power, electric energy, and the cost of electric energy:

1. Complete a data table listing information both given and implied, for example, the current, voltage, resistance, and time the device is used.

2. Determine the power dissipated in watts and kilowatts.
3. Determine the number of kWh of energy used and multiply the number of kWh by the cost per kWh.

For problems involving alternating current:

1. Complete a data table listing information both given and implied, for example, the rms current and voltage, peak current and voltage, and resistance.
2. Use the data table to solve for the average and instantaneous power dissipated in the resistor.

SOLUTIONS TO SELECTED TEXTBOOK PROBLEMS

TEXTBOOK PROBLEM 9. (Sections 18-2 and 18-3) A hair dryer draws 13.5 A when plugged into a 120-V line. (*a*) What is its resistance? (*b*) How much charge passes through it in 15 min? (Assume direct current.)

Part a. Step 1.

Determine the resistance of the hair dryer.

$$R = \frac{V}{R} = \frac{120\text{ V}}{13.5\text{ A}}$$

$$R = 8.89\ \Omega$$

Part a. Step 2.

Determine the total charge in coulombs that passes in 15 minutes.

$$\Delta Q = I\ \Delta t$$

$$= (13.5\text{ A})\left[(15\text{ min})\left(\frac{60\text{ s}}{1\text{ min}}\right)\right]$$

$$\Delta Q = 1.22 \times 10^4\text{ C}$$

TEXTBOOK PROBLEM 37. (Section 18-6) How many kWh of energy does a 550-W toaster use in the morning if it is in operation for a total of 5.0 min? At a cost of 9.0 cents/kWh, estimate how much this would add to your monthly electric energy bill if you made toast four mornings per week.

Part a. Step 1.

Determine the number of kWh used each morning.

$$W = Pt$$

$$= (550\text{ W})\left(\frac{1.0\text{ kW}}{1000\text{ W}}\right)\left(\frac{5.0\text{ min}}{\text{day}}\right)\left(\frac{1\text{ h}}{60\text{ min}}\right)$$

$$W = 0.046\text{ kWh}$$

Part a. Step 2.

Determine the number of kWh used each month.

Assume an average month of 30 days:

$$W = (0.046 \text{ kWh})\left(\frac{4 \text{ mornings}}{\text{week}}\right)(4 \text{ weeks})$$

$$W = 0.74 \text{ kWh}$$

Part a. Step 3.

Calculate the monthly cost.

$$\text{Cost} = (0.74 \text{ kWh})\left(\frac{9.0 \text{ cents}}{1.0 \text{ kWh}}\right)$$

$$\text{Cost} = 6.7 \text{ cents}$$

TEXTBOOK PROBLEM 45. (Section 18-6) A small immersion heater can be used in a car to heat a cup of water for coffee or tea. If the heater can heat 120 mL of water from 25°C to 95°C in 8.0 min, (*a*) approximately how much current does it draw from the car's 12-V battery, and (*b*) what is its resistance? Assume the manufacturer's claim of 85% efficiency.

Part a. Step 1.

Determine the amount of heat required to raise the temperature of the system from 25°C to 95°C.

$$Q = m_{H_2O}\, c_{H_2O}\, \Delta T_{H_2O}$$

$$Q = (120 \text{ mL})\left(\frac{1 \text{ g}}{1 \text{ mL}}\right)\left(\frac{1.0 \text{ cal}}{\text{g°C}}\right)(95°\text{C} - 25°\text{C})$$

$$Q = 8.4 \times 10^3 \text{ cal}$$

Part a. Step 2.

Convert the energy from calories to joules.

Heat energy is measured in calories, while electrical energy is measured in joules, where $4.186 \text{ J} = 1.00$ calories:

$$(8.4 \times 10^3 \text{ cal})\left(\frac{4.186 \text{ J}}{\text{cal}}\right) = 3.5 \times 10^4 \text{ J}$$

Part a. Step 3.

Write a formula for the electrical energy used to heat water.

The immersion heater is 85% efficient:

$$(0.85)(\text{electrical energy}) = (0.85)(\text{power})(\text{time}) = (0.85)IVt$$

$$(0.85)(\text{electrical energy}) = (0.85)(I)(12 \text{ V})\left[(8.0 \text{ min})\left(\frac{60 \text{ s}}{1 \text{ min}}\right)\right]$$

Part a. Step 4.

Use the law of conservation of energy to determine the electric current.

(0.85)(electrical energy) = heat energy

$$(0.85)(I)(12\text{ V})\left[(8.0\text{ min})\left(\frac{60\text{ s}}{1\text{ min}}\right)\right] = 3.5\times10^4\text{ J}$$

$I = 7.15$ A

Part a. Step 5.

Determine the resistance of the heating coil.

$$R = \frac{V}{R}$$

$$R = \frac{12\text{ V}}{7.15\text{ A}} = 1.7\ \Omega$$

TEXTBOOK PROBLEM 51. (Section 18-8) An 1800-W arc welder is connected to a 660-V_{rms} ac line. Calculate (*a*) the peak voltage and (*b*) the peak current.

Part a. Step 1.

Determine the peak voltage.

$V_{rms} = 0.707V_0$, where V_0 represents the peak voltage;

$660\text{ V} = 0.707V_0$

$V_0 = 933\text{ V} \approx 930\text{ V}$

Part a. Step 2.

Determine the rms current.

$\overline{P} = I_{rms}V_{rms}$

$1800\text{ W} = I_{rms}(660\text{ V})$

$I_{rms} = 2.7$ A

Part a. Step 3.

Determine the peak current.

$I_{rms} = 0.707I_0$, where I_0 represents the peak current;

$2.7\text{ A} = 0.707I_0$

$I_0 = 3.9$ A

TEXTBOOK PROBLEM 61. (Section 18-5) A person accidentally leaves a car with the lights on. If each of the two headlights uses 40 W and each of the two tail lights 6 W, for a total of 92 W, how long will a fresh 12-V battery last if it is rated at $75\ \text{A}\cdot\text{h}$? Assume the full 12 V appears across each bulb.

Part a. Step 1.

Determine the number of coulombs in $75\ \text{A}\cdot\text{h}$.

$$(75\ \text{A}\cdot\text{h})\left[\left(\frac{1\ \text{C/s}}{1\ \text{A}}\right)\left(\frac{3600\ \text{s}}{1\ \text{h}}\right)\right] = 2.7\times10^5\ \text{C}$$

Part a. Step 2.

Determine the total energy supplied by the battery.

$$W = QV = (2.7\times10^5\ \text{C})(12\ \text{V}) = 3.2\times10^6\ \text{J}$$

Part a. Step 3.

Determine the total power used by the lights.

$$2(40\ \text{W}) + 2(6\ \text{W}) = 92\ \text{W} = 92\ \text{J/s}$$

Part a. Step 4.

Determine the time until all of the battery energy is converted to heat.

$$P = \frac{W}{t} \quad \text{or} \quad t = \frac{W}{P}$$

$$t = \left(\frac{3.2\times10^6\ \text{J}}{92\ \text{J/s}}\right) \approx 3.5\times10^4\ \text{s}$$

$$t \approx 9.7\ \text{hours}$$

TEXTBOOK PROBLEM 77. (Section 18-5) A proposed electric vehicle makes use of storage batteries as its source of energy. It is powered by 24 batteries, each 12 V, $95\ \text{A}\cdot\text{h}$. Assume that the car is driven on level roads at an average speed of 45 km/h, and the average friction force is 440 N. Assume 100% efficiency and neglect energy used for acceleration. No energy is consumed when the vehicle is stopped, since the engine doesn't need to idle. (*a*) Determine the horsepower required. (*b*) After approximately how many kilometers must the batteries be recharged?

Part a. Step 1.

Convert the car's speed to m/s.

$$\left(\frac{45\ \text{km}}{\text{h}}\right)\left(\frac{1000\ \text{m}}{1\ \text{km}}\right)\left(\frac{1\ \text{h}}{3600\ \text{s}}\right) = 12.5\ \text{m/s}$$

Part a. Step 2.

Determine the horsepower produced by the engine to maintain the average.

The average speed is 12.5 m/s and 1 horsepower = 746 watts. The average force produced by the engine is equal to but opposite the frictional force:

$$P = \frac{W}{t} = Fv \cos\theta$$

$$P = (440 \text{ N})(12.5 \text{ m/s}) \cos 0° = 5500 \text{ W}$$

$$\frac{(5500 \text{ W})(1 \text{ hp})}{746 \text{ W}} \approx 7.4 \text{ hp}$$

Part b. Step 1.

Determine the total charge in coulombs provided by the 24 batteries.

$$(24)(95 \text{ A}\cdot\text{h})\left(\frac{1 \text{ C/s}}{1 \text{ A}}\right)\left(\frac{3600 \text{ s}}{1 \text{ h}}\right) = 8.2 \times 10^6 \text{ C}$$

Part b. Step 2.

Determine the total energy supplied by the batteries.

$$W = QV = (8.2 \times 10^6 \text{ C})(12 \text{ V}) = 9.9 \times 10^7 \text{ J}$$

Part b. Step 3.

Determine the time until all of the battery energy is dissipated.

$$P = \frac{W}{t}, \quad \text{or} \quad t = \frac{W}{P}$$

$$t = \left(\frac{9.8 \times 10^7 \text{ J}}{5500 \text{ J/s}}\right) \approx 1.8 \times 10^4 \text{ s}$$

Part b. Step 4.

Determine the maximum distance the car travels.

Distance = (speed)(time)

$$= (12.5 \text{ m/s})(1.8 \times 10^4 \text{ s})$$

Distance $\approx 2.2 \times 10^5$ m $\approx$ 220 km

TEXTBOOK PROBLEM 82. (Section 18-4) A tungsten filament used in a flashlight bulb operates at 0.20 A and 3.0 V. If its resistance at 20°C is 1.5 Ω, what is the temperature of the filament when the flashlight is on?

Part a. Step 1.

Determine the resistance of the filament when the flashlight is on.

$$V = IR$$

$$3.0\text{ V} = (0.20\text{ A})R$$

$$R = 15\ \Omega$$

Part a. Step 2.

Determine the temperature of the hot filament.

From Table 18-1, the temperature coefficient of resistivity of tungsten is $\alpha = 0.0045(\text{C}°)^{-1}$:

$$R = R_0(1 + \alpha\ \Delta T)$$

$$15\ \Omega = (1.5\ \Omega)[1 + (0.0045/\text{C}°)(T_\text{F} - 20.0°\text{C})]$$

$$10 = 1 + (0.0045/\text{C}°)(T_\text{F} - 20.0°\text{C})$$

$$9.0 = (0.0045/\text{C}°)(T_\text{F} - 20.0°\text{C})$$

$$T_\text{F} - 20.0°\text{C} = \frac{9.0}{0.0045/\text{C}°}$$

$$T_\text{F} = 20.0°\text{C} + 2000°\text{C}$$

$$T_\text{F} = 2020°\text{C} \approx 2000°\text{C}$$

DC CIRCUITS

OBJECTIVES

After studying the material of this chapter, the student should be able to:

- define emf and write the symbol for emf.
- determine the equivalent resistance of resistors arranged in series, in parallel, or in a series–parallel combination.
- use Ohm's law and Kirchhoff's rules to determine the current through each resistor and the voltage drop across each resistor in dc circuits with one or more loops.
- distinguish between the emf and the terminal voltage of a battery and calculate the terminal voltage given the emf, internal resistance of the battery, and external resistance in the circuit.
- know the symbols used to represent a source of emf, a resistor, a voltmeter, and an ammeter and how to interpret a simple circuit diagram.
- determine the equivalent capacitance of capacitors arranged in series or in parallel or the equivalent capacitance of a series–parallel combination.
- determine the charge on each capacitor and the voltage drop across each capacitor in a circuit where capacitors are arranged in series, in parallel, or in a series–parallel combination.
- calculate the time constant of an *RC* circuit. Determine the charge on the capacitor and the potential difference across the capacitor at a particular moment of time and the current through the resistor at a particular moment in time.
- describe the basic operation of a galvanometer and calculate the resistance that must be added to convert a galvanometer into an ammeter or a voltmeter.

KEY TERMS AND PHRASES

source of electromotive force (emf) is a device that transforms one form of energy (chemical, mechanical, etc.) to electric energy. Examples of sources of emf are a chemical battery and an electric generator.

electromotive force, or **emf**, is a measure of the potential difference across the source of voltage (that is, a battery or generator).

series circuit is an electric circuit with only a single path for electric current to travel. The current through each circuit element is the same.

parallel circuit is an electric circuit with more than one path or branch for electric current to travel. The current is divided among the branches of the circuit. The voltage drop is the same across each branch.

equivalent resistance (R_{eq}) is the resistance of a single resistor that is equivalent to the total resistance of a network of resistors.

internal resistance refers to the resistance to electric current inside the voltage source. The internal resistance of the source of emf is always considered to be in a series with the external resistance present in the electric circuit.

terminal voltage is the potential difference available to the circuit outside the source of emf. The terminal voltage equals the difference between the emf and the voltage drop across the internal resistance.

Kirchhoff's first rule, or **junction rule**, states that the sum of all currents entering any junction point equals the sum of all currents leaving the junction point. This rule is based on the law of conservation of electric charge.

Kirchhoff's second rule, or **loop rule**, states that the algebraic sum of all the gains and losses of potential around any closed path must equal zero. This law is based on the law of conservation of energy.

***RC* circuit** consists of a resistor and a capacitor connected in series to a dc power source.

time constant (τ) of an *RC* circuit equals the product of the resistance and the capacitance and is measured in seconds.

galvanometer consists of a moving coil placed in a magnetic field. When an electrical current flows through a galvanometer, the interaction of the current in the coil and the magnetic field causes the coil to deflect. The galvanometer is used to detect and measure very low currents.

ammeter measures the amount of electric current passing a particular point in a circuit. An ammeter is placed in series in the circuit and consists of a galvanometer and a resistor of very low value, called the **shunt resistor**, placed in parallel with the galvanometer.

voltmeter measures the potential difference across a circuit element. A voltmeter consists of a galvanometer of internal resistance r and a resistor of high resistance R_{ser} placed in series with the galvanometer.

SUMMARY OF MATHEMATICAL FORMULAS

resistors arranged in series	$I = I_1 = I_2 = I_3$	The current (I) at every point in a series circuit equals the current leaving the battery.
	$V = V_1 + V_2 + V_3$	The potential difference between the terminals of the battery (V) equals the sum of the potential differences across the resistors.
	$R_{eq} = R_1 + R_2 + R_3$	The equivalent electrical resistance (R_{eq}) for a series combination equals the sum of the individual resistances.
resistors arranged in parallel	$I = I_1 + I_2 + I_3$	The battery current (I) equals the sum of the currents in the branches.
	$V = V_1 = V_2 = V_3$	The potential difference across each resistor in the arrangement is the same.

	$\frac{1}{R_{eq}} = \frac{1}{R_1} + \frac{1}{R_2} + \frac{1}{R_3}$	The equivalent resistance (R_{eq}) of a parallel combination is always less than the smallest of the individual resistances.
terminal voltage of a source of emf	$V = \xi - Ir$	The terminal voltage (V) equals the difference between the emf of the source (ξ) and the drop in potential due to internal resistance (Ir).
Kirchhoff's first rule, or junction rule	$\sum I = 0$	The sum of all currents entering any junction point equals the sum of all currents leaving the junction point. This rule is based on the law of conservation of electric charge.
Kirchhoff's second rule, or loop rule	$\sum V = 0$	The algebraic sum of all the gains and losses of potential around any closed path must equal zero. This rule is based on the law of conservation of energy.
capacitors arranged in parallel	$V = V_1 = V_2 = V_3$	The potential difference across each capacitor equals the potential difference (V) of the source of emf.
	$Q = Q_1 + Q_2 + Q_3$	The total charge stored on the capacitor plates (Q) equals the sum of the charges stored on the individual capacitances.
	$C_{eq} = C_1 + C_2 + C_3$	The equivalent capacitance (C_{eq}) equals the sum of the individual capacitances.
capacitors arranged in series	$Q = Q_1 = Q_2 = Q_3$	The charge (Q) that leaves the source of emf equals the charge that forms on each capacitor.
	$V = V_1 + V_2 + V_3$	The potential difference across the source of emf (V) equals the sum of the potential differences across the capacitors.
	$\frac{1}{C_{eq}} = \frac{1}{C_1} + \frac{1}{C_2} + \frac{1}{C_3}$	The equivalent capacitance (C_{eq}) of a parallel combination is always less than the smallest of the individual capacitances.

charging an *RC* circuit	$I = I_0 e^{-t/RC}$	The current *I* in the circuit at time *t* after the switch is closed depends on the initial current (I_0), the resistance (*R*), and the capacitance (*C*).
	$V_C = \xi(1 - e^{-t/RC})$	The potential difference (V_C) across the capacitor as it is being charged
	$Q = Q_0(1 - e^{-t/RC})$	The amount of charge (*Q*) accumulated on the capacitor as it is being charged
discharging an *RC* circuit	$I = I_0 e^{-t/RC}$	The current in an *RC* discharge circuit as a function of time
	$V_C = V_0 e^{-t/RC}$	The voltage across resistance (*R*) in an *RC* discharge circuit as a function of time
	$Q = Q_0 e^{-t/RC}$	The charge (*Q*) on a capacitor in an *RC* discharge circuit as a function of time

CONCEPT SUMMARY

Electromotive Force (emf)

The potential difference between the terminals of a battery in which no internal energy losses occur is referred to as the **electromotive force**, or **emf**, and is measured in volts. The symbol for electromotive force is ξ.

A **source** of emf is a device that transforms one form of energy (chemical, mechanical, etc.) to electric energy. Examples of sources of emf are a chemical battery and an electric generator.

Resistors in Series and Parallel

A simple dc circuit may contain resistors arranged in series, in parallel, or in a series–parallel combination. A simple **series** circuit is shown in the diagram. The current (*I*) at every point in a series circuit equals the current leaving the battery:

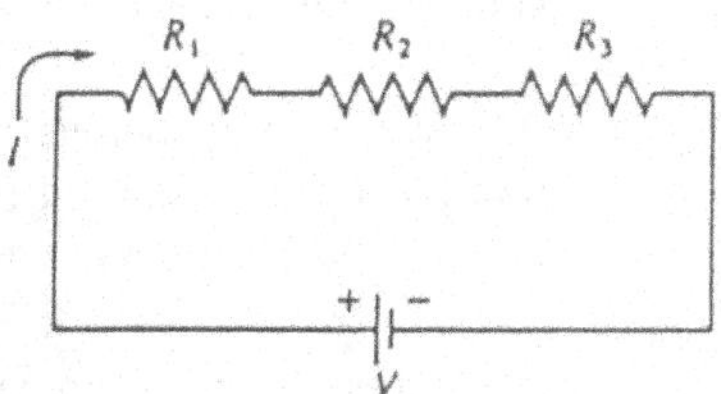

$$I = I_1 = I_2 = I_3$$

Assuming that the connecting wires offer no resistance to current flow, the potential difference between the terminals of the battery (*V*) equals the sum of the potential differences across the resistors:

$$V = V_1 + V_2 + V_3$$

The equivalent electrical resistance (R_{eq}) for this combination is equal to the sum of the individual resistances:

$$R_{eq} = R_1 + R_2 + R_3$$

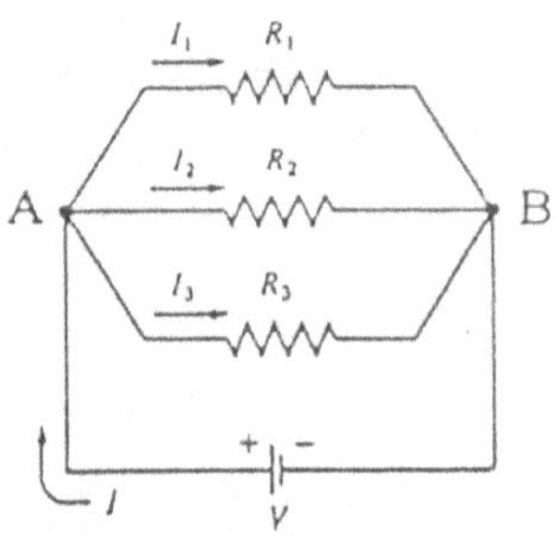

The next diagram represents a simple **parallel** circuit. The current leaving the battery divides at junction point A and recombines at point B. The battery current (I) equals the sum of the currents in the branches. In general,

$$I = I_1 + I_2 + I_3$$

The potential difference across each resistor in the arrangement is the same:

$$V = V_1 = V_2 = V_3$$

If no other resistance is present, then the potential difference across each resistor equals the potential difference across the terminals of the battery.

The equivalent resistance (R) of a parallel combination is always less than the smallest of the individual resistances. The formula for the equivalent resistance is as follows:

$$\frac{1}{R_{eq}} = \frac{1}{R_1} + \frac{1}{R_2} + \frac{1}{R_3}$$

TEXTBOOK QUESTION 1. (Section 19-7) Explain why birds can sit on power lines safely, even though the wires have no insulation around them, whereas leaning a metal ladder up against a power line is extremely dangerous.

ANSWER: When a bird sits on a power line, its feet are at the same voltage, and the electrical potential difference between the bird's two feet is zero. As a result, no current flows through the bird's body, so the bird is safe.

If a ladder is placed against a power line, the potential difference between the top of the ladder (touching the power line) and the bottom of the ladder (touching the ground) is very large. A short circuit occurs, and a very large current flows from the power line through the ladder to the ground. As a result, a person touching the ladder may receive a potentially fatal electrical shock.

TEXTBOOK QUESTION 5. (Section 19-2) Household outlets are often double outlets. Are these connected in series or parallel? How do you know?

ANSWER: If the outlets are connected in series, then both outlets would have to be used at the same time in order for current to flow. But when a lamp is connected to one outlet, it can be turned on without the second outlet being used. Therefore, the outlets must be connected in parallel.

Outlets connected in parallel operate independently of each other. A device connected to one outlet can be used whether or not the other outlet is being used. Since one outlet can be used independently of the other outlet, the outlets are connected in parallel.

TEXTBOOK QUESTION 8. (Section 19-2) You have a single 60-W lightbulb lit in your room. How does the overall resistance of your room's electric circuit change when you turn on an additional 100-W bulb? Explain.

ANSWER: The 60-W bulb and 100-W bulb are on the same circuit and are therefore connected in parallel. As a result, the voltage across each bulb is the same, while the total current flowing through the circuit is increased when both bulbs are turned on. According to Eq. 19-4 of the textbook, the resistance to current flow through the circuit decreases when resistors are connected in parallel.

EXAMPLE PROBLEM 1. (Section 19-2) Resistors $R_1 = 2.0\ \Omega$, $R_2 = 6.0\ \Omega$, and $R_3 = 12.0\ \Omega$ are connected in parallel as shown in the diagram. Determine the (*a*) equivalent resistance of the arrangement, (*b*) total current leaving the battery, (*c*) potential drop across each resistor, and (*d*) current through each resistor.

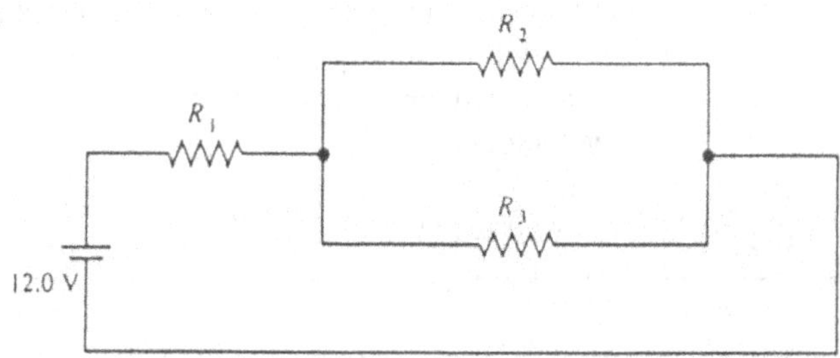

Part a. Step 1.

Determine the equivalent resistance of the two resistors in parallel.

$$\frac{1}{R_{eq}} = \frac{1}{6.0\ \Omega} + \frac{1}{12.0\ \Omega} = \frac{2+1}{12.0\ \Omega}$$

$$R_{eq} = \frac{12.0\ \Omega}{3} = 4.0\ \Omega$$

Part a. Step 2.

Determine the total resistance.

The parallel combination is in series with the 2.0-Ω resistor.

Redraw the circuit showing the 2.0-Ω resistor and R_{eq} and solve for R:

$$R = 2.0\ \Omega + 4.0\ \Omega = 6.0\ \Omega$$

The dashed line drawn around the 4.0-Ω resistor indicates that this resistor is equivalent to the parallel combination.

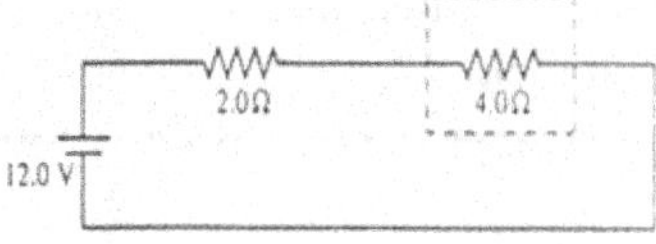

Part b. Step 1.

Determine the current leaving the battery.

Using Ohm's law:

$$I = \frac{V}{R} = \frac{12.0\ V}{6.0\ \Omega} = 2.0\ \text{A}$$

Part c. Step 1.

Determine the drop in potential across each resistor.	The current that leaves the battery passes through the 2.0-Ω resistor. The potential drop across the 2.0-Ω resistor is $V_1 = IR_1 = (2.0\text{ A})(2.0\ \Omega) = 4.0$ volts The potential drop across the parallel combination is the product of the total current that enters the combination at the junction and the equivalent resistance of the parallel combination: $V = (2.0\text{ A})(4.0\ \Omega) = 8.0$ volts

Part d. Step 1.

Determine the current through each resistor.	It was shown in Part b that the current in the 2.0-Ω resistor is 2.0 A. The potential difference across the 6.0-Ω and 12.0-Ω resistors is 8.0 V. Use Ohm's law is to determine each current: $I_2 = \frac{8.0\ V}{6.0\ \Omega} = 1.3\text{ A}$ $I_3 = \frac{8.0\ V}{12.0\ \Omega} = 0.67\text{ A}$

TEXTBOOK QUESTION 18. (Section 19-7) Why is it more dangerous to turn on an electric appliance when you are standing outside in bare feet than when you are inside wearing shoes with thick soles?

ANSWER: The soles of your shoes are made of a material that has high resistance. If the resistance is high, then the current flow through your body will be low. The resistance of bare feet to current flow is significantly lower, and as a result a much larger current can flow through your body when you are barefoot.

EMF and Terminal Voltage

All sources of emf have what is known as **internal resistance** (r) to the flow of electric current. The internal resistance of a fresh battery is usually small but increases with use. Thus, the voltage across the terminals of a battery is less than the emf of the battery. The **terminal voltage** (V) is given by the equation $V = \xi - Ir$, where ξ represents the emf of the source of potential in volts, I is the current leaving the source of emf in amperes, and r is the internal resistance in ohms. The internal resistance of the source of emf is always considered to be in series with the external resistance present in the electric circuit.

Kirchhoff's Rules

Kirchhoff's rules are used in conjunction with Ohm's law in solving problems involving complex circuits:

Kirchhoff's first rule, or the **junction rule**: The sum of all currents entering any junction point equals the sum of all currents leaving the junction point. This rule is based on the law of conservation of electric charge.

Kirchhoff's second rule, or the **loop rule**: The algebraic sum of all the gains and losses of potential around any closed path must equal zero. This law is based on the law of conservation of energy.

Suggestions for Using Kirchhoff's Laws

Diagram (a) is an example of a complex circuit that can be solved using Kirchhoff's rules.

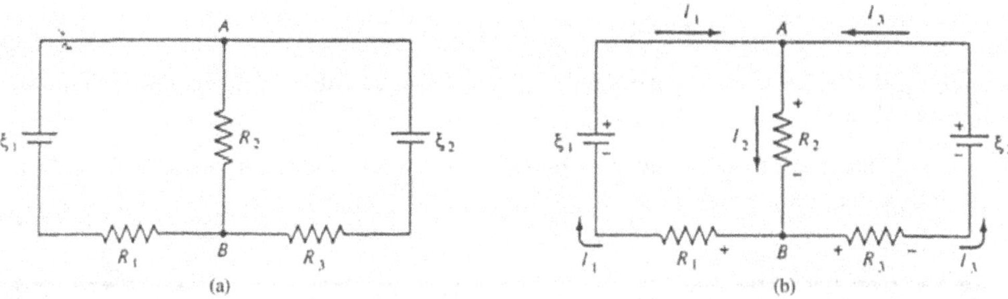

Step 1. As shown in diagram (b), assign a direction to the current in each independent branch of the circuit. Assume that the current flows away from the positive side of the battery and toward the negative side. Because of the directions of I_1 and I_3, in the diagram, current I_2 is directed from point A to point B.

Step 2. Place a plus $(+)$ sign on the side of each resistor where the current enters and a minus $(-)$ sign on the side where the current exits. This indicates that a drop in potential occurs as the current passes through the resistor. Place a $+$ sign next to the long line of the battery symbol and a $-$ sign next to the short line. If you choose the wrong direction for the flow of current in a particular branch, then your final answer for the current in that branch will be negative. The negative answer indicates that the current actually flows in the opposite direction.

Step 2. Select a junction point and apply the junction rule, for example, at point A in the diagram:

The junction rule may be applied at more than one junction point. In general, apply the junction rule to enough junction points so that each branch current appears in at least one equation.

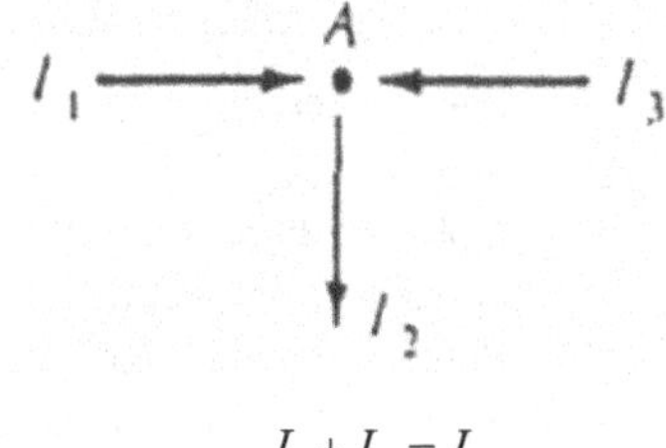

$$I_1 + I_3 = I_2$$

Step 3. Apply Kirchhoff's loop rule by first taking note whether there is a gain or loss of potential at each resistor and source of emf as you trace the closed loop. Remember that the sum of the gains and losses of potential must add to zero. For example, for the left loop of the sample circuit, start at point B and travel clockwise around the loop. Because the direction chosen for the loop is also the direction assigned for the current, there is a gain in potential across the battery $(-\text{ to }+)$, but a loss of potential across each resistor $(+\text{ to }-)$. For the next diagram (a), using the loop rule, we can write the following equation: $-I_1R_1 + \xi_1 - I_2R_2 = 0$.

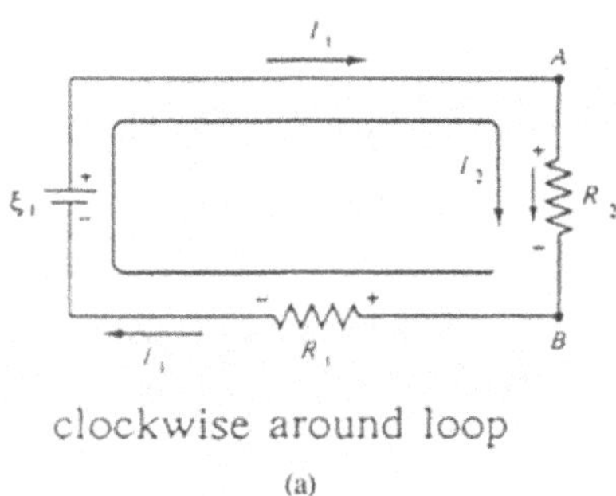

clockwise around loop

(a)

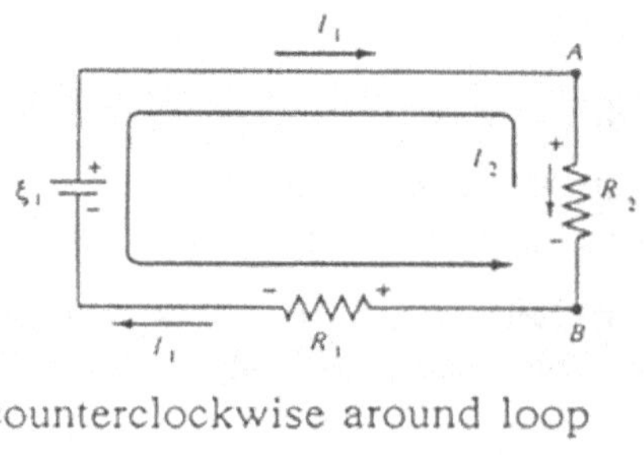

counterclockwise around loop

(b)

The direction taken around the loop is arbitrary. As shown in diagram (b), when tracing a counterclockwise path around the circuit starting at B, there is a gain in potential across each resistor (– to +) and a drop in potential across the battery (+ to –). The loop equation is $+I_2R_2 - \xi_1 + I_1R_1 = 0$.

By multiplying both sides of the above equation by –1 and algebraically rearranging, we can show that the two equations are equivalent.

Be sure to apply the loop rule to enough closed loops so that each branch current appears in at least one loop equation. Use standard algebraic methods to solve for each branch current.

EXAMPLE PROBLEM 2. (Sections 19-3 and 19-4) Determine the current in each branch of the complex circuit shown in the previous diagram, given $\xi_1 = 4.00\text{ V}$, $\xi_2 = 2.00\text{ V}$, $R_1 = 1.00\ \Omega$, $R_2 = 2.00\ \Omega$, and $R_3 = 3.00\ \Omega$.

Part a. Step 1.

Apply the junction rule at point A and write an equation for the current.

At point A: $I_1 + I_3 = I_2$

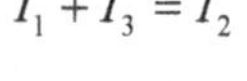

Rearranging gives

$I_1 - I_2 + I_3 = 0$ (Eq. 1)

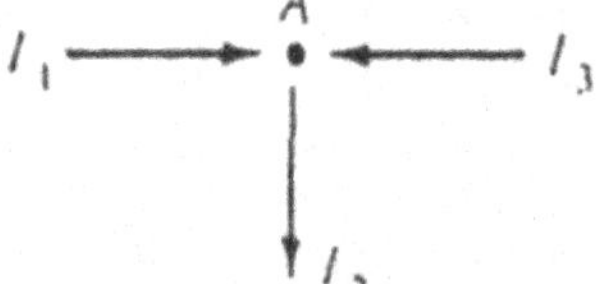

Part a. Step 2.

Note the direction of the current through each resistor. Place + and – signs on the appropriate sides of each resistor. Apply the loop rule. Start at point B and proceed clockwise around the left loop.

Note that at each resistor there will be a potential drop (+ to –), while at the source of emf (ξ_1) there will be a gain:

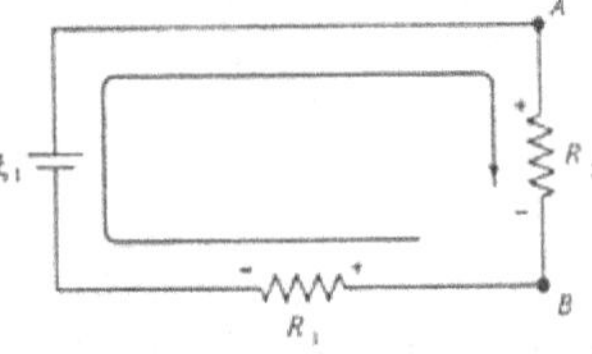

$-I_1R_1 + \xi_1 - I_2R_2 = 0$

Rearranging and simplifying give

$+I_1R_1 + I_2R_2 = \xi_1$

Substitute the values given in the statement of the problem and arrange the resulting equation in the form of Eq. 1:

$(1.00\ \Omega)I_1 + (2.00\ \Omega)I_2 = 4.00\ \text{V}$

$1.00\ I_1 + 2.00\ I_2 = 4.00\ \text{A}$ (Eq. 2)

Part a. Step 3.

Again apply the loop rule. Start at point B and proceed clockwise around the right loop.

By proceeding clockwise, there will be a potential gain $(-\text{ to }+)$ at each resistor, while at the source of emf ξ_2 there will be a potential drop $(+\text{ to }-)$:

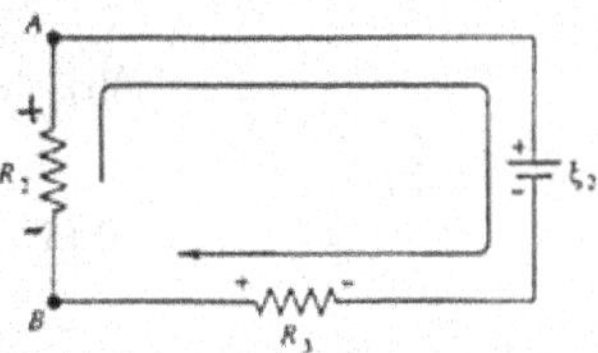

$+I_2R_2 - \xi_2 + I_3R_3 = 0$ and rearranging gives

$+I_2R_2 + I_3R_3 = \xi_2$

Substitute the values given in the statement of the problem and arrange as in Eq. 1.

$(2.00\ \Omega)I_2 + (3.00\ \Omega)I_3 = 2.00\ \text{V}$

$2.00I_2 + 3.00I_3 = 2.00\ \text{A}$ (Eq. 3)

Part a. Step 4.

Solve the three equations, using algebra.

The following method is known as substitution:

$$\begin{aligned} I_1 - I_2 + I_3 &= 0 && \text{(Eq. 1)} \\ 1.00I_1 + 2.00I_2 &= 4.00 && \text{(Eq. 2)} \\ 2.00I_2 + 3.00I_3 &= 2.00 && \text{(Eq. 3)} \end{aligned}$$

Subtract Eq. 2 from Eq. 1 in order to eliminate I_1:

$$\begin{aligned} I_1 - I_2 + I_3 &= 0 && \text{(Eq. 1)} \\ 1.00I_1 + 2.00I_2 &= 4.00 && \text{(Eq. 2)} \\ \hline -3.00I_2 + 1.00I_3 &= -4.00 && \text{(Eq. 4)} \end{aligned}$$

Multiply both sides of Eq. 4 by 3 and then subtract Eq. 3 from Eq. 4:

$$\begin{array}{llll} -9.00I_2 & +3.00I_3 & = & -12.0 \quad (\text{Eq. }4) \\ 2.00I_2 & +3.00I_3 & = & 2.00 \quad (\text{Eq. }3) \\ \hline -11.0I_2 & & = & -14.0 \end{array}$$

$$I_2 = \frac{-14.0}{-11.0} = 1.27\text{ A}$$

Substitute $I_2 = 1.27\text{ A}$ into Eq. 3 and determine I_3:

$$2.00I_2 + 3.00I_3 = 2.00 \quad (\text{Eq.}3)$$

$$(2.0)(1.27) + 3.00I_3 = 2.00 \quad \text{and} \quad I_3 = -0.182\text{ A}$$

The negative sign indicates that the actual direction of the current is opposite from the direction initially assumed. In this instance, the direction of current I_3 is away from point A.

Substitute the values of I_2 and I_3 into Eq. 1 to solve for I_1:

$$I_1 - I_2 + I_3 = 0 \quad (\text{Eq. }1)$$

$$I_1 - 1.27\text{ A} + -0.182\text{ A} = 0$$

$$I_1 = 1.45\text{ A}$$

Capacitors in Series and Parallel

Examples of circuits with **capacitors in series** and **in parallel** are shown.

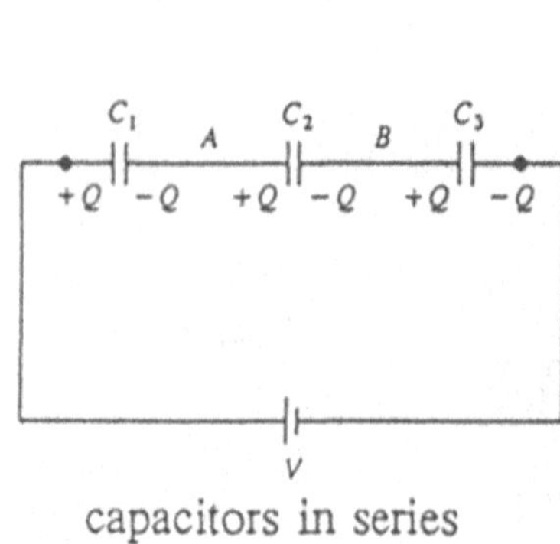

capacitors in series

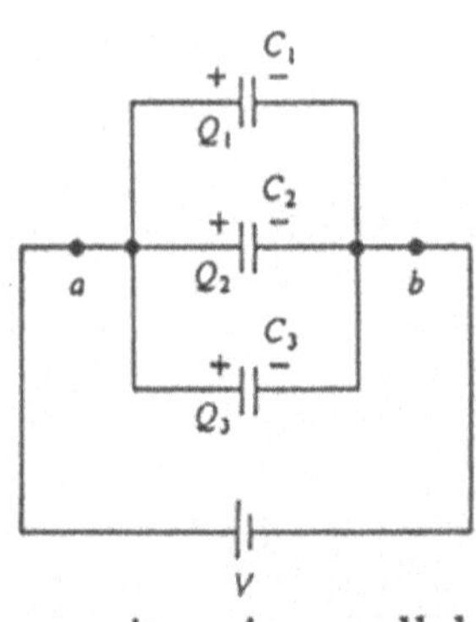

capacitors in parallel

For capacitors in parallel, according to Kirchhoff's loop rule, the potential difference (V) of the source of emf is as follows:

$$V = V_1 = V_2 = V_3$$

The total charge stored on the capacitor plates (Q) equals the amount of charge that left the source of emf:

$$Q = Q_1 + Q_2 + Q_3$$

Since $Q = CV$,

$$CV = C_1V_1 + C_2V_2 + C_3V_3 \quad \text{and} \quad C = C_1 + C_2 + C_3$$

For capacitors in series, the amount of charge (Q) that leaves the source of emf equals the amount of charge that forms on each capacitor:

$$Q = Q_1 = Q_2 = Q_3$$

From Kirchhoff's loop rule, the potential difference across the source of emf (V) equals the sum of the potential differences across the individual capacitors:

$$V = V_1 + V_2 + V_3$$

Since $V = \dfrac{Q}{C}$,

$$\frac{Q}{C} = \frac{Q_1}{C_1} + \frac{Q_2}{C_2} + \frac{Q_3}{C_3} \quad \text{and} \quad \frac{1}{C_{\text{eq}}} = \frac{1}{C_1} + \frac{1}{C_2} + \frac{1}{C_3}$$

EXAMPLE PROBLEM 3. (Section 19-5) Three capacitors are arranged as shown in the diagram. Determine the (*a*) equivalent capacitance of the combination, (*b*) potential difference across each capacitor, and (*c*) charge on each capacitor.

$C_1 = 1.0\ \mu\text{F}$

$C_2 = 2.0\ \mu\text{F}$

$C_3 = 3.0\ \mu\text{F}$

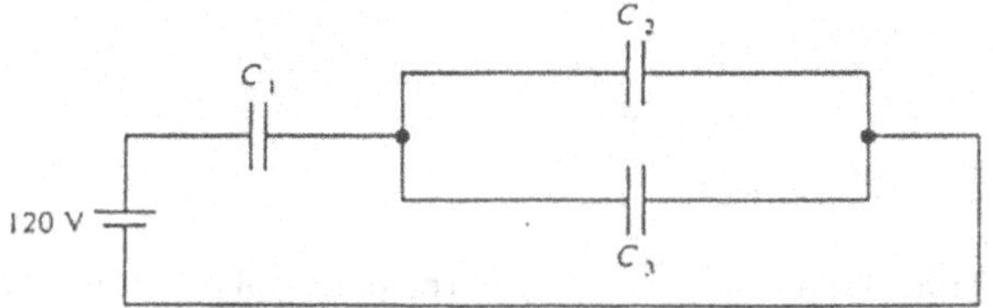

Part a. Step 1.

Determine the equivalent capacitance of the two capacitors arranged in parallel.	$C_{\text{eq}} = C_2 + C_3$ $C_{\text{eq}} = 2.0\ \mu\text{F} + 3.0\ \mu\text{F} = 5.0\ \mu\text{F}$

Part a. Step 2.

Redraw the diagram and solve for the total capacitance.

C_1 and C_{eq} are in series; therefore,

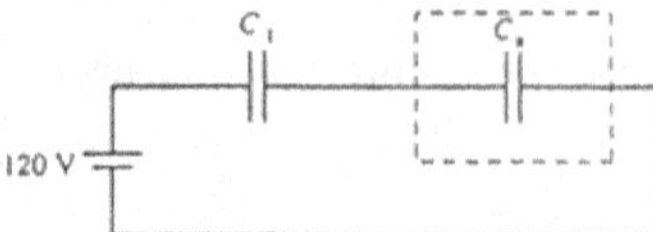

$$\frac{1}{C}=\frac{1}{C_1}+\frac{1}{C_{eq}}$$

$$=\frac{1}{1.0\ \mu\text{F}}+\frac{1}{5.0\ \mu\text{F}}=\frac{6}{5.0\ \mu\text{F}}$$ and

$C = 0.83\ \mu\text{F}$

Part b. Step 1.

Determine the total charge that leaves the source of emf and the charge that forms on C_1.

$Q = C_{total}V = (0.83\ \mu\text{F})(120\ \text{V}) = 100\ \mu\text{C}$

$Q = 1.0 \times 10^{-4}\ \text{C}$

C_1 is in series with the source of emf. The charge that forms on C_1 equals the total charge that left the source of emf, $Q_1 = 100\ \mu\text{C}$.

Part b. Step 2.

Determine the potential difference across C_1.

$$V_1 = \frac{Q_1}{C_1}$$

$$V_1 = \frac{(100\ \mu\text{C})}{(10\ \mu\text{F})} = 100\ \text{V}$$

Part b. Step 3.

Determine the potential difference across the two capacitors arranged in parallel.

The potential difference across the capacitors in parallel can be determined by applying Kirchhoff's loop rule. Starting at the source of emf and following the loop through C_1 and C_2:

$120\ \text{V} - V_1 - V_2 = 0$ but $V_1 = 100$ volts

Therefore, $V_2 = 20\ \text{V}$.

The potential difference is the same across capacitors in parallel; therefore, $V_3 = 20\ \text{V}$.

Part c. Step 1.

Determine the charge stored on each of the capacitors in the parallel combination.

The charge stored on C_1 was determined to be 1000 C. The charge stored on C_2 and C_3 can be determined as follows:

$$Q_2 = C_2V_2 = (2.0\ \mu\text{F})(20\ \text{V}) = 40\ \mu\text{C}$$

$$Q_3 = C_3V_3 = (3.0\ \mu\text{F})(20\ \text{V}) = 60\ \mu\text{C}$$

Note that the total charge stored on the parallel combination equals 100 μC and is equal to the amount stored on C_1.

Circuits Containing Resistor and Capacitor

An ***RC* circuit** consists of a resistor and a capacitor connected in series to a dc power source. When switch 1 (S_1), shown in the accompanying diagram, is closed, the current will begin to flow from the source of emf and charge will begin to accumulate on the capacitor.

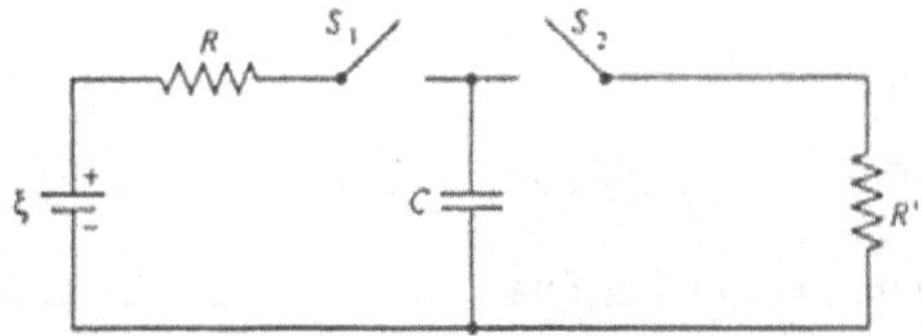

Using Kirchhoff's loop rule, we can show that

$$\xi - IR - \frac{Q}{C} = 0$$

where *IR* refers to the drop in potential across the resistor and $\frac{Q}{C}$ refers to the drop in potential across the capacitor. As charge accumulates on the capacitor, the potential difference increases until it equals ξ, and at that point the current ceases to flow.

By using the methods of calculus, it can be shown that the current *I* through the circuit at time *t* after switch S_1 is closed is given by

$$I = I_0 e^{-t/RC}$$

where I_0 is the initial value of the current just after switch S_1 is closed. *RC* is the product of the resistance and the capacitance and is referred to as the **time constant** (τ) of the circuit; it is measured in seconds. The potential difference across the capacitor as it is being charged is

$$V_C = \xi(1 - e^{-t/RC})$$

The amount of charge (Q) accumulated on the capacitor as it is being charged is

$$Q = Q_0(1 - e^{-t/RC})$$

where Q_0 is the charge on the capacitor when it is fully charged.

The variation of current, voltage, and charge is shown graphically. Note that the time is marked off in terms of the time constant (τ).

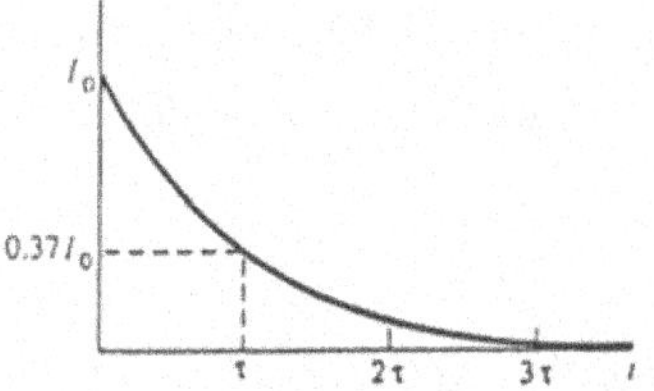

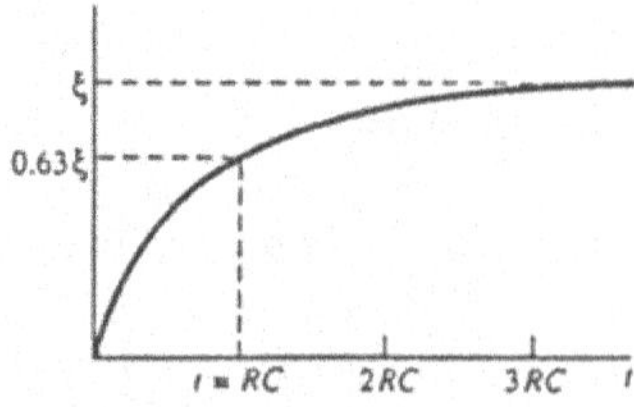

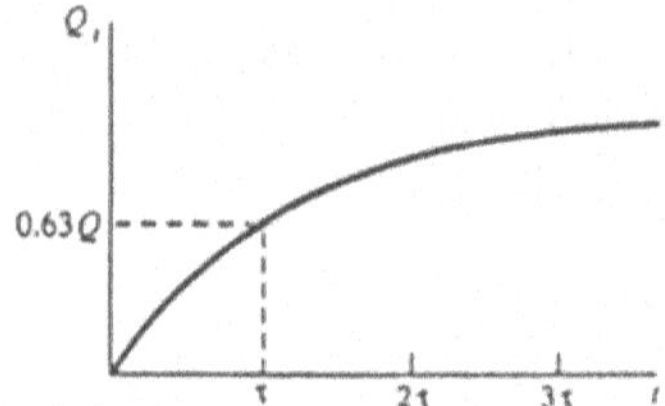

If switch S_1 is open and switch S_2 is closed, then the capacitor will **discharge** through resistor R. The charge on the capacitor (Q) decreases exponentially with time as follows:

$$Q = Q_0 e^{-t/RC}$$

where Q_0 is the charge on the capacitor when it is fully charged.

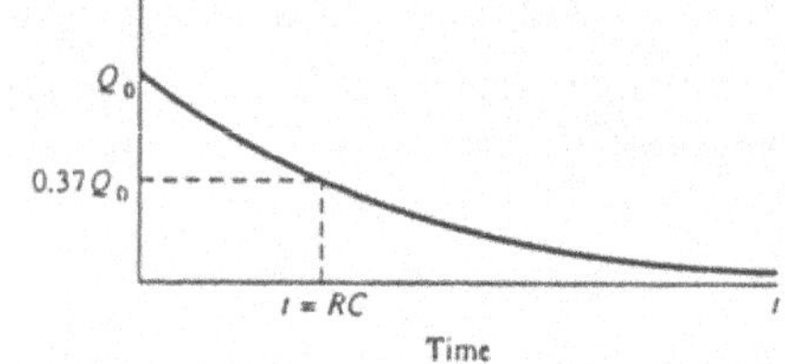

The diagram shows the variation of charge with time as the capacitor discharges. The time is marked off in time constants. The current in a discharge circuit as a function of time is

$$I = I_0 e^{-t/RC}$$

where I_0 represents the initial current as the discharge begins.

The voltage across the resistance (R) in a discharging circuit as a function of time is

$$V_C = V_0 e^{-t/RC}$$

where V_0 represents the initial voltage across the capacitor as the discharge begins.

TEXTBOOK QUESTION 15. (Section 19-6) In an *RC* circuit, current flows from the battery until the capacitor is completely charged. Is the total energy supplied by the battery equal to the total energy stored by the capacitor? If not, where does the extra energy go?

ANSWER: As shown in Fig. 19-20 of the textbook, when switch S is closed, current will flow through the resistor R as the capacitor is being charged. Therefore, some of the energy supplied by the battery is dissipated in the form of heat in the resistor while the rest is stored by the capacitor. As a result, the energy stored by the capacitor is less than the energy supplied by the battery.

Galvanometers, Ammeters, and Voltmeters

A **galvanometer** is a device used to detect and measure very low currents, usually in the range of 1 milliampere or less. Because the coil of a galvanometer consists of wires, the galvanometer presents resistance to the flow of current through it. This internal resistance is usually low.

An **ammeter** is a device designed to measure the amount of electric current passing a particular point in a circuit. The ammeter is placed in series in the circuit and consists of (1) a galvanometer and (2) a resistor of very low value, called the **shunt resistor**, placed in parallel with the galvanometer. Since a potential drop occurs across it, an ammeter is designed to have low overall resistance. Therefore, if it is properly designed, the ammeter will have minimal effect on the external circuit.

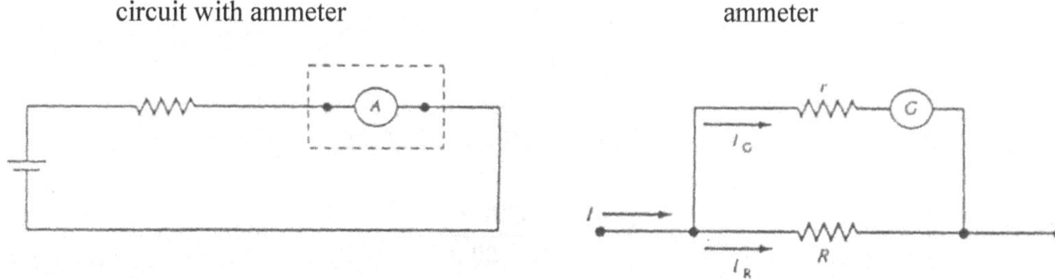

A **voltmeter** measures the potential difference across a circuit element. The voltmeter is placed in parallel with the circuit element and therefore the potential difference across the voltmeter is the same as that across the circuit element. A voltmeter consists of (1) a galvanometer of internal resistance r and (2) a resistor of high resistance R placed in series with the galvanometer.

The resistance of the voltmeter is very large compared with the circuit element. As a result, very little current flows through the voltmeter, and it has a minimal effect on the circuit.

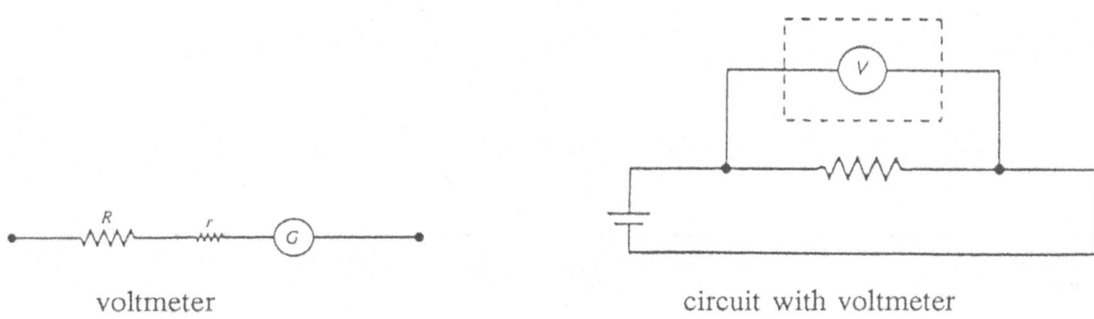

EXAMPLE PROBLEM 4. (Section 19-8) A 50.0-mA current causes a galvanometer of internal resistance of 100 ohms to deflect full scale. Determine the resistance that must be added in order to convert the galvanometer into (*a*) a voltmeter that measures potential differences from 0 to 50.0 V and (*b*) an ammeter that measures currents from 0 to 10.0 A.

Part a. Step 1.

Draw a diagram indicating the location of the resistance that must be added to convert the galvanometer into a voltmeter.

To convert the galvanometer into a voltmeter, a resistance R must be placed in series with the galvanometer.

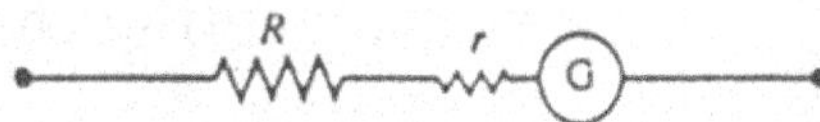

Part a. Step 2.

Determine the magnitude of the resistance that must be added to convert the galvanometer into a voltmeter.

The voltage drop across the circuit must be 50.0 V when a current of 50.0 mA = 0.0500 A passes through the galvanometer:

$$V = I(r + R)$$

$$50.0 \text{ V} = (0.0500 \text{ A})(100\ \Omega + R)$$

$$100\ \Omega + R = \frac{50.0 \text{ V}}{0.0500 \text{ A}}$$

$$100\ \Omega + R = 1000\ \Omega \quad \text{and} \quad R = 900\ \Omega$$

Part b. Step 1.

Draw a diagram showing the resistance that must be added to convert the galvanometer into an ammeter.

In order to convert the galvanometer into an ammeter that reads from 0 to 10.0 A, it is necessary to add a shunt resistor (R_{sh}) in parallel with the galvanometer.

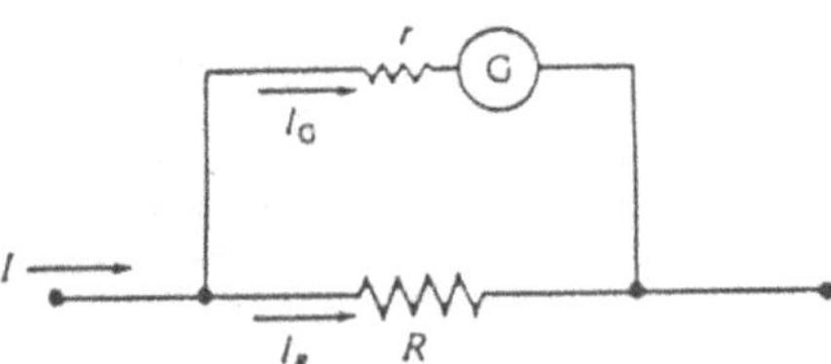

Part b. Step 2.

Determine the magnitude of the current that passes through the shunt.

Only 50.0 mA may pass through the galvanometer when it reads full-scale deflection. Therefore, if the ammeter is to measure 10.0 A when it is fully deflected, then the shunt current is

$$I_{sh} = 10.0 \text{ A} - 0.0500 \text{ A} = 9.95 \text{ A}$$

Part b. Step 3.

Determine the voltage drop across the galvanometer and across the shunt.

The galvanometer and shunt resistor are in parallel. The potential difference across each will be equal; that is, $V_G = V_{sh}$:

$$V_G = I_G r = (0.0500 \text{ A})(100\ \Omega)$$

$$V_G = 5.0 \text{ V}$$

$$\text{but} \quad V_{sh} = V_G = 5.0 \text{ V}$$

Part b. Step 4.

Determine the shunt resistance. $V_{sh} = I_{sh}R_{sh}$

$5.0\text{ V} = (0.995\text{ A})R_{sh}$

$R_{sh} = 5.03\ \Omega$

PROBLEM SOLVING SKILLS

For problems involving equivalent resistance:

1. Determine whether the resistors are arranged in series or in parallel.
2. If necessary, simplify the circuit step by step until it is reduced to a simple series or parallel combination. Use the appropriate formula to determine the equivalent resistance.
3. Use Ohm's law and Kirchhoff's laws to solve for the current through each resistor and the potential difference across each resistor.

For problems involving a complex circuit:

1. Assign a direction to the current in each branch of the circuit. Place a + sign on the side of each resistor where the current enters and a – sign where the current exits.
2. Place a + sign at the positive terminal of each battery and a – sign at the negative terminal.
3. Select a junction point and use Kirchhoff's junction rule to write an equation for the currents entering the point.
4. Apply Kirchhoff's loop rule and write equations based on the gains and losses of potential around selected loops.
5. Use algebraic techniques to solve for the unknown currents. Recall that it is necessary to have as many equations as there are unknowns in order to solve the problem.

For problems involving equivalent capacitance:

1. Determine whether the capacitors are arranged in series or in parallel.
2. If necessary, simplify the circuit step by step until it is reduced to a simple series or parallel combination. Use the appropriate formula to determine the equivalent capacitance.
3. Use $C = \frac{Q}{V}$ and apply Kirchhoff's rules to solve for the charge stored on each capacitor and the potential difference across each capacitor.

For problems involving the construction of an ammeter or voltmeter from a galvanometer and an added resistor:

1. Recall that for an ammeter the shunt resistor is added in parallel with the galvanometer and that for a voltmeter the additional resistor is added in series with the galvanometer.
2. Apply Ohm's law and Kirchhoff's rules and solve for the magnitude of the resistance necessary for the conversion.

SOLUTIONS TO SELECTED TEXTBOOK PROBLEMS

TEXTBOOK PROBLEM 3. (Section 19-1) What is the internal resistance of a 12.0-V car battery whose terminal voltage drops to 8.8 V when the starter motor draws 95 A? What is the resistance of the starter?

Part a. Step 1.

Use Ohm's law to determine the internal resistance of the battery.

As shown in Fig. 19-2 of the textbook, the starter motor is in series with the internal resistance of the battery:

The terminal voltage = V_{ab}

while the emf of the battery = ξ

and the drop in voltage due to the internal resistance = Ir

$$V_{ab} = \xi - Ir$$

$$8.8\text{ V} = 12.0\text{ V} - (95\text{ A})r$$

$$r = \frac{12.0\text{ V} - 8.8\text{ V}}{95\text{ A}} = 0.0.34\ \Omega$$

Part b. Step 1.

Use Ohm's law to determine the total resistance in the circuit.

$$\xi = IR$$

$$12.0\text{ V} = (95\text{ A})R$$

$$R = 0.126\ \Omega$$

Part b. Step 2.

Determine the resistance of the starter.

Let $R_{starter}$ = the resistance of the starter and

r = the internal resistance of the battery; then

$$R = R_{starter} + r$$

$$0.126\ \Omega = R_{starter} + 0.034\ \Omega$$

$$R_{starter} = 0.092\ \Omega$$

TEXTBOOK PROBLEM 6. (Section 19-2) Suppose you have a 580-Ω, a 790-Ω and a 1.20-kΩ resistor. What is (*a*) the maximum, and (*b*) the minimum resistance you can obtain by combining these?

Part a. Step 1.

Determine the maximum resistance.

The maximum resistance is obtained when the resistors are connected in series:

$$R_{eq} = R_1 + R_2 + R_3$$

$$R_{eq} = 580\ \Omega + 790\ \Omega\ + 1200\ \Omega$$

$$R_{eq} = 2570\ \Omega = 2.57\ \text{k}\Omega$$

Part b. Step 1.

Determine the minimum resistance.

The minimum resistance is obtained when the resistors are connected in parallel:

$$\frac{1}{R_{eq}} = \frac{1}{R_1} + \frac{1}{R_2} + \frac{1}{R_3}$$

$$\frac{1}{R_{eq}} = \frac{1}{580\ \Omega} + \frac{1}{790\ \Omega} + \frac{1}{1200\ \Omega}$$

$$= 1.72 \times 10^{-3}\ \Omega^{-1} + 1.27 \times 10^{-3}\ \Omega^{-1} + 8.33 \times 10^{-4}\ \Omega^{-1}$$

$$= 3.82 \times 10^{-3}\ \Omega^{-1}$$

$$R_{eq} = 262\ \Omega$$

TEXTBOOK PROBLEM 11. (Section 19-2) A battery with an emf of 12.0 V shows a terminal voltage of 11.8 V when operating in a circuit with two lightbulbs, each rated at 4.0 W (at 12.0 V), which are connected in parallel. What is the battery's internal resistance?

Part a. Step 1.

Draw a diagram of the circuit described in the Problem.

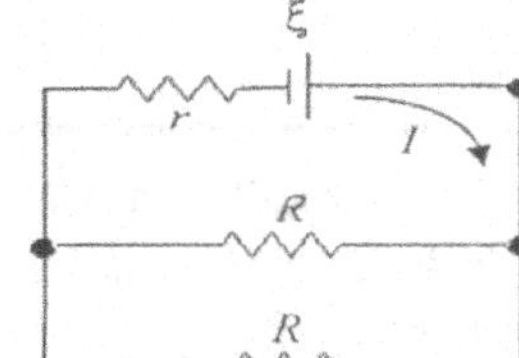

Part a. Step 2.

Determine the resistance of each bulb.

$$P = \frac{V^2}{R}$$

$$R = \frac{V^2}{P} = \frac{(12.0\ \text{V})^2}{4.0\ \text{W}} = 36\ \Omega$$

Part a. Step 3.

Determine the equivalent resistance of the two bulbs.

The two bulbs are connected in parallel:

$$\frac{1}{R_{\text{eq}}} = \frac{1}{R_1} + \frac{1}{R_2} = \frac{1}{36\ \Omega} + \frac{1}{36\ \Omega}$$

$$\frac{1}{R_{\text{eq}}} = \frac{2}{36\ \Omega} \quad \text{and}$$

$$R_{\text{eq}} = 18\ \Omega$$

Part a. Step 4.

Determine the current through the equivalent resistance.

$$V_{\text{ab}} = IR_{\text{eq}}$$

$$11.8\ \text{V} = I(18\ \Omega)$$

$$I = 0.66\ \text{A}$$

Part a. Step 5.

Determine the battery's internal resistance (r).

$$V_{\text{ab}} = \xi - Ir$$

$$11.8\ \text{V} = 12.0\ \text{V} - (0.66\ \text{A})r$$

$$r = \frac{(11.8\ \text{V} - 12.0\ \text{V})}{(-0.66\ \text{A})}$$

$$r = 0.30\ \Omega$$

TEXTBOOK PROBLEM 28(*a*). (Sections 19-3 and 19-4) Determine the magnitudes and directions of the currents in each resistor shown in Fig. 19-56. The batteries have emfs of $\xi_1 = 9.0\ \text{V}$ and $\xi_2 = 12.0\ \text{V}$ and the resistors have values of $R_1 = 25\ \Omega$, $R_2 = 68\ \Omega$, and $R_3 = 35\ \Omega$. Ignore internal resistance of the batteries.

Part a. Step 1.

Choose a direction for the current through each branch.

The choice of direction for the current in each branch is arbitrary. However, a reasonable assumption would be that the current flows away from the batteries, as shown in the diagram, and toward resistor R_2.

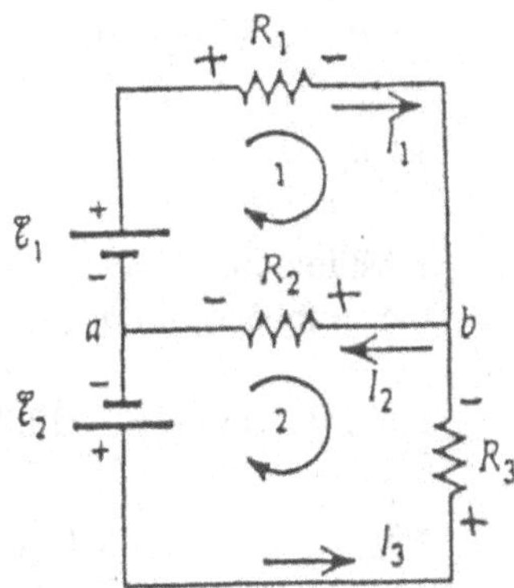

Part a. Step 2.

Apply the junction rule at point b and write an equation for the current.

At point b: $I_1 + I_3 = I_2$ and rearranging gives

$I_1 - I_2 + I_3 = 0$ (Eq. 1)

Part a. Step 3.

Apply the loop rule to loop 1 shown in the diagram. Note: Loop 1 is the top loop in the diagram. Start at point a and proceed clockwise around the loop.

Note the direction of the current through each resistor. A plus sign is placed on the side of the resistor where the current enters. A minus sign is placed on the side where the current exits. A drop in potential occurs when the current flows from the plus $(+)$ side of a resistor to the negative $(-)$ side.

Traveling clockwise starting at point a, there is a gain of potential across the source of emf (ξ_1) and a drop of potential across each resistor:

$$+\xi_1 - I_1R_1 - I_2R_2 = 0$$

Rearranging and simplifying gives

$$+I_1R_1 + I_2R_2 = \xi_1$$

Substitute the values given in the statement of the problem and arrange the resulting equation in the form of Eq. 1:

$$(25\ \Omega)I_1 + (68\ \Omega)I_2 = 9.0\text{ V}$$

$$2.5I_1 + 6.8I_2 = 0.90\text{ A (Eq. 2)}$$

Part a. Step 4.

Again apply the loop rule. Start at point a and proceed clockwise around loop 2. Loop 2 is the bottom loop in the diagram.

By proceeding clockwise, there will be a gain of potential across each resistor and a drop in potential across the source of emf (ξ_2):

$+I_2R_2 + I_3R_3 - \xi_2 = 0$ and rearranging gives

$+I_2R_2 + I_3R_3 = \xi_2$

Substitute the values given in the statement of the problem and arrange as in Eq. 1:

$(68\ \Omega)I_2 + (35\ \Omega)I_3 = 12.0\ \text{V}$

$6.8I_2 + 3.5I_3 = 1.2\ \text{A}$ (Eq. 3)

Part a. Step 5.

Solve the three equations by using algebra.

Use the substitution method to solve for each current:

$$\begin{array}{lllll} I_1 & - I_2 & + I_3 & = 0 & \text{(Eq. 1)} \\ 2.5I_1 & +6.8I_2 & & = 0.90\ \text{A} & \text{(Eq. 2)} \\ & 6.8I_2 & +3.5I_3 & = 1.2\ \text{A} & \text{(Eq. 3)} \end{array}$$

Multiply Eq. 3 by -1.0 and add Eqs. 2 and 3:

$$\begin{array}{lllll} 2.5I_1 & +6.8I_2 & & = 0.90\ \text{A} & \text{(Eq. 2)} \\ & -6.8I_2 & -3.5I_3 & = -1.2\ \text{A} & \text{(Eq. 3)} \\ \hline 2.5I_1 & & -3.5I_3 & = -0.30\ \text{A} & \text{(Eq. 4)} \end{array}$$

Multiply Eq. 1 by 6.8 and add Eqs. 1 and 2:

$$\begin{array}{lllll} 6.8I_1 & -6.8I_2 & +6.8I_3 & = 0 & \text{(Eq. 1)} \\ 2.5I_1 & +6.8I_2 & & = 0.90\ \text{A} & \text{(Eq. 2)} \\ \hline 9.3I_1 & & +6.8I_3 & = 0.90\ \text{A} & \text{(Eq. 5)} \end{array}$$

Multiply Eq. 4 by -3.72 and add Eqs. 4 and 5:

$$\begin{array}{llll} -9.3I_1 & +13.0I_3 & = 1.12\ \text{A} & \text{(Eq. 4)} \\ 9.3I_1 & +6.8I_3 & = 0.90\ \text{A} & \text{(Eq. 5)} \\ \hline & 19.8I_3 & = 2.02\ \text{A} & \end{array}$$

$I_3 = (2.02\ \text{A})/(19.8) = +0.102\ \text{A}$

The positive sign indicates that the direction of the current is in the direction initially assumed. A negative value would indicate that the direction of current I_3 is opposite from the direction arbitrarily chosen.

Substitute $I_3 = 0.102\text{ A}$ into Eq. 3 and determine I_2:

$$6.8I_2 + 3.5(0.102\text{ A}) = 1.2\text{ A}$$
$$6.8I_2 + 0.357 = 1.2\text{ A}$$
$$6.8I_2 = 0.843$$

$$I_2 = 0.124\text{ A}$$

Substitute the values of I_2 and I_3 into Eq. 1 to solve for I_1:

$$I_1 - I_2 + I_3 = 0 \quad \text{(Eq. 1)}$$
$$I_1 - 0.124\text{ A} + 0.102\text{ A} = 0 \quad \text{and} \quad I_1 + 0.022\text{ A}$$

TEXTBOOK PROBLEM 40. (Section 19-5) If 21.0 V is applied across the whole network of Fig. 19-63, calculate (*a*) the voltage across each capacitor and (*b*) the charge on each capacitor.

Part a. Step 1.

Determine the voltage across the $2.00\text{-}\mu\text{F}$ capacitor.

The $2.00\text{-}\mu\text{F}$ capacitor is in parallel with the series combination. Based on Kirchhoff's loop rule, the voltage drop across the $2.00\text{-}\mu\text{F}$ capacitor equals the battery voltage; that is, $V_3 = 21.0\text{ V}$.

Part a. Step 2.

Determine the voltage across each capacitor in series.

For capacitors in series, the amount of charge stored on each capacitor is the same:

$$Q_1 = Q_2, \text{ where } Q = CV$$

$$C_1V_1 = C_2V_2 \quad \text{and} \quad V_1\left(\frac{C_2}{C_1}\right)V_2$$

The sum of the voltage drops across C_1 and C_2 equals the battery voltage; that is,

$$V_1 + V_2 = 21.0\text{ V}$$

$$\left(\frac{C_2}{C_1}\right)V_2 + V_2 = 21.0\text{ V}$$

$$\left(\frac{C_2}{C_1}+1\right)V_2 = 21.0\text{ V} \quad \text{and} \quad \left(\frac{C_2+C_1}{C_1}\right)V_2 = 21.0\text{ V}$$

$$\left(\frac{4.00\ \mu\text{F}+3.00\ \mu\text{F}}{3.00\ \mu\text{F}}\right)V_2 = 21.0\text{ V}$$

$$V_2 = \tfrac{3}{7}(21.0\text{ V}) = 9.0\text{ V}$$

$$V_1 + V_2 = 21.0\text{ V}$$

$$V_1 + 9.0\text{ V} = 21.0\text{ V}$$

$$V_1 = 12.0\text{ V}$$

Part b. Step 1.

Determine the charge stored on C_1, C_2, and C_3.

$$Q_1 = C_1 - V_1 = (3.00\ \mu\text{F})(12.0\text{ V}) = 36.0\ \mu C$$

$$Q_2 = C_2 - V_2 = (4.00\ \mu\text{F})(9.0\text{ V}) = 36.0\ \mu C$$

$$Q_3 = C_3 - V_3 = (2.00\ \mu\text{F})(21.0\text{ V}) = 42.0\ \mu C$$

TEXTBOOK PROBLEM 62. (Section 19-10) A galvanometer has an internal resistance of 32 Ω and deflects full scale for a 55-μA current. Describe how to use this galvanometer to make (*a*) an ammeter to read currents up to 25 A, and (*b*) a voltmeter to give a full-scale deflection of 250 V.

Part a. Step 1.

Draw a diagram locating the resistance which must be added to convert the galvanometer into an ammeter.

In order to convert the galvanometer into an ammeter which reads from 0 to 25 A, it is necessary to add a shunt resistor (R_{sh}) in parallel with the galvanometer.

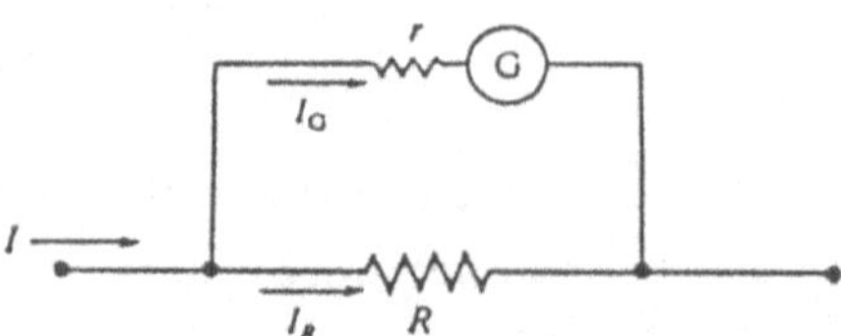

Part a. Step 2.

Determine the magnitude of the current which passes through the shunt.

Only 55 μA—that is, 5.5×10^{-5} A—may pass through the galvanometer when it reads full-scale deflection. Therefore, if the ammeter is to measure 25 A when it is fully deflected, then the shunt current is

$$I_{sh} = 25\text{ A} - 5.0 \times 10^{-5}\text{A} \approx 25\text{ A}$$

Part a. Step 3.

Determine the voltage drop across the galvanometer and across the shunt.

The galvanometer and shunt resistance are in parallel. The potential difference across each will be equal; that is, $V_G = V_{sh}$:

$$V_G = I_G r = (5.5 \times 10^{-5}\text{A})(32\ \Omega)$$

$$V_G = 1.76 \times 10^{-3}\text{ V}$$

but $V_{sh} = V_G = 1.76 \times 10^{-3}$ V

Part a. Step 4.

Determine the shunt resistance.

$$V_{sh} = I_{sh} R_{sh}$$

$$1.76 \times 10^{-3}\text{ V} = (25\text{ A})R_{sh}$$

$$R_{sh} = 7.0 \times 10^{-5}\ \Omega$$

Part b. Step 1.

Draw a diagram locating the resistance which must be added to convert the galvanometer into a voltmeter.

In order to convert the galvanometer into a voltmeter, a resistance R must be placed in series with the galvanometer.

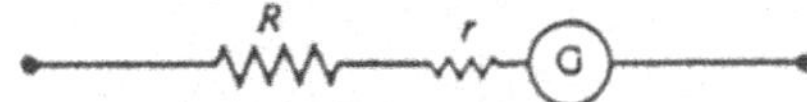

Part b. Step 2.

Determine the magnitude of the resistance that must be added in order to convert the galvanometer into a voltmeter.

The voltage drop across the circuit must be 250 volts when a current of 55 μA $(5.5 \times 10^{-5}\text{ A})$ passes through the galvanometer.

$$V = I(r + R)$$

$$250\text{ V} = (5.5 \times 10^{-5}\text{A})(32\ \Omega + R)$$

$$32\ \Omega + R = \frac{250\ \text{V}}{5.5\times 10^{-5}\ \text{A}}$$

$$32\ \Omega + R = 4.5\times 10^{6}\ \Omega$$

$$R = 4.5\times 10^{6}\ \Omega - 32\ \Omega \approx 4.5\times 10^{6}\ \Omega,\ \text{or } 4.5\ \text{M}\Omega$$

TEXTBOOK PROBLEM 71. (Section 19-6) A heart pacemaker is designed to operate at 72 beats/min using a 6.5-μF capacitor in a simple *RC* circuit. What value of resistance should be used if the pacemaker is to fire (capacitor discharge) when the voltage reaches 75% of maximum and then drops to 0 V (72 times a minute)?

Part a. Step 1.

Determine the time in seconds between beats.

$$t_{\text{beat}} = \left(\frac{1\ \text{min}}{72\ \text{beats}}\right)\left(\frac{60\ \text{s}}{1\ \text{min}}\right) = 0.833\ \text{s}$$

Part a. Step 2.

The resistance can be determined from the formula for the voltage increase across a capacitor which is being charged.

$$V_C = \xi[1 - e^{-t/(RC)}],\quad \text{where } V_C = 0.75$$

$$0.75\xi = \xi[1 - e^{-t/(RC)}]\quad \text{but } \xi \text{ cancels; so}$$

$$0.75 = 1 - e^{-t/(RC)}$$

$$0.75 - 1 = e^{-t/(RC)}$$

$$\ln 0.25 = \ln e^{-t/(RC)}$$

$$-1.386 = -t/(RC)\ \ln e\quad \text{but}\quad \ln e = 1;\ \text{so}$$

$$t = 1.386RC\quad \text{and}\quad C = 6.5\times 10^{-6}\ \text{F}$$

$$0.833\ \text{s} = 1.386R(6.5\times 10^{-6}\ \text{F})$$

$$R = 9.25\times 10^{4}\ \Omega$$

20

MAGNETISM

OBJECTIVES

After studying the material of this chapter, the student should be able to:

- determine the magnitude of the magnetic field produced by both a long, straight, current-carrying wire and a current loop. Use the right-hand rule to determine the direction of the magnetic field produced by the current.
- explain what is meant by ferromagnetism and include in the explanation the concept of domains and the Curie temperature.
- state the conventions adopted to represent the direction of a magnetic field, the current in a current-carrying wire, and the direction of motion of a charged particle moving through a magnetic field.
- apply the right-hand rule to determine the direction of the force on either a charged particle traveling through a magnetic field or a current-carrying wire placed in a magnetic field.
- determine the torque on a current loop arranged in a magnetic field and explain galvanometer movement.
- state Ampere's law and understand when to apply it.
- explain how a mass spectrometer can be used to determine the mass of an ion and how the device can be used to separate isotopes of the same element.

KEY TERMS AND PHRASES

north pole, or "north-seeking pole," of a bar magnet tends to align with the Earth's magnetic field and point toward magnetic north. The **south pole** of a bar magnet tends to point toward magnetic south.

magnetic field surrounds every magnet and is also produced by a charged particles in motion relative to some reference point. The direction of the field at any point is indicated by the north pole of a compass needle placed at that point. The SI unit of magnetic field strength is the tesla (T).

right-hand rule is used to predict the direction of the magnetic field produced by a current-carrying wire. The thumb of the right hand points in the direction of the conventional current in the wire. The fingers encircle the wire in the direction of the magnetic field.

magnetic force acts on a charged particle traveling through a magnetic field. The magnetic force always acts perpendicular to the direction of the magnetic field and the velocity vector. A second right-hand rule is used to predict the direction of the force on the wire.

Ampere's law relates the current in a wire of any shape (not just a long straight wire) and the magnetic field it produces.

velocity selector allows only charged particles which have a particular velocity to pass undeflected. The velocity is the same regardless of the magnitude of the charge or the mass of the particle.

mass spectrometer uses charged particles traveling through magnetic fields to determine the relative mass of the particle.

galvanometer is a device that forms the basis of most meters (ammeters, voltmeters, and ohmmeters) as well as electric motors. Galvanometer movement is the result of interaction of the magnetic field of a permanent magnet, which is directed perpendicular to a current carried by a loop or coil of wire.

electric motor is essentially a galvanometer that is arranged so that a coil of wire, referred to as the **armature**, runs continuously. The electric current through the armature interacts with the external field and the resulting torque causes the armature to rotate. Depending on the design of the motor, the motor can be arranged to run on direct current (dc) or alternating current (ac) electricity.

ferromagnetic materials, such as iron, can be permanently magnetized. Each atom has a net magnetic effect, and the atoms tend to align their magnetic fields in arrangements known as **domains**. Each domain contributes to the overall magnetic field of the piece of iron. In an ordinary piece of iron or other ferromagnetic material, the magnetic fields produced by the individual domains cancel out so that the object is not a magnet. In a magnet, the domains are preferentially aligned in one direction and a net magnetic effect is produced.

Curie temperature refers to the temperature where it is no longer possible to magnetize an object and a permanent magnet loses its magnetic effect.

SUMMARY OF MATHEMATICAL FORMULAS

magnetic field due to a long straight wire	$B = \dfrac{\mu_0 I}{2\pi r}$	The magnitude of the magnetic field strength (B) a perpendicular distance (r) from a long, straight wire carrying a current (I). The SI unit of B is the tesla (T) and $\mu_0 = 4\pi \times 10^{-7}\ \text{T} \cdot \text{m/A}$.
magnetic field due to a loop of wire	$B = \dfrac{\mu_0 NI}{\ell}$	The magnitude of the magnetic field strength (B) at the center of a loop of N turns of wire in a length (ℓ), each carrying a current (I)
force on a current- carrying wire in a magnetic field	$F = I\ell B \sin\theta$	The force (F) on a wire carrying a current depends on the magnitude of the current (I), the length of the wire (ℓ), the magnetic field strength (B), and the angle (θ) between the direction of the current in the wire and the direction of the magnetic field.
force on a charged particle traveling through a magnetic field	$F = qvB\sin\theta$	The force (F) on a charged particle traveling through a magnetic field depends on the magnitude of the charge (q), the velocity of the charge (v), the magnetic field strength (B), and the angle (θ) between the direction of motion of the particle and the direction of the magnetic field.

Ampere's law	$\Sigma B_{\parallel} \Delta \ell = \mu_0 I_{encl}$	Around any closed loop path, the sum of each path segment $\Delta \ell$ times the component of $\vec{\mathbf{B}}$ parallel to the segment equals μ_0 times the current I enclosed by the closed path.
torque on a current loop (galvanometer)	$\tau = NIAB \sin\theta$	When a current flows in a closed loop of wire in an external magnetic field, a torque is produced that depends on the number of turns of wire in the coil (N), the current (I), the area of the coil (A), the magnetic field strength (B), and the angle between the magnetic field and the perpendicular to the face of the coil. The quantity NIA is the magnetic dipole moment (M) of the coil.
force between two parallel conducting wires	$\frac{F}{\ell} = \frac{\mu_0 I_1 I_2}{2\pi d}$	Two parallel, current-carrying conductors produce magnetic fields that result in a force between the conductors. The force per unit length (F/ℓ) depends on the product of the currents (I_1) and (I_2) and is inversely related to the distance (d) between the wires.

CONCEPT SUMMARY

Magnets and Magnetic Fields

Two **magnets** exert a force on one another. If two **north poles** (or two **south poles**) are brought near, then a repulsive force is produced. If a north pole and a south pole are brought near, then a force of attraction results. Thus, "like poles repel, unlike poles attract."

The concept of a field is applied to magnetism as well as to gravity and electricity. A **magnetic field** surrounds every magnet and is also produced by a charged particle in motion relative to some reference point. The presence of the magnetic field about a bar magnet can be seen by placing a piece of paper over the bar magnet and sprinkling the paper with iron filings. The magnetic field produced by certain arrangements of bar magnets are represented in the following diagrams.

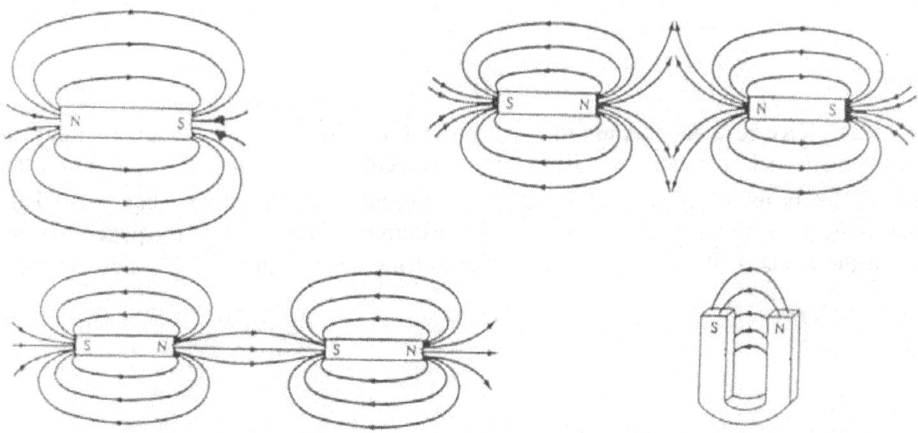

Electric Currents Produce Magnetism

A wire carrying a current (I) produces a magnetic field. The magnitude of the magnetic field strength (B) a perpendicular distance r from a long straight wire is given by

$$B = \frac{\mu_0 I}{2\pi r},$$

where μ_0 is known as the **permeability of free space**.

As shown in diagram (a), the direction of the magnetic field produced by a current-carrying wire can be predicted by using a **right-hand rule**. The thumb of the right hand points in the direction of the conventional current in the wire, and the fingers encircle the wire in the direction of the magnetic field.

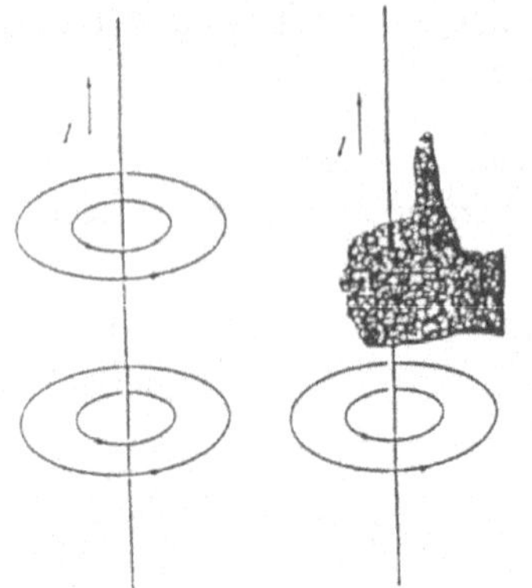

(a) magnetic field produced by a straight wire

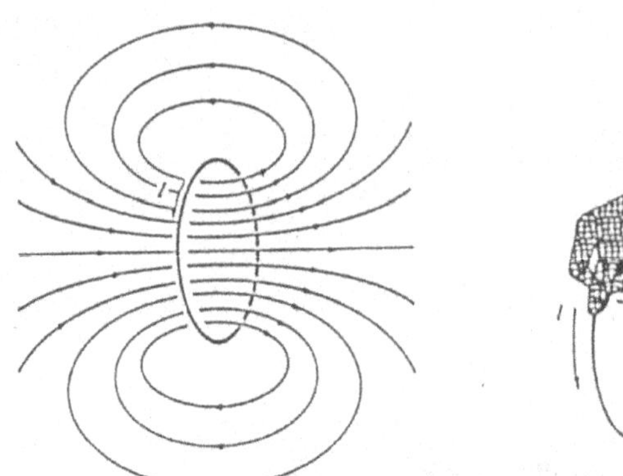

(b) magnetic field produced by a loop of wire

The magnitude of the strength of the magnetic field (B) at the center of a **loop of wire** of N turns of wire in a length (ℓ), each carrying a current I, is

$$B = \frac{\mu_0 NI}{\ell}$$

As diagrams (a) and (b) show, the direction of the magnetic field at the center of the loop can again be predicted by using the right-hand rule. The thumb is placed tangent to a point on the loop and is pointed in the same direction as the current in the loop at that point. The fingers encircle the wire in the same direction as the magnetic field.

Conventions

Certain conventions have been adopted in order to represent the direction of the magnetic field and the current in a wire. A magnetic field directed into the page is represented by a group of ×'s, while a magnetic field directed out of the page is represented by a group of dots. A current-carrying wire that is arranged perpendicular to the page is represented by a circle. If the current is directed into the page, then an × is placed in the center of the circle. If the current is directed out of the page, then a dot is placed in the center of the circle.

× × × × × × × × × × × × × × × *B* field into the page	· · · · · · · · · · · · · · · *B* field out of the page	⊙ wire with current out of the page	⊗ wire with current into the page

---

EXAMPLE PROBLEM 1. (Section 20-5) Determine the magnetic field strength (*a*) at a point 0.25 meters from a long straight wire that carries a 2.0-A current and (*b*) at the center of a loop of wire 0.050 m in radius that contains 300 turns and carries a 4.0-A current. (*c*) Use the right-hand rule and the established conventions to indicate the direction of the magnetic field produced in each of the situations shown.

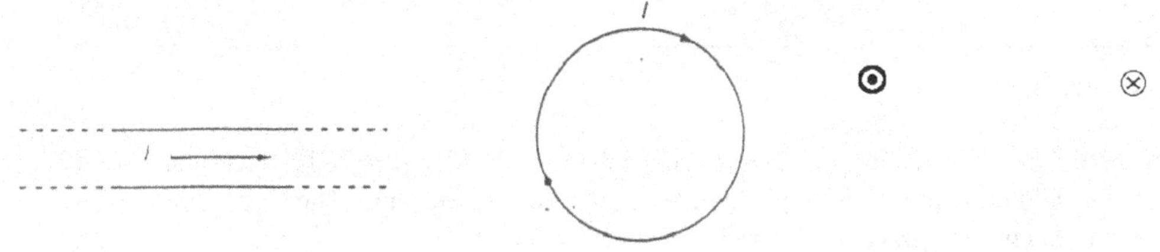

Part a. Step 1.

Determine *B* by using the formula for the magnetic field produced by a current-carrying wire.	$B = \dfrac{\mu_0 I}{2\pi r} = \dfrac{(4\pi \times 10^{-7}\ \mathrm{T \cdot m/A})(2.0\ \mathrm{A})}{2\pi(0.25\ \mathrm{m})}$ $B = 1.6 \times 10^{-6}\mathrm{T}$

Part b. Step 1.

Determine *B* by using the formula for the magnetic field produced by a current-carrying loop of wire.	$B = \dfrac{\mu_0 NI}{2r}$ $= \dfrac{(4\pi \times 10^{-7}\ \mathrm{T \cdot m/A})(300)(4.0\ \mathrm{A})}{2(0.05\ \mathrm{m})}$ $B = 1.5 \times 10^{-2}\mathrm{T}$

Part c. Step 1.

Indicate the direction of the magnetic field about the long straight current-carrying wire.	For the long straight wire, the thumb points toward the right while the fingers of the right hand tend to pass into the page below the wire and out of the page above the wire.

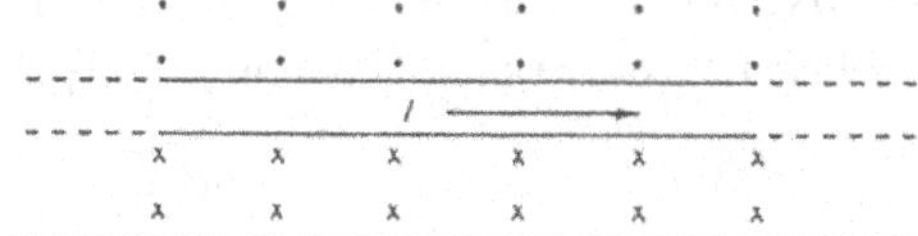

Part c. Step 2.

Indicate the direction of the current-carrying loop of wire.

For the current-carrying loop, the thumb is placed tangent to the loop at arbitrarily selected points. However, it must point in the direction of the current at that location. The fingers of the field produced by the right hand tend to pass into the page inside the loop and out of the page outside the loop.

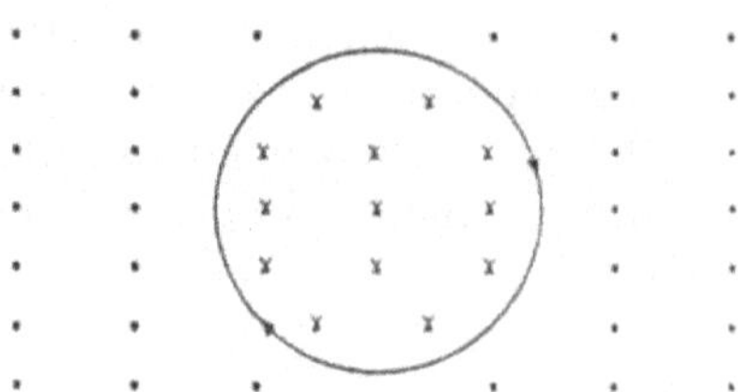

Part c. Step 3.

Indicate the direction of the magnetic field produced by a long straight current-carrying wire if the current is directed into the page.

The thumb of the right hand is directed perpendicular to the plane of the page and into the page. The fingers encircle the wire in a clockwise direction.

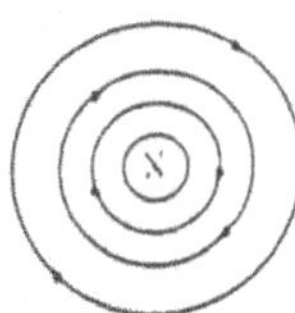

Part c. Step 4.

Indicate the direction of the magnetic field produced by a long straight current-carrying wire if the current is directed out of the page.

The thumb of the right hand is directed perpendicular to the plane of the page and out of the page. The fingers encircle the wire in a counterclockwise direction.

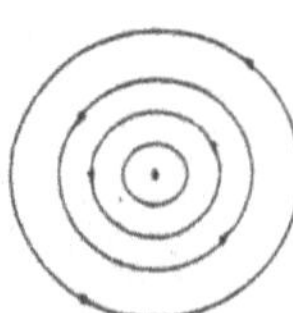

Force on a Current-Carrying Wire in a Magnetic Field

A current (I) in a wire consists of moving electrical charges, and a force (F) may be produced when a current-carrying wire of length ℓ is placed in a magnetic field. The magnitude of the force is given by the equation

$$F = I\ell B \sin\theta,$$

where B is the **magnetic field strength** in teslas (T). Other units for magnetic field strength include newtons per ampere meter $(\text{N/A}\cdot\text{m})$, webers per square meter (Wb/m^2), and gauss (G):

$$1\ \text{T} = 1\ \text{N/A}\cdot\text{m} = 1\ \text{Wb/m}^2 = 10^4\ \text{G}$$

Here θ is the angle between the direction of the current in the wire and the magnetic field. The force on the wire is zero if $\theta = 0°$ $(\sin 0° = 0)$ and is a maximum if $\theta = 90°$ $(\sin 90° = 1.0)$.

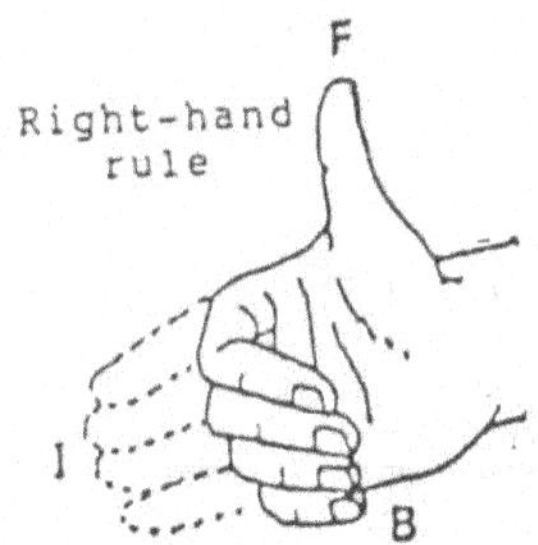

As described in the textbook and shown in the diagram, a second **right-hand rule** is used to predict the direction of the force on the wire: First you orient your right hand so that the outstretched fingers point in the direction of the (conventional) current; from this position, when you bend your fingers they should point in the direction of the magnetic field lines. If they do not, then rotate your hand and arm about the wrist until they do, remembering that your straightened fingers must point in the direction of the current. When your hand is oriented in this way, the extended thumb points in the direction of the force on the wire.

TEXTBOOK QUESTION 4. (Section 20-3) A horseshoe magnet is held vertically with the north pole on the left and south pole on the right. A wire passing between the poles, equidistant from them, carries a current directly away from you. In what direction is the force on the wire? Explain.

ANSWER: See Fig. 20-11a of the textbook. The current in the wire is directed away from you into the page. The north pole (N) of the horseshoe magnet is to the left of the wire, while the south pole (S) is to the right of the wire. Using the right-hand rule, your fingers initially point into the page in the direction of the conventional current. From this position, you bend your fingers until they point toward the right side of the page from the north pole toward the south pole of the horseshoe magnet. The direction your thumb points is the direction of the force on the wire. Your thumb points downward toward the bottom of the page; therefore, the force on the wire is directed toward the bottom of the page.

EXAMPLE PROBLEM 2. (Section 20-3) In the diagram shown below, a wire carrying 3.0 A of current is placed in a uniform magnetic field of strength 4.0 T. Determine the direction and magnitude of the force on a 0.020-meter section of the wire.

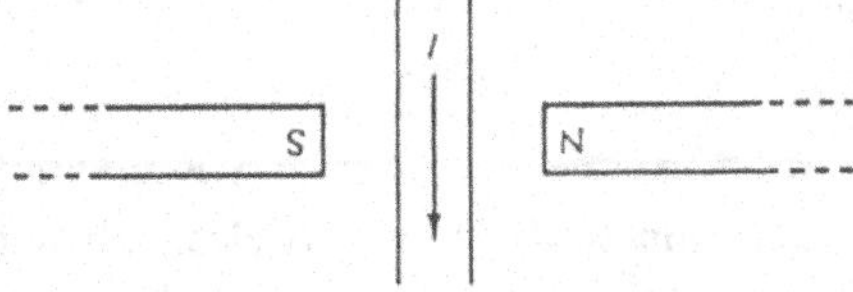

Part a. Step 1.

Use the right-hand rule to determine the direction of the force.	Based on the diagram, the current is directed toward the bottom of the page. Using the right-hand rule, your fingers point in the direction of the current. Your hand is then arranged so that when you bend your fingers, they bend toward the direction of B, which is toward the left side of the page. Your thumb is directed into the page, which means that the force is directed into the page.

Part a. Step 2.

Determine the magnitude of the force.	The current is at right angles to the direction of the magnetic field, $\theta = 90°$. Also, $1.0\ \text{T} = 1.0\ \text{N/A}\cdot\text{m}$. Then

$$F = I\ell B\sin\theta$$

$$= (3.0\ \text{A})(0.020\ \text{m})(4.0\ \text{T})(\sin 90°) \quad \text{but } \sin 90° = 1.0$$

$$F = 0.24\ \text{N}$$

Force on a Charged Particle Moving in a Magnetic Field

An electrically charged particle (q) moving through a magnetic field (B) at speed v may be acted upon by a force (F). The magnitude of the force (F) on the particle is given by

$$F = qvB\sin\theta,$$

where θ is the angle between the direction of motion of the particle and the direction of the magnetic field. If $\theta = 0°$, then the particle is traveling parallel to the field and no force exists on the particle ($\sin 0° = 0$). If $\theta = 90°$, then $\sin 90° = 1$, the particle is traveling perpendicular to the magnetic field, and the force is a maximum.

The direction of the force can be predicted by again using the right-hand rule: the outstretched fingers of your right hand point along the direction of motion of the positively charged particle (v), and when you bend your fingers they must point along the direction of B; then your thumb points in the direction of the force. If the particle is negatively charged, then the force is directed opposite from the direction of the thumb.

TEXTBOOK QUESTION 18. (Section 20-4) If a moving charge is deflected sideways in some region of space, can we conclude, for certain, that $\vec{\mathbf{B}} \neq 0$ in that region? Explain.

ANSWER: A magnetic field is not required for a moving charge to be deflected. As shown in Figs. 17-29 and 17-50 of the textbook, in the absence of a magnetic field, an electric field directed perpendicular to the direction of motion of the moving charge will cause a sidewise deflection.

TEXTBOOK QUESTION 10. (Section 20-4) Three particles, a, b, and c, enter a magnetic field as shown in Fig. 20-48. What can you say about the charge on each particle? Explain.

ANSWER: All three particles are shown moving into the magnetic field. Based on the formula $F = qvB\sin\theta$, it can be said that particle b has no excess electric charge (that is, $q = 0$), because it is not deflected. Particles a and c are deflected; therefore, they must carry an excess charge.

Using the right-hand rule: let the outstretched fingers of your right hand point toward the right in the direction of motion of particles a and b. Your bent fingers point in the direction of b, and your thumb points in the direction of the force. By this rule, it can be said that particle a is charged positively while particle c is charged negatively.

EXAMPLE PROBLEM 3. (Section 20-4) An alpha (α) particle has a mass 6.68×10^{-27} kg and charge $+3.2\times10^{-19}$ C. The particle is traveling at 3.00×10^{7} m/s and enters a magnetic field of magnitude 0.400 T. As shown in the diagram, the magnetic field is directed at right angles to the direction of motion of the particle. Determine the (*a*) radius of the circle in which the particle travels and (*b*) period of its motion. (*c*) Using the right-hand rule, determine the direction of motion in which the α particle travels—that is, either clockwise or counterclockwise as viewed from above.

$\alpha\rightarrow$

Part a. Step 1.

Determine the radius of the circle in which the particle travels.

The alpha particle is deflected by a magnetic force and travels in a circular path. The magnetic force provides the centripetal acceleration (a_R), thus

$$\text{net } F = ma_R \quad \text{but} \quad a_R = \frac{v^2}{r}$$

$$F_{\text{magnetic}} = m\frac{v^2}{r}$$

$$qVB\sin\theta = m\frac{v^2}{r} \quad \text{but} \quad \theta = 90° \text{ and } \sin 90° = 1$$

Solving for *r*:

$$r = \frac{mv}{qB} = \frac{(6.68\times10^{-27}\ \text{kg})(3.00\times10^{7}\ \text{m/s})}{(3.2\times10^{-19}\ \text{C})(0.400\ \text{N/A}\cdot\text{m})}$$

$$r = 1.57\ \text{m}$$

Part b. Step 1.

Determine the period of the motion.

Hint: The period (T) of the motion refers to the time required for the alpha particle to complete one revolution.

The distance traveled in one revolution equals the circumference ($2\pi r$) of the circle. The period can be determined by dividing the circumference of the circle by the velocity (v):

$$T = \frac{2\pi r}{v} = \frac{2\pi(1.57\ \text{m})}{3.00\times10^{7}\ \text{m/s}}$$

$$T = 3.29\times10^{-7}\ \text{s}$$

Part c. Step 1.

Use the right-hand rule to determine the direction of motion of the particle.

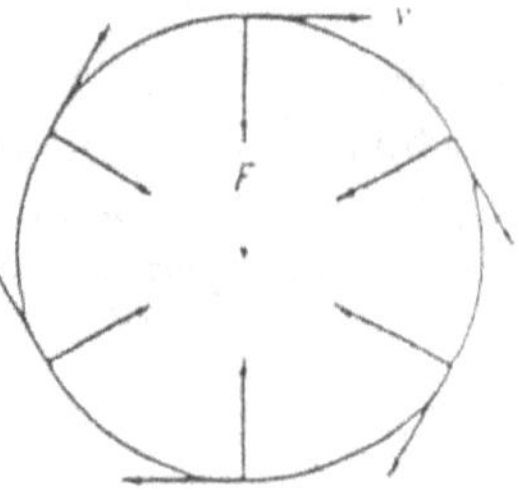

The fingers point toward the right side of the page. Your hand is then arranged so that when you bend your fingers, they point out of the page in the direction of B. Your thumb points toward the bottom on the page. At the point where it enters the magnetic field, the particle is deflected downward. Applying the right-hand rule at several more points indicates that the particle travels in a clockwise circle as viewed from above.

EXAMPLE PROBLEM 4. (Section 20-4) An electron is accelerated from rest through a potential difference of 200 volts into a magnetic field of 3.0×10^{-4} T which is directed perpendicular to the electron's path. Determine the (*a*) velocity of the electron as it enters the field, (*b*) radius of the circle in which it travels, and (*c*) period of its motion. (*d*) Based on the diagram shown below, determine the direction of motion of the electron in the magnetic field—that is, either clockwise or counterclockwise as viewed from above. Assume that the initial velocity of the electron is zero. Note: $m_e = 9.1\times10^{-31}$ kg and $q_e = 1.6\times10^{-19}$ C.

Part a. Step 1.

Determine the electron's velocity as it enters the magnetic field.

The electron is accelerated through a potential difference and gains kinetic energy. The work done on the particle is equal to the product of the charge on the particle and the potential difference:

$$W = qV = \tfrac{1}{2}mv^2 - \tfrac{1}{2}mv_0^2 \quad \text{but } v_0 = 0$$

$$qV = \tfrac{1}{2}mv^2$$

Rearranging gives

$$v = \sqrt{\frac{2qV}{m}}$$

$$v = \sqrt{\frac{2(1.6\times10^{-19}\ \text{C})(200\ \text{V})}{9.1\times10^{-3}\ \text{kg}}}$$

$$v = 8.39\times10^{6}\ \text{m/s}$$

Part b. Step 1.

Determine the radius of the circle in which the electron travels.

The magnetic force provides the centripetal acceleration and causes the electrons to follow a circular path.

$$F_{\text{magnetic}} = ma_{\text{centripetal}}$$

$$qVB\sin\theta = \frac{mv^2}{r} \quad \text{but} \quad \theta = 90° \quad \text{and} \quad \sin 90° = 1$$

Solving for r gives

$$r = \frac{mv}{qB} = \frac{(9.1\times10^{-31}\ \text{kg})(8.39\times10^{6}\ \text{m/s})}{(1.6\times10^{-19}\ \text{C})(3.0\times10^{-4}\ \text{T})}$$

$$r = 0.159\ \text{m}$$

Part c. Step 1.

Determine the period of the motion.

Hint: The period (T) of the motion refers to the time required for the electron to complete one revolution.

The distance traveled in one revolution equals the circumference $(2\pi r)$ of the circle. The period can be determined by dividing the circumference of the circle by the velocity (v):

$$T = \frac{2\pi r}{v} = \frac{2\pi(0.159\ \text{m})}{8.39\times10^{6}\ \text{m/s}}$$

$$T = 1.19\times10^{-7}\,\text{s}$$

Part d. Step 1.

Determine the direction of motion of the particle. Note: Remember that an electron carries a negative charge.

Based on the diagram, your fingers point toward the top of the page. Your hand is then arranged so that when you bend your fingers, they point into the page in the direction of B. Your thumb points toward the left. However, since the particle carries a negative charge, the force is in the opposite direction. Therefore, the force particle is toward the right. Arbitrarily selecting a few more points along the path shows that the electron will travel in a counterclockwise circle as viewed from above.

Ampere's Law

Equation 20–6 of the textbook, $B = \mu_0 I / 2\pi r$, applies to long straight wires only. **Ampere's law** is useful because it relates the current in a wire of any shape, not just a long straight wire, and the magnetic field it produces. Around any closed loop path, the sum of each path segment $\Delta\ell$ times the component of $\vec{\mathbf{B}}$ parallel to the segment equals μ_0 times the current I enclosed by the closed path:

$$\Sigma B_{\parallel}\, \Delta\ell = \mu_0 I_{\text{encl}}$$

The sum must be made over a closed path. Here I_{encl} is the total net current that path encloses.

The Mass Spectrometer

A **velocity selector** allows only charged particles that have a particular velocity to pass undeflected regardless of the magnitude of their charge or their mass. As a result, a velocity selector can be used along with a second magnetic field (B') to determine the mass of an ion. As shown in the diagram, such a particle passes through the velocity selector into magnetic field B' arranged perpendicular to its path. The mass of the particle is given by

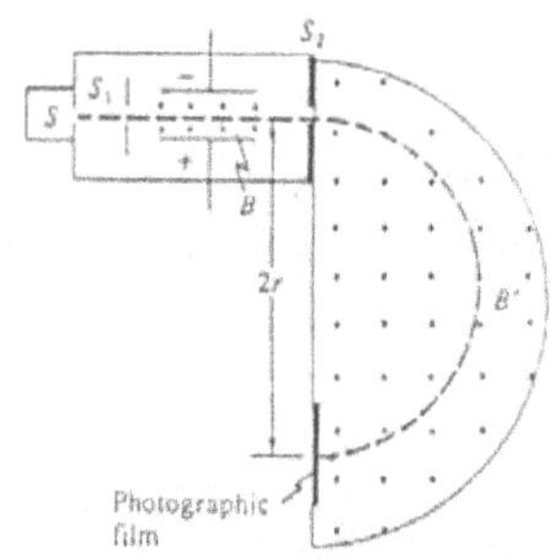

$$m = \frac{qB'r}{v} \quad \text{and since} \quad v = \frac{E}{B}, \quad m = \frac{qBB'r}{E}$$

Ions of different elements will not follow the same path because they differ in mass. But even if a pure substance is used, particles with different masses are found. This is because elements contain **isotopes**, particles that have the same chemical and physical properties but a different number of neutrons in the nucleus. Two isotopes of the same element do not have the same mass; thus the ions will follow paths of different radius.

EXAMPLE PROBLEM 5. (Section 20-11) In an experiment using a mass spectrometer, two singly charged isotopes of neon travel undeflected when passing through an electric field of magnitude 8.0×10^{3} V/m and a magnetic field of magnitude 4.0×10^{-1} T. As shown in the above diagram, the electric and magnetic fields are directed perpendicular to each other. (*a*) Determine the velocity of each isotope. (*b*) The isotopes then enter a section where only the magnetic field is directed perpendicular to their path. Determine the radius of curvature of the path followed by each particle. Note: The approximate mass of Ne-20 is 3.32×10^{-26} kg, while the mass of Ne-22 is approximately 3.65×10^{-26} kg. The charge on each isotope is 1.6×10^{-19} C.

Part a. Step 1.

Determine the velocity of each isotope as they pass through the crossed electric and magnetic fields.

The isotopes travel at constant speed and are undeflected:

net $F = 0$

$F_{\text{electric}} - F_{\text{magnetic}} = 0$

$qE - qvB\sin 90° = 0$

$F_{\text{electric}} \uparrow$ $\oplus \rightarrow$ $F_{\text{magnetic}} \downarrow$

$$v = \frac{E}{B} = \frac{8.0\times10^{3}\ \text{V/m}}{4.0\times10^{-1}\ \text{T}}$$

$$= 2.0\times10^{4}\,\text{m/s}$$

Part b. Step 1.

Determine the radius of the circle in which the electron travels.

The magnetic force provides the centripetal acceleration and causes the isotopes to follow a circular path:

$F_{\text{magnetic}} = ma_{\text{centripetal}}$

$qvB'\sin\theta = \dfrac{mv^2}{r}$ but $\theta = 90°$ and $\sin 90° = 1$

Solving for r for the Ne-20 isotope:

$$r = \frac{mv}{qB} = \frac{(3.32\times10^{-26}\ \text{kg})(2.0\times10^{4}\ \text{m/s})}{(1.6\times10^{-19}\,\text{C})(4.0\times10^{-1}\ \text{T})}$$

$r = 1.04\times10^{-2}\,\text{m}$

Solving for r for the Ne-22 isotope:

$$r = \frac{mv}{qB} = \frac{(3.65\times10^{-26}\ \text{kg})(2.0\times10^{4}\ \text{m/s})}{(1.6\times10^{-19}\,\text{C})(4.0\times10^{-1}\ \text{T})}$$

$r = 1.14\times10^{-2}\,\text{m}$

Galvanometer

A **galvanometer** is a device that forms the basis of most meters (ammeters, voltmeters, and ohmmeters) as well as electric motors. Galvanometer movement is the result of interaction of the magnetic field of a permanent magnet, which is directed perpendicular to a current carried by a loop or coil of wire.

Based on the diagram, the force on each vertical segment of length a is

$$F = NIaB \sin 90°,$$

where N refers to the number of turns of wire in the coil. If the coil is pivoted in the center, then a torque is produced that equals

$$\tau = NIabB,$$

where b is the distance between the vertical segments. Since ab equals the area (A) of the coil, $\tau = NIAB$; the quantity NIA is the **magnetic moment** of the coil.

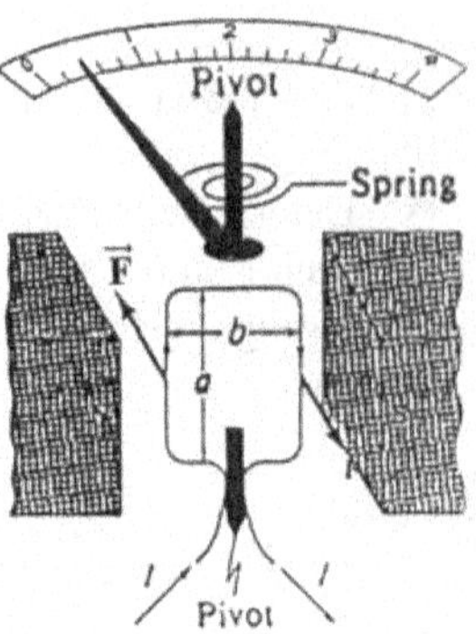

Electric Motor

The electric motor is essentially a galvanometer that is arranged so that a coil of wire, referred to as the **armature**, runs continuously. The electric current through the armature interacts with the external field, and the resulting torque causes the armature to rotate. Depending on the design of the motor, the motor can be arranged to run on direct current (dc) or alternating current (ac) electricity.

Force between Two Parallel Conductors

Two parallel current-carrying conductors produce magnetic fields that result in a force between the conductors. If the currents are in the same direction, then the force is of one attraction. A repulsive force results if the currents are in opposite directions. The force per unit length (F/ℓ) exerted by conductor 1 on conductor 2 and vice versa is given by

$$\frac{F}{\ell} = \frac{\mu_0 I_1 I_2}{2\pi d}$$

where d is the distance between the wires in meters, and I_1 and I_2 represent the magnitude of the current in each conductor.

EXAMPLE PROBLEM 6. (Section 20-5) Two long parallel wires carry currents of 2.0 A and 5.0 A, respectively. As shown in the diagram, the wires are 0.20 m apart and the currents flow in the same direction. Determine the magnitude and direction of the (*a*) force on a 0.50-m segment of the wire carrying the 2.0-A current and (*b*) force on a 0.50-m segment of the wire carrying the 5.0-A current.

Part a. Step 1.

Determine the magnitude of the magnetic field acting on the wire carrying the 2.0-A current.

The magnetic field acting on the wire is due to the magnetic field produced by the wire carrying the 5.0-A current. Therefore,

$$B = \frac{(4\pi \times 10^{-7}\ \text{T}\cdot\text{m/A})(5.0\ \text{A})}{(2\pi)(0.20\ \text{m})}$$

$$= 5.0 \times 10^{-6}\ \text{T}$$

Part a. Step 2.

Determine the magnitude of the force acting on the wire carrying the 2.0-A current.

The force on a wire carrying the 2.0-A current is given by the following:

$$F = I\ell B \sin\theta$$

$$F = (2.0\ \text{A})(0.50\ \text{m})(5.0 \times 10^{-6}\ \text{N/A}\cdot\text{m})\sin 90°$$

$$F = 5.0 \times 10^{-6}\ \text{N}$$

Part a. Step 3.

Determine the direction of the force acing on the 2.0-A current.

Based on the diagram and from the right-hand rule, the magnetic field produced by the 5.0-A current encircles the wire such that it is vertically into the page at the position of the 2.0-A current.

Using the right-hand rule for the force produced by a magnetic field acting on a current-carrying wire, the fingers point toward the bottom of the page in the direction of the 2.0-A current. The magnetic field is inward, perpendicular to the plane of the page. The fingers bend until they point inward. The thumb points to the right. The force is therefore toward the right.

Part b. Step 1.

Determine the magnitude of the magnetic field produced by the 2.0-A current.

The magnetic field acting on the 5.0-A current is due to the magnetic field produced by the 2.0-A current.

$$B = \frac{(4\pi \times 10^{-7}\ \text{T}\cdot\text{m/A})(2.0\ \text{A})}{(2\pi)(0.20\ \text{m})}$$

$$B = 2.0 \times 10^{-6}\ \text{N/A}\cdot\text{m}$$

Part b. Step 2.

Determine the force acting on the wire carrying the 5.0-A current.

$$F = I\ell B \sin\theta \quad \text{but} \quad \theta = 90°$$

$$F = (5.0\ \text{A})(0.50\ \text{m})(2.0 \times 10^{-6}\ \text{N/A}\cdot\text{m})(\sin 90°)$$

$$F = 5.0 \times 10^{-6}\ \text{N}$$

Part b. Step 3.

Determine the direction of the force acting on wire carrying the current.

The magnetic field produced by the 2.0-A current is directed perpendicularly out of the plane of the page at the position of the 5.0-A current. The fingers point along the direction of the 5.0-A current. The fingers bend to point in the direction of the magnetic field. The thumb points toward the left in the direction of the wire carrying the 2.0-A current. Therefore, the two wires are attracted to one another.

Note: The answer to Part b can be predicted by using Newton's third law of motion. The magnetic field produced by one wire interacts with the current in the other wire. Each wire is acted upon by a force due to the other wire. The forces should be equal in magnitude but opposite in direction.

Alternate solution: As described in Section 20-5 of the textbook, the force per unit length between two long parallel current-carrying wires is given by

$$\frac{F}{\ell} = \frac{\mu_0 I_1 I_2}{2\pi d}$$

where d is the distance between the wires and I_1 and I_2 are the currents.

Substituting the data given in the problem:

$$\frac{F}{0.50\text{ m}} = \frac{(4\pi\times10^{-7}\text{ T}\cdot\text{m/A})(2.0\text{ A})(5.0\text{ A})}{2\pi(0.20\text{ m})}$$

$$F = 5.0\times10^{-6}\text{ N}$$

Ferromagnetism; Domains

If an object is made from a **ferromagnetic** material, such as iron, each atom has a net magnetic effect. As shown in Fig. 20-42a of the textbook, the atoms tend to align their magnetic fields in arrangements known as **domains**; each domain contributes to the overall magnetic field of the piece of iron. In an ordinary piece of iron or other ferromagnetic material, the magnetic field produced by the individual domains cancels out so that the object is not a magnet. In a magnet, the domains are preferentially aligned in one direction, and a net magnetic effect is produced.

As shown in Fig 20-42b, if an unmagnetized ferromagnetic object is placed in a magnetic field, the domains which are aligned with the direction of the external magnetic field tend to grow. The increase in the size of the domains comes at the expense of neighboring domains that are not aligned with the external field. The result is that the unmagnetized object is attracted to the external magnetic field, and the object is said to exhibit **ferromagnetism**.

When the object is removed from the external field, the domains may remain aligned and the object retain a net magnetic effect. The object may lose this net magnetic effect if it is struck, dropped, or heated. Above a certain temperature, known as the **Curie temperature**, it is not possible to magnetize an object, and a permanent magnet loses its magnetic effect.

TEXTBOOK QUESTION 25. (Section 20-12) An unmagnetized nail will not attract an unmagnetized paper clip. However, if one end of the nail is in contact with a magnet, the other end *will* attract a paper clip. Explain.

ANSWER: Assume that the unmagnetized nail is brought near the north pole of a bar magnet. The domains in the nail grow and become aligned with the direction of the external magnetic field of the magnet. The end of the nail close to the magnet becomes a south pole, while the far end becomes a north pole and the nail is attracted to the magnet.

When the paper clip comes close to the nail, the domains in the paper clip grow in the direction of the magnetic field of the nail. The end of the paper clip close to the nail becomes a south pole and the paper clip is attracted to the nail.

PROBLEM SOLVING SKILLS

For problems involving a current-carrying wire in a magnetic field:

1. Draw an accurate diagram showing the orientation of the wire in the magnetic field. Use the adopted conventions to indicate the direction of the current and the magnetic field.
2. Use the right-hand rule to determine the direction of the force on the wire.
3. Use $F = I\ell B \sin \theta$ to determine the magnitude of the force.

For problems involving a charged particle traveling through a magnetic field:

1. Draw an accurate diagram showing the motion of the particle through the magnetic field. Use the adopted conventions to indicate the direction of the magnetic field and the direction of motion of the particle.
2. Take note of whether the particle is positively charged or negatively charged and then use the right-hand rule to determine the direction of the force on the particle.
3. If the particle is deflected into circular motion, then the magnetic force produces a centripetal acceleration. Use Newton's second law to determine the magnitude of the acceleration.
4. If the speed of the particle as it enters the magnetic field is given or can be calculated, it is possible to determine the radius of the circle in which the particle travels.

For problems involving the magnetic field produced by a current-carrying wire:

1. Draw a diagram showing the orientation of the straight wire or the loop of wire and the direction of the current in the wire.
2. Use the second right-hand rule to determine the direction of the magnetic field. Use the adopted conventions to indicate the direction of the magnetic field.
3. Use the appropriate formula to determine the magnitude of the magnetic field.

SOLUTIONS TO SELECTED TEXTBOOK PROBLEMS

TEXTBOOK PROBLEM 6. (Section 20-3) The force on a wire carrying 6.45 A is a maximum of 1.28 N when placed between the pole faces of a magnet. If the pole faces are 55.5 cm in diameter, what is the approximate strength of the magnetic field?

Part a. Step 1.

Determine the angle between the direction of the current in the wire and the direction of the magnetic field.	Since the force is a maximum, the angle between the direction of the direction of the current and the direction of the magnetic field is 90°.

Part a. Step 2.

Determine the strength of the magnetic field.	The length of the wire in the magnetic field equals the diameter of the pole faces; that is, $\ell = 55.5\text{ cm} = 0.555\text{ m}$:

$$F = I\ell B \sin 90°$$

$$1.28\text{ N} = (6.45\text{ A})(0.555\text{ m})B \sin 90°$$

$$B = 0.358\text{ N/A}\cdot\text{m} = 0.358\text{ T}$$

TEXTBOOK PROBLEM 8. (Section 20-3) Suppose a straight 1.00-mm-diameter copper wire could just "float" horizontally in air because of the force due to the Earth's magnetic field $\vec{\mathbf{B}}$, which is horizontal, perpendicular to the wire, and of magnitude 5.0×10^{-5} T. What current would the wire carry? Does the answer seem feasible? Explain briefly.

Part a. Step 1.

Determine the density of copper and the radius of the wire in meters.	From Table 10-1, density of copper $\rho_{\text{cu}} = 8.9\times10^3\text{ kg/m}^3$:

$$\text{radius} = (\tfrac{1}{2})\text{diameter} = (\tfrac{1}{2})(1.00\text{ mm}) = 0.50\text{ mm} = 5.0\times10^{-4}\text{ m}$$

Part a. Step 2.

Derive a formula for the wire's mass in terms of its density and volume.

$$\rho = \frac{m}{V}, \quad \text{where } V = \pi r^2 \ell$$

$$m = \rho V = \rho\pi r^2 \ell$$

Part a. Step 3.

The magnetic force balances gravity. Use $F = I\ell B \sin\theta$ to determine the magnitude of the current.

The wire is suspended at rest; therefore, net $F = 0$:

$$F_{\text{magnetic}} - F_{\text{gravity}} = 0$$

$$I\ell B \sin 90° - mg = 0$$

$$I = \frac{mg}{\ell B}\frac{\rho\pi r^2 \ell g}{\ell B} = \frac{\rho\pi r^2 g}{B}$$

$$I = \frac{(8.9\times10^3\ \text{kg/m}^3)(\pi)(5.0\times10^{-4}\ \text{m})^2(9.8\ \text{m/s})}{5.0\times10^{-5}\ \text{T}}$$

$$I = 1370\ \text{A}$$

The answer is not feasible. A current of this magnitude would quickly melt the wire.

TEXTBOOK PROBLEM 17. (Section 20-4) A 1.5-MeV (kinetic energy) proton enters a 0.30-T field, in a plane perpendicular to the field. What is the radius of its path? See Section 17-4.

Part a. Step 1.

Express the proton's kinetic energy in joules.

$$\text{KE} = (1.5\ \text{MeV})\left(\frac{1.6\times10^{-13}\ \text{J}}{1\ \text{MeV}}\right) = 2.4\times10^{-13}\ \text{J}$$

Part a. Step 2.

Determine the proton's velocity.

$$\text{KE} = \tfrac{1}{2}mv^2$$

$$2.4\times10^{-13}\ \text{J} = \tfrac{1}{2}(1.67\times10^{-27}\ \text{kg})v^2$$

$$v^2 = 2.87\times10^{14}\ \text{m}^2/\text{s}^2$$

$$v = 1.7\times10^7\ \text{m/s}$$

Part a. Step 3.

Determine the radius of the proton's path.

The proton is deflected by a magnetic force and travels in a circular path. The magnetic force provides the centripetal acceleration (a_R). Therefore,

$$\text{net } F = ma_\text{R} \quad \text{but} \quad a_\text{R} = \frac{v_2}{r}$$

$$F_\text{magnetic} = \frac{mv^2}{r}$$

$$qvB \sin\theta = \frac{mv^2}{r} \quad \text{but} \quad \theta = 90° \quad \text{and} \quad 90° = 1$$

Solving for r:

$$r = \frac{mv}{qB} = \frac{(1.67\times10^{-27}\ \text{kg})(1.7\times10^7\ \text{m/s})}{(1.6\times10^{-19}\ \text{C})(0.30\ \text{T})}$$

$$r = 0.59\ \text{m}$$

TEXTBOOK PROBLEM 30. (Sections 20-5 and 20-6) An experiment on the Earth's magnetic field is being carried out 1.00 m from an electric cable. What is the maximum allowable current in the cable if the experiment is to be accurate to $\pm 3\%$?

Part a. Step 1.

Determine the strength of the magnetic field produced by the cable.

The Earth's magnetic field is approximately 5.0×10^{-5} T.

$$B_{\text{cable}} \approx (\pm 0.03)(5 \times 10^{-5}\ \text{T}) \approx \pm 1.5 \times 10^{-6}\ \text{T}$$

Part a. Step 2.

Determine the maximum current that the cable can carry.

Assume that the cable is in the form of a long straight wire.

$$B_{\text{cable}} = \frac{\mu_0 I_{\text{cable}}}{2\pi r}$$

$$I = \frac{(B_{\text{cable}})(2\pi r)}{\mu_0}$$

$$I \approx \frac{(\pm 1.5\times 10^{-6}\ T)(2\pi)(1.0\ \text{m})}{(4\pi \times 10^{-7}\ \text{T}\cdot\text{m/A})} \approx 7.5\ \text{A}$$

TEXTBOOK PROBLEM 35. (Sections 20-5 and 20-6) Determine the magnetic field midway between two long straight wires 2.0 cm apart in terms of the current I in one when the other carries 25 A. Assume these currents are (*a*) in the same direction, and (*b*) in opposite directions.

Part a. Step 1.

Determine the magnitude of the magnetic field produced by each wire at the midpoint between the wires.

$$B_1 = \frac{\mu_0 I_1}{2\pi r}$$

$$B_1 = \frac{(4\pi \times 10^{-7}\ \text{T}\cdot\text{m/A})I}{2\pi(0.010\ \text{m})}$$

$$= (2.0\times 10^{-5})I\ \text{T/A}$$

$$B_2 = \frac{\mu_0 I_2}{2\pi r} = \frac{(4\pi \times 10^{-7}\ \text{T}\cdot\text{m/A})(25\ \text{A})}{2\pi(0.010\ \text{m})}$$

$$B_2 = 5.0 \times 10^{-4}\ \text{T}$$

Part a. Step 2.

Determine the magnitude of the resultant magnetic field when the currents are in the same direction.

Using the right-hand rule, it can be shown that when the currents are in the same direction, the magnetic fields oppose each other.

$$B_{\text{total}} = B_1 - B_2 = (2.0\times10^{-5} I\ \text{T}) - 5.0\times10^{-4}\ \text{T}$$

$$B_{\text{total}} = (2.0\times10^{-5} I - 5.0\times10^{-4})\ \text{T}$$

$\downarrow B_2$ $\uparrow B_2$

$\otimes$. $\otimes$

I 15 A

The direction of the resultant magnetic field is in the direction of the field that has the greater magnitude.

Part b. Step 1.

Determine the magnitude of the resultant magnetic field when the currents are in the opposite direction.

Using the right-hand rule, it can be shown that when the currents are in the opposite direction, the magnetic fields are in the same direction.

$$B_{\text{total}} = B_1 + B_2 = (2.0\times10^{-5} I\ \text{T}) + 5.0\times10^{-4}\ \text{T}$$

$$B_{\text{total}} = (2.0\times10^{-5} I + 5.0\times10^{-4})\ \text{T}$$

$B_1 \uparrow\uparrow B_2$

$\odot$. $\otimes$

I 15 A

Based on the diagram, the direction of the resultant magnetic field is vertically upward.

TEXTBOOK PROBLEM 59. (Section 20-11) An unknown particle moves in a straight line through crossed electric and magnetic fields with $E = 1.5\ \text{kV/m}$ and $B = 0.034\ \text{T}$. If the electric field is turned off, the particle moves in a circular path of radius $r = 2.7\ \text{cm}$. What might the particle be?

Part a. Step 1.

Determine the particle's speed.

In order for the particle to travel in a straight line, the force exerted by the electric field must be equal but opposite to the force exerted by the magnetic field.

$$F_{\text{electric}} = F_{\text{magnetic}}$$

$$qE = qvB$$

$$v = \frac{E}{B} = \frac{1.5\times10^3 \text{ V/m}}{0.034 \text{ T}}$$

$$v = 4.4\times10^4 \text{ m/s}$$

Part a. Step 2.

Determine the mass of the particle. Note: Assume the particle is a singly charged ion:

$q = 1.6\times10^{-19}$ C.

The magnetic force provides the centripetal acceleration and causes the particle to follow a circular path.

$$F_{\text{magnetic}} = ma_{\text{centripetal}}$$

$qvB \sin\theta = mv2/r$ but $\theta = 90°$ and $\sin 90° = 1$

Solving for m:

$$m = \frac{qBr}{v} = \frac{(1.6\times10^{-19} \text{ C})(0.034 \text{ T})(0.027 \text{ m})}{4.4\times10^4 \text{ m/s}}$$

$$m = 3.33\times10^{-27} \text{ kg}$$

Part a. Step 3.

Express the mass of the particle in atomic mass units.

$$(3.33\times10^{-27} \text{ kg})\left(\frac{1 \text{ amu}}{1.66\times10^{-27} \text{ kg}}\right) = 2.011 \text{ amu}$$

Part a. Step 4.

Use Appendix B to determine the identity of the particle.

From Appendix B, a deuterium nucleus has a mass of 2.014 amu. Therefore, the particle is mostly likely a deuterium nucleus.

TEXTBOOK PROBLEM 72. (Section 20-4) A doubly charged helium atom, whose mass is 6.6×10^{-27} kg, is accelerated by a voltage of 3200 V. (*a*) What will be its radius of curvature in a uniform 0.240-T field? (*b*) What is its period of revolution?

Part a. Step 1.

Determine the ion's velocity as it enters the magnetic field.

The ion is accelerated through the 3200-V potential difference and gains kinetic energy. The work done on the particle is equal to the product of the charge on the particle and the potential difference:

$W = qV = \text{KE}_\text{f} - \text{KE}_\text{i}$, where $q = 2(1.6\times10^{-19} \text{ C}) = 3.2\times10^{-19}$ C

$qV = \frac{1}{2}mv_\text{f}^2 - \frac{1}{2}mv_\text{i}^2$ but $v_\text{i} = 0$; therefore,

$$qV = \tfrac{1}{2}mv_\text{f}^2$$

Rearranging gives

$$v = \sqrt{\frac{2qV}{m}} = \sqrt{\frac{2(3.2\times10^{-19}\ \text{C})(3200\ \text{V})}{6.6\times10^{-27}\ \text{kg}}}$$

$v = 5.57\times10^{5}$ m/s

Part a. Step 2.

Determine the radius of the circle in which the electron travels.

The magnetic force provides the centripetal acceleration and causes the electrons to follow a circular path.

$$F_{\text{magnetic}} = ma_{\text{centripetal}}$$

$$qvB\sin\theta = m\frac{v^2}{r} \quad \text{but} \quad \theta = 90° \quad \text{and} \quad \sin 90° = 1$$

Solving for r:

$$r = \frac{mv}{qB} = \frac{(6.6\times10^{-27}\ \text{kg})(5.57\times10^{5}\ \text{m/s})}{(3.2\times10^{-19}\ \text{C})(0.240\ \text{N/A}\cdot\text{m})}$$

$r = 4.79\times10^{-2}$ m

Part b. Step 1.

Determine the period of the motion.

Hint: The period (T) of the motion refers to the time required for the electron to complete one revolution.

The distance traveled in one revolution equals the circumference $(2\pi r)$ of the circle. The period can be determined by dividing the circumference of the circle by the velocity (v).

$$T = \frac{2\pi r}{v} = \frac{(2\pi)(4.79\times10^{-2}\ \text{m})}{5.57\times10^{5}\ \text{m/s}}$$

$T = 5.4\times10^{-7}$ s

21

ELECTROMAGNETIC INDUCTION AND FARADAY'S LAW

OBJECTIVES

After studying the material of this chapter, the student should be able to:

- determine the magnitude of the magnetic flux through a surface of known area, given the strength of the magnetic field and the angle between the direction of the magnetic field and the surface.
- write a statement of Faraday's law in terms of changing magnetic flux. Use Faraday's law to determine the magnitude of the induced emf in a closed loop due to a change in the magnetic flux through the loop.
- use Faraday's law to determine the magnitude of the induced emf in a straight wire moving through a magnetic field.
- state Lenz's law and use Ohm's law and Lenz's law to determine the magnitude and direction of the induced current.
- explain the basic principle of the electric generator. Determine the magnitude of the maximum value of the induced emf in a loop that is rotating at a constant rate in a uniform magnetic field.
- explain how an eddy current can be produced in a piece of metal. Describe situations in which eddy currents are beneficial and situations when they must be eliminated.
- explain how a transformer can be used to step up or step down the voltage. Apply the equations that relate number of turns, voltages, and currents in the primary and secondary coils to solve transformer problems.
- explain what is meant by mutual inductance and self-inductance. List the factors that determine the self-inductance of a solenoid. State the SI unit for inductance.
- write the equations for the average induced emf in a solenoid in which the current is changing at a known rate.
- write the equations for the energy stored in an inductor's magnetic field and for the energy stored per unit volume—that is, the energy density.
- write the equation for the voltage across the inductor as a function of time after the inductor is connected to the source of emf in an *LR* circuit. Graph the current as a function of time after the initial connection is completed.
- write the equation for the voltage across the inductor as a function of time if the inductor is disconnected from the source of emf and discharged. Graph the current as a function of time after the discharge is initiated.
- distinguish among resistance, capacitive reactance, inductive reactance, and impedance in an *LR, LC*, or *LRC* circuit. Calculate the reactance of a capacitor and/or inductor which is connected to a source of known frequency.

- use a phasor diagram to determine the phase angle and total impedance for an *LR*, *LC*, or *LRC* circuit.
- determine the rms current and power dissipated in an *LRC* circuit. Determine the voltage drop across each circuit element and the resonant frequency of the circuit.

KEY TERMS AND PHRASES

magnetic flux is the product of the magnetic field strength, the area of the plane of the loop through which the magnetic field passes, and the cosine of the angle that the magnetic field makes with a line drawn normal to the plane of the loop.

Faraday's law states that a voltage is produced in a conductor when the magnetic flux through the conductor changes. The voltage produced is called the induced emf.

Lenz's law states that an induced emf produces a current whose magnetic field always opposes the change in magnetic flux that caused it.

electric generator uses mechanical work to produce electric energy. The **armature** of the generator is turned by an external torque and an emf is induced as the coil passes through an external magnetic field.

back emf, or **counter emf**, is produced as the armature of a motor rotates in the external magnetic field. This induced emf produces a torque that opposes the motion of the armature.

eddy current is an induced current produced by a changing magnetic flux in a piece of metal. The direction of the induced current is such as to oppose the magnetic field that caused the current.

mutual induction occurs when a changing current in one circuit induces a current in another circuit.

self-induction occurs when a changing current in a circuit induces a back emf in the circuit.

transformer is a device used to increase or decrease an ac voltage.

***LRC* series circuit** consists of an inductor, a resistor, and a capacitor, all connected in series with an alternating current source of voltage.

phasor diagram is used to represent either current or voltage in an ac circuit. Vector algebra is used to analyze an *LRC* circuit.

inductive reactance is a measure of the effect an inductor has on the current through an ac circuit. It is analogous to the effect a resistor has to the flow of current through a dc circuit.

capacitive reactance is a measure of the effect a capacitor has on the current through an ac circuit. It is analogous to the effect a resistor has to the flow of current through a dc circuit.

impedance in an ac circuit is analogous to resistance in a dc circuit. Impedance results from the combination of inductive reactance, capacitive reactance, and resistance.

resonant frequency is the frequency at which an *LRC* circuit has minimum impedance to current flow. At this frequency current flow is a maximum and the energy transferred to the system is a maximum.

SUMMARY OF MATHEMATICAL FORMULAS

magnetic flux	$\Phi_B = BA\cos\theta$	Magnetic flux (Φ_B) is the product of the magnetic field strength (B), the area (A) of the plane of the loop through which the magnetic field passes, and the cosine of the angle ($\cos\theta$) that the magnetic field makes with a line drawn normal to the plane of the loop.
Faraday's law	$\xi = -N\dfrac{\Delta\Phi_B}{\Delta t}$	The magnitude of the induced emf (ξ) in a coil depends on the number of turns (N) in the loop and the rate of change of flux $\left(\dfrac{\Delta\Phi_B}{\Delta t}\right)$ through the loop.
emf induced in a moving conductor	$\xi = B\ell v\sin\theta$	The emf (ξ) induced in a conducting rod or wire which moves through a magnetic field depends on the velocity (v) of the rod relative to the magnetic field, the length (ℓ) of the rod, and the angle (θ) between the direction of motion of the rod and the direction of the magnetic field.
emf produced by an electric generator	$\xi = NB\omega A\sin\omega t$ $\xi_0 = NB\omega A$	The magnitude of the emf (ξ) produced by a generator rotating continuously with a constant angular velocity (ω) depends on the number of turns in the armature (N), the area of the loop (A), the magnitude of the magnetic field (B), and the angle (ωt) that the face of the loop makes with the direction of the magnetic field at time t. ξ_0 represents the peak voltage of the generator.
transformer	$\dfrac{V_S}{V_P} = \dfrac{N_S}{N_P}$	The voltage in the secondary coil (V_S) depends on the voltage in the primary coil (V_P) and the number of turns in each coil, N_S and N_P.
	$I_P V_P = I_S V_S$	In a transformer (assuming that energy losses can be ignored), the power in the primary ($P_{input} = I_P V_P$) equals the power in the secondary ($P_{output} = I_S V_S$).
	$\dfrac{I_S}{I_P} = \dfrac{N_P}{N_S}$	The ratio of the currents is related to the ratio of the turns.

induced voltage due to mutual inductance	$\xi = -M\dfrac{\Delta I}{\Delta t}$	The induced voltage (ξ) in one coil depends on the mutual inductance (M) and the rate of change of the current in a second coil $\dfrac{\Delta I}{\Delta t}$.
induced voltage due to self-inductance	$\xi = -L\dfrac{\Delta I}{\Delta t}$	The induced voltage in a single coil depends on the self-inductance (L) in a single coil and the rate of change of the current in the coil $\dfrac{\Delta I}{\Delta t}$.
	$L = \dfrac{\mu_0 N^2 A}{\ell}$	The self inductance (L) of a long coil, called a solenoid, of length ℓ, cross-sectional area A, and N turns.
energy (U) stored in an inductor	$U = \frac{1}{2}LI^2$	The energy stored in a coil is related to the self-inductance (L) and current (I).
	$U = \frac{1}{2}\dfrac{B^2}{\mu_0}A\ell$	The energy stored in the inductor's magnetic field is related to the magnetic field strength (B) and the volume enclosed by the windings $(A\ell)$.
energy density (u)	$u = \frac{1}{2}\dfrac{B^2}{\mu_0}$	The energy density (u) refers to the energy stored per unit volume.
current rise in an LR circuit	$I = \dfrac{V_0}{R}(1 - e^{-t/\tau})$	The current (I) in an LR circuit is related to the voltage of the source (V_0), the resistance (R), the time constant (τ), and time (t).
time constant	$\tau = \dfrac{L}{R}$	τ represents the inductive time constant of an LR circuit.
current decay in an LR circuit	$I = I_{\max}e^{-t/\tau}$	The current (I) in an LR decay circuit depends on the voltage of the source (V_0), the resistance (R), the time constant (τ), and time (t).
inductive reactance	$X_L = 2\pi fL = \omega L$	The inductive reactance (X_L) is related to both the frequency of the source (f) and the self-inductance of the coil (L).

capacitive reactance	$X_C = \frac{1}{2\pi fC} = \frac{1}{\omega C}$	The capacitive reactance (X_C) is inversely proportional to the frequency of the source and the capacitance of the capacitor.
LRC series circuit peak voltage	$V_0 = \sqrt{V_{R0}^2 + (V_{L0} - V_{C0})^2}$	The peak voltage (V_0) across the source is related to the peak voltage across the resistor (V_{R0}), the inductor (V_{L0}), and the capacitor (V_{C0}).
impedance	$Z = \sqrt{R^2 + (X_L - X_C)^2}$	The impedance (*Z*) is related to the inductive reactance (X_L), the capacitive reactance (X_C), and the resistance (*R*).
phase angle	$\tan\phi = \frac{X_L - X_C}{R}$	The phase angle (ϕ) between the voltage and current equals the ratio of the difference between the inductive and capacitive reactance $(X_L - X_C)$ and the resistance (*R*).
power dissipated by impedance	$\overline{P} = I_{\text{rms}} V_{\text{rms}} \cos\phi$ $\overline{P} = I_{\text{rms}}^2 Z \cos\phi$	Power dissipated in the form of heat is related to the current (I_{rms}), voltage (V_{rms}), impedance (*Z*), and power factor $\cos\phi$.
power factor	$\cos\phi = \frac{R}{Z}$	$\cos\phi$ is defined as the power factor of the *LRC* circuit. For a resistor, $\phi = 0°$ and $\cos 0° = 1$. Therefore, all of the power is dissipated in the form of heat. For a capacitor or an inductor, $\phi = 90°$ and $\cos 90° = 0$. Therefore, no heat is dissipated by a capacitor (or inductor); all of the energy is stored in the electric (or magnetic) field.
resonant frequency for an *LRC* circuit	$f_0 = \frac{1}{2\pi}\sqrt{\frac{1}{LC}}$	Resonance for an *LRC* circuit occurs when the impedance is a minimum. At resonance, the energy transferred to the system from the source of emf is a maximum. The resonant frequency (f_0) is inversely related to the square root of the inductance and the capacitance.

CONCEPT SUMMARY

Faraday's Law of Induction

If a bar magnet is moved toward a coil of wire, an emf will be induced in the wire. The magnitude of the induced emf depends on the magnetic field strength, the area of the loop, and the time required for the change in **magnetic flux** through the area of the loop to occur. The product of the magnetic field strength (*B*) and the area of the plane of the loop through which it passes is known as the magnetic flux (Φ_B).

If the magnetic field is given in teslas $\text{T} = \text{Wb/m}^2$ and the area of the plane of the loop in m^2, then the flux has the unit of webers (Wb). The formula for the magnetic flux is

$$\Phi_B = BA\cos\theta$$

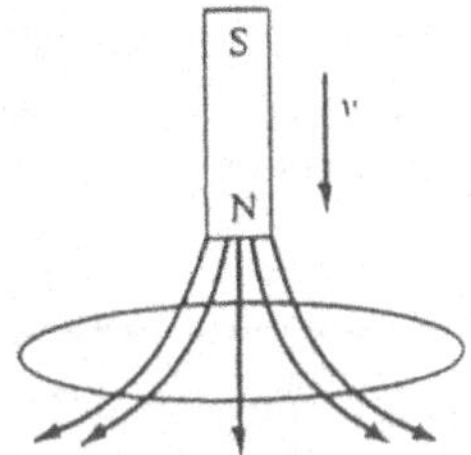

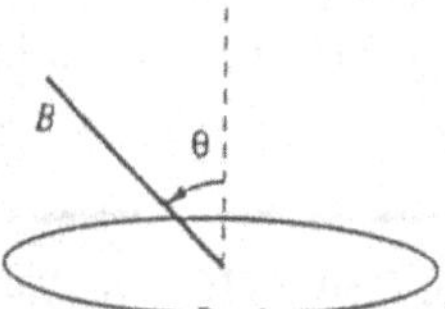

Here θ is the angle between the direction of *B* and a line drawn perpendicular to the plane of the loop. If the angle is 0°, then the flux passing through the loop is a maximum, as $\cos 0° = 1.0$. If the angle is 90°, then no flux passes through the loop, as $\cos 90° = 0$.

The magnitude of the induced emf (ξ) in the coil is given by **Faraday's law**:

$$\xi = -N\frac{\Delta\Phi_B}{\Delta t}$$

N is the number of turns in the loop, and $\frac{\Delta\Phi_B}{\Delta t}$ is the rate of change of flux through the loop.

TEXTBOOK QUESTION 2. (Section 21-2) What is the difference between magnetic flux and magnetic field?

ANSWER: A magnetic field surrounds every magnet and current-carrying wire. The magnetic field strength is represented by $\vec{\mathbf{B}}$, and because it has both magnitude and direction, it is a vector quantity. The unit used for the magnetic field strength is the tesla (T), where $1\ \text{T} = 1\ \text{N/A}\cdot\text{m} = 1\ \text{weber/m}^2$. Magnetic flux equals the product of the perpendicular component of the magnetic field (*B*) as it passes through a loop of area *A*. The magnetic flux (Φ_B) is given by Eq. 21-1 of the textbook, $\Phi_B = B_\perp A = BA\cos\theta$. Magnetic flux is a scalar quantity. The unit used for magnetic flux is the weber, where $1\ \text{Wb} = 1\ \text{T}\cdot\text{m}^2$.

Lenz's Law

The minus sign in the Faraday's law formula indicates that the induced emf opposes the change in flux through the loop. This opposition is described by **Lenz's law**, which states that an induced emf produces a current whose magnetic field always opposes the change in magnetic flux that caused it.

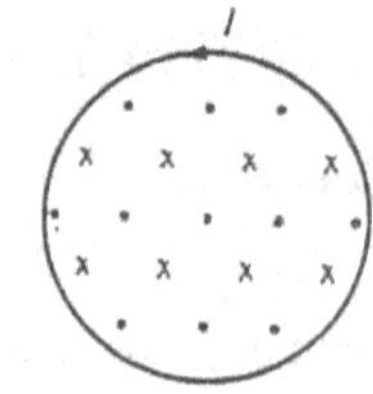

An example of Lenz's law is a north pole of a bar magnet being inserted into a coil of wire, as shown in the figure to the right, in which the ×'s represent the magnetic field of the bar magnet. According to Lenz's law, an induced current will be produced in the counterclockwise direction through the coil. Using the right-hand rule, it can be determined that this induced current produces a magnetic field (dots in the figure) which opposes the magnetic field of the bar magnet.

If the bar magnet is held motionless, then the flux is no longer changing—that is, $\frac{\Delta \Phi_B}{\Delta t} = 0$, and the induced current disappears. However, if the magnet is withdrawn, then an induced current will be produced in the clockwise direction. This current produces a magnetic field which attempts to maintain the magnetic flux through the loop at the same magnitude as when the bar magnet was motionless in the loop. Since the loop contains electrical resistance, this current will quickly be reduced to zero.

It is the relative motion between the magnetic field and the loop that causes the induced emf. Thus, it is possible to induce an emf in a loop by (1) holding the magnetic field constant and changing the area of the loop through which the magnetic flux passes, (2) holding the loop motionless and changing the magnitude of the flux through the loop, or (3) changing the orientation of the loop in the magnetic field by rotating the loop in the field.

TEXTBOOK QUESTION 6. (Section 21-2) Suppose you are looking along a line through the centers of two circular (but separate) wire loops, one behind the other. A battery is suddenly connected to the front loop, establishing a clockwise current. (*a*) Will a current be induced in the second loop? (*b*) If so, when does this current start? (*c*) When does it stop? (*d*) In what direction is this current? (*e*) Is there a force between the two loops? (*f*) If so, in what direction?

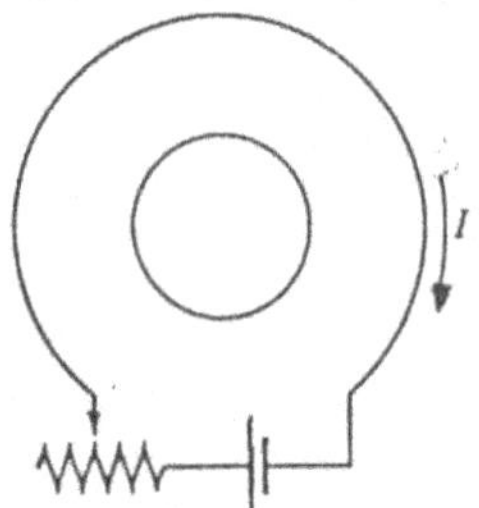

ANSWER: (*a*) and (*b*) According to Faraday's law, the changing magnetic field produced by the first loop passes through the second loop, inducing a voltage in the second loop. The induced voltage will produce a current in the second loop that begins immediately.

(*c*) The induced current in the second loop will stop when the magnetic field through the second loop stops changing. This occurs once the clockwise current in the first loop reaches a steady-state value.

(*d*) According to Lenz's law, the induced current in the second loop will oppose the changing magnetic field that created it. Therefore, a counterclockwise current will be produced in the second loop.

(*e*) and (*f*) Yes. As long as the current in the first loop is increasing, the induced current in the second loop creates a magnetic field that opposes the magnetic field in the first loop. The two loops will repel each other.

TEXTBOOK QUESTION 14. (Sections 21-1 to 21-4) A bar magnet falling inside a vertical metal tube reaches a terminal velocity even if the tube is evacuated so that there is no air resistance. Explain.

ANSWER: The falling magnet causes a change in flux in the region of the tube through which it is passing. According to Faraday's law, this change of flux induces an emf. The induced emf produces a current, which in turn produces a magnetic field that opposes the motion of the falling magnet. The speed of the falling magnet increases until the fields are equal and opposite. The bar magnet stops accelerating and falls at a constant speed which is called the terminal velocity.

EXAMPLE PROBLEM 1. (Section 21-2) The magnitude of the magnetic field through a 500-turn loop of wire changes from 1.00 Wb/m^2 to 4.00 Wb/m^2 in 0.150 s. The radius of the loop is 0.0300 m and its electrical resistance is 5.00 Ω. (*a*) Determine the magnitude of the average induced emf and the current in the loop. (*b*) Determine the direction of the induced current in the loop if the loop is in the plane of the page and the magnetic field is directed out of the page.

Part a. Step 1.

Determine the area of the loop of wire.

$$A = \pi r^2 = \pi(0.0300\ \text{m})^2$$

$$A = 2.83\times10^{-3}\ \text{m}^2$$

Part a. Step 2.

Determine the change in magnetic flux through the loop.

The magnetic field is directed perpendicular to the coil. Using the convention discussed in the textbook, the angle θ is 0°:

$$\Delta\Phi_B = \Phi_\text{f} - \Phi_\text{i}$$

$$= B_\text{f} A \cos\theta - B_\text{i} A \cos\theta$$

$$= (4.00\ \text{Wb/m}^2)(2.83\times10^{-3}\ \text{m}^2)(\cos 0°) - (1.00\ \text{Wb/m}^2)(2.83\times10^{-3}\ \text{m}^2)(\cos 0°)$$

$$\Delta\Phi_B = 8.49\times10^{-3}\ \text{Wb}$$

Part a. Step 3.

Use Faraday's law to determine the magnitude of the induced emf and Ohm's law to determine the indeed current.

$$\xi = -N\frac{\Delta\Phi_B}{\Delta t} = -\frac{(500)(8.49\times10^{-3}\ \text{Wb})}{0.150\ \text{s}}$$

$$\xi = -28.3\ \text{V}$$

$$I = \xi/R = \frac{-28.3\ \text{V}}{5.00\ \Omega} = -5.66\ \text{A}$$

Part b. Step 1.

Draw a diagram showing the loop and the external magnetic field.

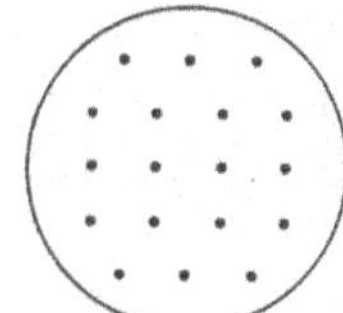

• = external magnetic field

Part b. Step 2.

Use Lenz's law to predict the direction of the induced current.

Based on Lenz's law, the direction of the induced current must oppose the external magnetic field that caused it. Since the external magnetic field is directed out of the page, the current in the loop must produce a magnetic field which is directed into the page. The right-hand rule indicates that a clockwise current in the loop will produce a magnetic field that is directed into the page.

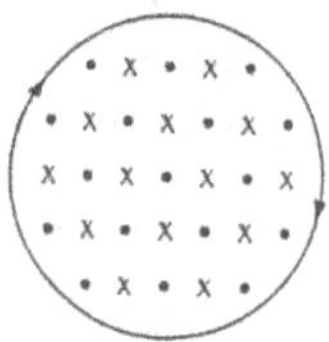

• = external magnetic field

× = magnetic field produced by the induced current

EXAMPLE PROBLEM 2. (Section 21-2) A 5.30-Wb/m^2 magnetic field is directed perpendicular to the plane of a 300-turn loop that has radius 0.250 m. The loop is rotated through $180°$ in 0.130 s. Determine the magnitude of the induced emf in the loop.

Part a. Step 1.

Determine the change in flux through the loop.

The area of the loop and the strength of the external magnetic field are not changing. However, the angle between the plane of the loop and the magnetic field changes from $0°$ to $180°$ as the coil is rotated. An emf will be induced in the loop because as the loop rotates, it cuts through lines of magnetic flux. The change in flux can be determined as follows:

$$\Delta\Phi_B = \Phi_f - \Phi_i$$

$$= B_f A_f \cos 180° - B_i A_i \cos 0°$$

$$= BA\,(\cos 180° - \cos 0°)$$

but $\cos 180° = -1$ and $\cos 0° = +1$

$$= (5.30 \text{ Wb/m}^2)[\pi(0.250 \text{ m})^2](-1.0 - +1.0)$$

$$\Delta\Phi_B = -2.08 \text{ Wb}$$

Part a. Step 2.

Use Faraday's law to determine the magnitude of the induced emf.

$$\xi = -N\frac{\Delta\Phi_B}{\Delta t}$$

$$= -(300)\left(\frac{-2.08 \text{ Wb}}{0.130 \text{ s}}\right)$$

$$\xi = +4800 \text{ V}$$

EMF Induced in a Moving Conductor

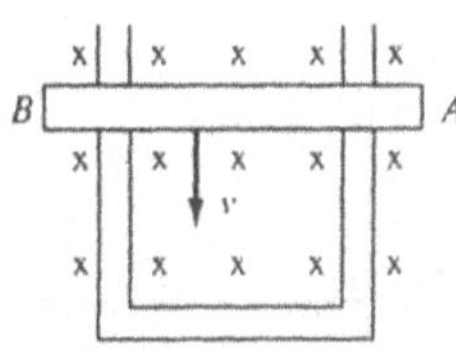

An emf is induced in a conducting rod or wire that moves through a magnetic field. This can be predicted by the right-hand rule. In the diagram, a conducting rod is moved along a U-shaped conductor in a uniform magnetic field that points into the page. The positive charges in the rod are moving with the rod as the rod moves through the magnetic field. Using the right-hand rule, the outstretched fingers point in the direction that the rod is moving. When the fingers bend, they must point in the direction of the magnetic field—that is, into the page. The thumb points toward point A, indicating that the positive charges tend to move away from B and toward A.

A difference in potential exists between the ends of the rod. This difference in potential is the induced emf. The induced emf in the wire is given by the formula

$$\xi = B\ell v \sin\theta$$

where ℓ is the length in meters of the segment of the rod, v is the velocity of the rod relative to the magnetic field, and θ is the angle between the direction of motion of the rod and the direction of the magnetic field.

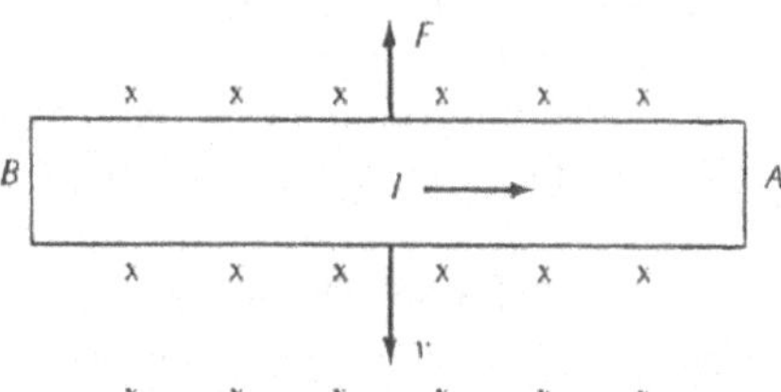

Since the rod is part of a closed circuit, a current will flow through the circuit. As shown in the diagram, the current flows from B toward A and then clockwise through the circuit. The magnitude of this induced current is determined from Ohm's law, $I = \xi/R$.

The direction of the induced current is such that a force is produced that opposes the external force that causes the rod to move through the magnetic field. This also can be predicted by using the right-hand rule. The outstretched fingers point in the direction of the induced current. When the fingers bend to point in the direction of the magnetic field, the thumb of the right hand points toward the top of the page. The force F, which is produced by the interaction of the induced current and the magnetic field, opposes the motion of the wire. This is another example of Lenz's law.

If the rod travels at constant speed, and friction between the rod and the U-shaped conductor is negligible, then the induced force is equal in magnitude but opposite in direction to the external force moving the rod through the field.

EXAMPLE PROBLEM 3. (Section 21-3) The metal rails of the U-shaped conductor shown in the preceding diagram are 1.00 m apart and the magnetic field is 0.200 Wb/m^2. The magnetic field is directed perpendicular to the page. The electrical resistance of the closed circuit is 1.00 Ω and is assumed to be constant. Determine the (*a*) magnitude and direction of the induced current in the circuit when the rod is moving at 3.00 m/s, (*b*) magnitude and direction of the external force required to keep the rod moving at a constant speed, and (*c*) power required to keep the rod moving at constant speed.

Part a. Step 1.

Determine the magnitude of the induced emf.

Since every point in the wire is moving at the same speed, use $\xi = B\ell v \sin\theta$:

$$\xi = (0.200\ \text{Wb/m}^2)(1.00\ \text{m})(3.00\ \text{m/s})\sin 90^\circ$$

$$= 0.600\ \text{V}$$

Part a. Step 2.

Determine the magnitude of the induced current.

$I = \xi/R$

$$I = \frac{0.600\ \text{V}}{1.00\ \Omega} = 0.600\ \text{A}$$

Part a. Step 3.

Use Lenz's law to determine the direction of the induced current.

Lenz's law predicts that the induced current must oppose the change that causes it. The current opposes the change by interacting with the external magnetic field to produce a force that opposes the rod's motion. Using the right-hand rule, the current can be shown to be flowing from B to A and therefore in a clockwise direction through the U-shaped conductor.

Part b. Step 1.

Determine the magnitude of the external force.

Note: $1\ \dfrac{\text{Wb}}{\text{m}^2} = 1\ \text{T} = 1\ \dfrac{\text{N}}{\text{A}\cdot\text{m}}$

$$F = I\ell B\sin\theta = (0.600\ \text{A})(1.00\ \text{m})(0.200\ \text{N/A}\cdot\text{m})\sin 90^\circ$$

$$F = 0.120\ \text{N}$$

Part c. Step 1.

Determine the mechanical power required to keep the rod moving at constant speed.

The external force is moving the wire through the magnetic field; the angle between the force and the direction of motion is 0°. The wire travels at constant speed; therefore,

$$P = Fv\cos 0^\circ$$

$$P = (0.120\ \text{N})(3.00\ \text{m/s})(\cos 0^\circ) = 0.36\ \text{W}$$

Generators

The **electric generator** uses mechanical work to produce electric energy. The **armature** is turned by an external torque and an emf is induced as the coil passes through an external magnetic field.

If the loop rotates continuously with a constant angular velocity (ω), then the magnitude of the induced emf is given by

$$\xi = NB\omega A \sin \omega t$$

where ωt is the angle that the face of the loop makes with the direction of the magnetic field at time t and A is the area of the loop.

The output emf of the generator is sinusoidally alternating. However, depending on the design, the generator can be made to produce either dc or ac current.

EXAMPLE PROBLEM 4. (Section 21-5) An ac generator consists of an 800-turn flat rectangular coil 0.0600 m long and 0.0400 m wide. The coil rotates at 1200 rpm in a magnetic field of 0.400 T. Determine the (*a*) angular velocity of the coil in radians per second and (*b*) peak output voltage of the generator.

Part a. Step 1.

Determine the angular velocity of the coil.	The angular velocity ω is related to the frequency (f) by the formula $$\omega = 2\pi f = \left(\frac{2\pi \text{ rad}}{1 \text{ rev}}\right)\left(\frac{1200 \text{ rev}}{1 \text{ min}}\right)\left(\frac{1 \text{ min}}{60 \text{ s}}\right) = 126 \text{ rad/s}$$

Part b. Step 1.

Determine the peak output voltage.	The peak voltage is given by $\xi_0 = NB\omega A,$ where $$A = (0.0600 \text{ m})(0.0400 \text{ m}) = 2.40\times10^{-3} \text{ m}^2$$ $$\xi_0 = (800)(0.400 \text{ T})(2.40\times10^{-3} \text{ m}^2)(126 \text{ rad/s})$$ $$\xi_0 = 96.8 \text{ V}$$

Counter EMF and Torque

As the armature of a motor rotates in the external magnetic field, an induced emf is produced. This induced emf, which is called a **back** or **counter emf**, produces a torque that opposes the motion of the armature (Lenz's law).

The magnitude of the back emf is proportional to the speed of the armature of the motor. When a motor is turned on, the magnitude of the back emf increases until a balance point is reached and the armature rotates at a constant speed. If the motor speed increases, then the back emf increases until a balance point is again achieved.

The above also applies to a generator. The external torque causing the armature to rotate is opposed by a counter torque produced by the counter emf. If this did not occur, it would violate the law of conservation of energy, since it would then be possible to start the armature rotating and produce electrical energy without expending mechanical energy.

TEXTBOOK QUESTION 11. (Section 21-6) Explain why, exactly, the lights may dim briefly when a refrigerator motor starts up. When an electric heater is turned on, the lights may stay dimmed as long as the heater is on. Explain the difference.

ANSWER: When a motor starts up, the current through the armature is large because the induced counter emf is small and the electrical resistance to current flow is small. Electrical outlets on the same circuit as the motor will experience a voltage drop, and any house lights on the same circuit with the motor will dim. As the armature speeds up, it produces an increasingly large counter emf that reduces the flow of current through the motor. The voltage at the electrical outlets rises, and the lights return to their former brightness.

When an electric heater is turned on, it also draws a large current. However, there is no induced counter emf to reduce the electric current to the heater. The heater has a large resistance, and the product of the electrical current and the resistance results in a large voltage drop in the heater. The result is that the voltage available to other electrical outlets remains reduced.

Eddy Currents

An induced current called an **eddy current** is produced by a changing magnetic flux in a piece of metal. The direction of the induced current is such as to oppose the magnetic field that caused the current (Lenz's law). Eddy currents can be produced by moving the metal through a magnetic field or by allowing a changing magnetic field to pass through a stationary piece of metal.

Eddy currents can be beneficial, for example, electromagnetic damping in certain analytical balances and the braking systems of some electric transit cars. However, eddy currents are often undesirable. For example, the coils of wire that make up the armature of a motor or generator as well as the primary and secondary of a transformer are often wound on an iron core. Eddy currents produced in the iron core dissipate energy in the form of heat ($P = I\xi = I^2R$). To avoid this problem, the core is laminated, which means that it is made of very thin sheets insulated from one another. This insulation causes a large electrical resistance along the path of the eddy current, and the magnitude of the eddy current is small. The result is negligible energy losses.

Transformers

A **transformer** is a device used to increase or decrease an ac voltage. It consists of two coils of wire, known as the **primary** and **secondary** coils. The primary is connected to a source of emf and the secondary to a device usually referred to as the load. A schematic of a simple transformer is shown at right. When the current changes in the primary, the magnetic field it produces changes. As the changing field produced by the primary passes through the secondary, an induced emf is produced. If the secondary is part of a closed circuit, then an induced current will pass through it. The advantage of a transformer is that it is possible to produce a higher or lower voltage in the secondary as compared to the primary. This is accomplished by having different numbers of turns of wire in the two coils. If the number of turns of wire in the primary is greater than in the secondary, then the voltage in the secondary is lower than in the primary and the transformer is

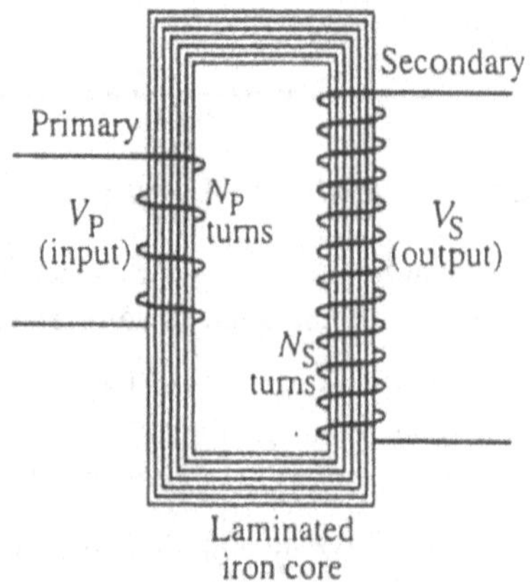

a **step-down** transformer. If the number of turns in the primary is less than in the secondary, then the voltage in the secondary is higher than in the primary and the transformer is a **step-up** transformer. We assume that (nearly) all the magnetic flux produced by the current in the primary coil also passes through the secondary coil and that we can ignore energy losses.

V_S and V_P are the voltages in the secondary and primary coils, and N_S and N_P are the number of turns in each coil. The following equation reflects the theoretical limit for a transformer:

$$\frac{V_S}{V_P} = \frac{N_S}{N_P}$$

Based on the law of conservation of energy, the power output in the secondary cannot be greater than the power input in the primary. In the "ideal" transformer they would be equal; thus

$$P_{\text{input}} = P_{\text{output}}$$

$$I_P V_P = I_S V_S \quad \text{and} \quad \frac{I_S}{I_P} = \frac{V_P}{V_S}$$

Power losses in transmission lines are due to heat, $P = I^2 R,$ so there is considerable advantage to transmitting power over long distances at high voltage but low current. Thus transformers can step the voltage up at the source and step the voltage down at the output. The ratio of the currents is related to the ratio of the turns as follows:

$$\frac{I_S}{I_P} = \frac{N_P}{N_S}$$

EXAMPLE PROBLEM 5. (Section 21-7) The primary winding of a transformer contains 300 turns and draws 4.0 A from a 120-V source of emf. If the secondary contains 600 turns, determine the emf and current in the secondary. Assume that the transformer is 100% efficient.

Part a. Step 1.

Complete a data table.	$V_P = 120$ volts	$V_S = ?$
	$N_P = 300$ turns	$N_S = 600$ turns
	$I_P = 4.0$ A	$I_S = ?$

Part a. Step 2.

Determine the emf in the secondary.

$$\frac{V_S}{V_P} = \frac{N_S}{N_P}$$

$$\frac{V_S}{120\text{ V}} = \frac{600}{300}$$

$$V_S = 240\text{ V}$$

Part a. Step 3.

Determine the current in the secondary.

The transformer is 100% efficient; therefore, the power output in the secondary equals the power input in the primary.

$$P_{\text{input}} = P_{\text{output}}$$

$$I_{\text{P}}V_{\text{P}} = I_{\text{S}}V_{\text{S}}$$

$$(4.0\text{ A})(120\text{ V}) = I_{\text{S}}(240\text{ V})$$

$$I_{\text{S}} = 2.0\text{ A}$$

Mutual Inductance

If two coils are placed near one another, then a changing current in coil 1 creates a changing magnetic field that passes through coil 2. According to Faraday's law, an induced emf is produced in coil 2. The induced emf in coil 2 is given by

$$\xi_2 = -M\frac{\Delta I_1}{\Delta t},$$

where M is a proportionality constant called the **mutual inductance** and $\Delta I_1/\Delta t$ is the rate of change of the current in the first coil. If the current is changing in coil 2 rather than coil 1, then the induced emf occurs in coil 1, and

$$\xi_1 = -M\frac{\Delta I_2}{\Delta t}$$

The value of M is the same whether the emf occurs in coil 1 or coil 2. The unit of mutual inductance is the **henry** (H), where $1\text{ H} = 1$ ohm-second. The value of M depends on whether or not iron is present, the size of the coils, the number of turns, and the distance between the coils.

Self-Inductance

Self-inductance appears in a single coil when the current in the coil changes. The changing current produces a changing magnetic field in the coil, and this induces a back emf in the coil. The back emf tends to retard the flow of current if the current in the coil is increasing and induces an emf in the same direction as the current flow if the current is decreasing. The average induced emf is

$$\xi = -L\frac{\Delta I}{\Delta t},$$

where $\frac{\Delta I}{\Delta t}$ is the rate of change of current in the coil and L is a proportionality constant measured in henrys (H).

The self-inductance (L) of a long coil, called a solenoid, of length ℓ and cross-sectional area A that contains N turns is $L = \frac{\mu_0 N^2 A}{\ell}$.

EXAMPLE PROBLEM 6. (Section 21-9) The current in a 100-mH coil changes from -40.0 mA to $+40.0\text{ mA}$ in 5.00 ms. Determine the magnitude of the induced emf.

Part a. Step 1.

Use the formula for the induced emf in a coil of self-inductance L.

The induced emf in a coil in which the current is changing is given by

$$\xi = -L\frac{\Delta I}{\Delta t}$$

but $\Delta I = I_f - I_i = 40.0\text{ mA} - (-40.0\text{ mA}) = +80.0\text{ mA}$

$$\xi = -\frac{(100\times10^{-3}\text{ H})(+80.0\times10^{-3}\text{ A})}{5.00\times10^{-3}\text{ s}}$$

$$\xi = -1.60\text{ V}$$

Energy Stored in a Magnetic Field

The energy stored in a coil of inductance L, carrying a current I, is given by energy $U = \frac{1}{2}LI^2$. The energy stored in the inductor's magnetic field is given by energy $U = \frac{1}{2}\frac{B^2}{\mu_0}A\ell$. The volume enclosed by the windings of the coil equals the product of A and ℓ. The energy stored per unit volume, or **energy density**, is given by the formula energy density $u = \frac{1}{2}\frac{B^2}{\mu_0}$.

EXAMPLE PROBLEM 7. (Sections 21-10 and 21-11) A 2000-turn, air-filled coil is 2.00 m long and 0.100 m in diameter and carries a current of 1.00 A. Determine the (*a*) self-inductance of the coil and (*b*) energy stored in the coil's magnetic field.

Part a. Step 1.

Determine the cross-sectional area of the coil.

$$A = \frac{\pi d^2}{4} = \frac{\pi(0.100\text{ m})^2}{4} = 7.85\times10^{-3}\text{ m}^2$$

Part a. Step 2.

Determine the self-inductance of the coil.

$$L = \frac{\mu_0 N^2 A}{\ell}$$

$$= \frac{(4\pi\times10^{-7}\text{ T}\cdot\text{m/A})(2000)^2(7.85\times10^{-3}\text{ m}^2)}{2.00\text{ m}}$$

$$L = 0.0197\text{ H} = 19.7\text{ mH}$$

Part b. Step 1.

Determine B if the magnitude of the field produced by a long coil is given by $B = \frac{\mu_0 NI}{\ell}$.

$$B = \frac{\mu_0 NI}{\ell}$$

$$= \frac{(4\pi \times 10^{-7}\ \text{T}\cdot\text{m/A})(2000)(1.00\ \text{A})}{2.00\ \text{m}}$$

$$B = 1.26\times10^{-3}\ \text{T}$$

Part b. Step 2.

Determine the energy stored in the magnetic field.

$$U = \text{energy} = \tfrac{1}{2}\frac{B^2}{\mu_0}A\ell$$

$$= \tfrac{1}{2}\left(\frac{(1.26\times10^{-3}\ \text{T})^2}{4\pi\times10^{-7}\ \text{T}\cdot\text{m/A}}\right)(7.85\times10^{-3}\ \text{m}^2)(2.00\ \text{m})$$

$$U = 9.86\times10^{-3}\ \text{J}$$

LR Circuits

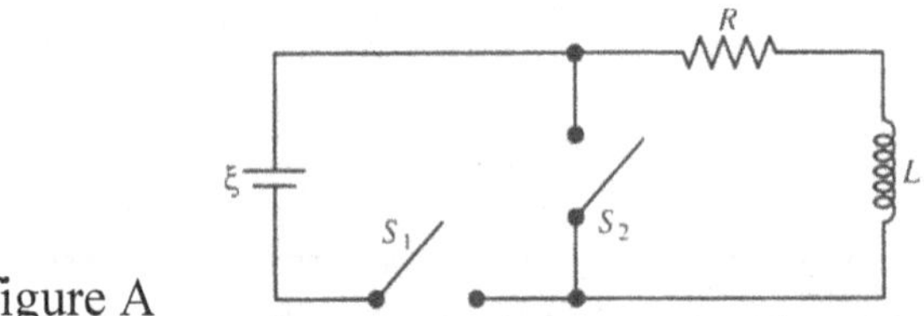

figure A

Figure A is a schematic drawing of an ***LR* circuit**. If switch 1 (S_1) is closed, then the source of emf ξ is connected to the inductor L, which has resistance R. The current through the circuit increases according to the formula

$$I = \frac{\xi}{R}(1 - e^{-t/\tau}),$$

where t is the time in seconds since the switch was closed, and τ is the inductive time constant of the circuit, $\tau = L/R$. It can be shown that when $t = \tau$, the current in the circuit has reached 63% of its maximum value; that is, $I = 0.63I_{\text{max}}$. Figure B shows the growth of current in the circuit as a function of time.

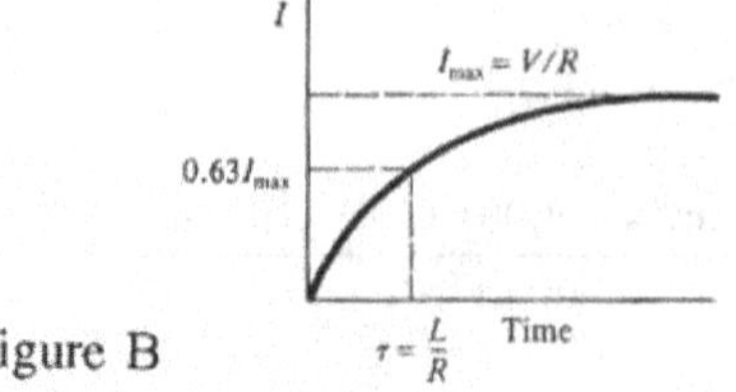

figure B

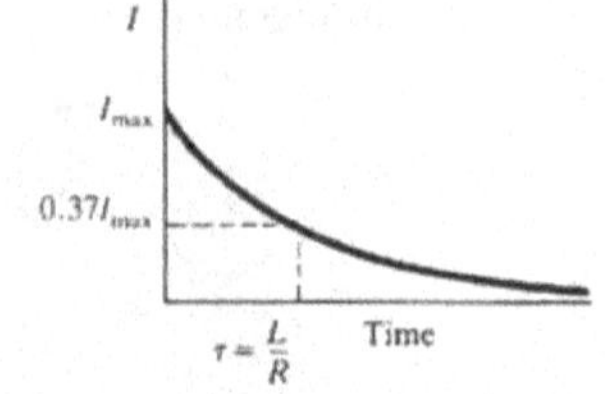

figure C

If switch 1 is disconnected after a steady current I is reached, then the battery is removed from the circuit and no current flows. If switch 2 is now connected, then the energy stored in the inductor L causes a current to flow as follows:

$$I = \frac{\xi}{R} e^{-t/\tau}$$

This decay current is represented graphically in figure C. It can be shown that for a decay current, the current reaches 37% of its initial value $(0.37I_{\text{max}})$ after one time constant $(t = \tau)$.

AC Circuits: Resistors

An ac source connected to a resistor produces a current (I) and a voltage drop (V) across the resistor given by the following equations:

$$I = I_0 \cos 2\pi ft \quad \text{and} \quad V = V_0 \cos 2\pi ft,$$

where I_0 and V_0 are the peak values, f is frequency of the source in hertz, and t is the time in seconds.

The instantaneous voltage and current are said to be in phase since both are zero at the same moment of time and both reach their maximum values in either direction at the same time. For the circuit shown in the diagram, the accompanying graphs represent the voltage and current through the resistor as a function of time.

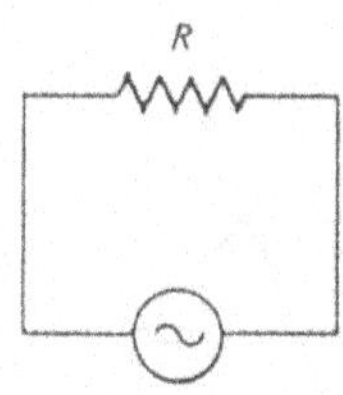

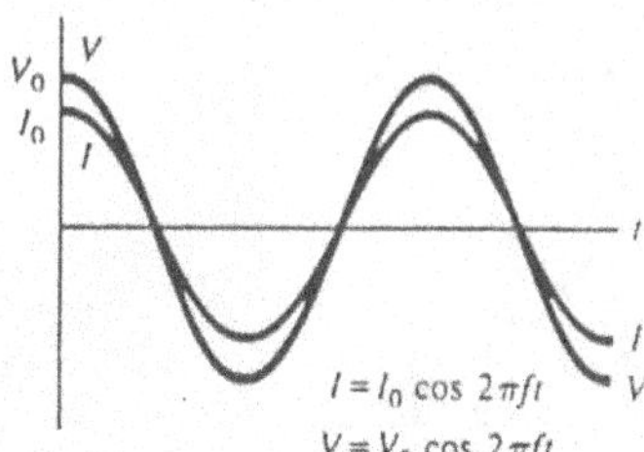

The relationship between V and I follows Ohm's law $I = V/R$, and the average power dissipated is given by

$$\overline{P} = \overline{IV} = I_{\text{rms}}{}^2 R = \frac{V_{\text{rms}}^2}{R}$$

where rms refers to root-mean-square and $I_{\text{rms}} = 0.707 I_0$ and $V_{\text{rms}} = 0.707 V_0$.

AC Circuits: Inductors

The current and voltage in an inductor connected to an ac source are given by

$$I = I_0 \cos 2\pi ft \quad \text{and} \quad V = V_0 \sin 2\pi ft$$

The current lags the voltage and is out of phase by 90°; that is, the current reaches its maximum and minimum values $\frac{1}{4}$ cycle after the voltage.

An inductor produces a back emf $\xi = -L\dfrac{\Delta I}{\Delta t}$ and therefore resists the flow of current through it. Energy from the source is momentarily stored in the magnetic field, and as the field decreases, the energy is transferred back to the source. Thus no power is dissipated in the inductor.

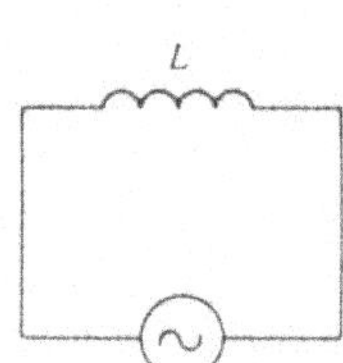

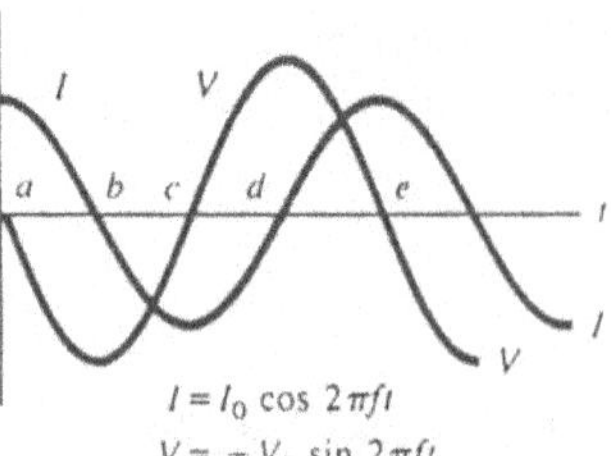

The magnitude of the current is related to the applied voltage at a given frequency by the equation

$$V_L = IX_L$$

where $X_L = 2\pi fL$. X_L is the **inductive reactance** of the inductor and has units of ohms. The inductive reactance is related to both the frequency of the source and the magnitude of the self-inductance of the coil.

The current and the voltage are not in phase; that is, the peak current and peak voltage in the inductor are not reached at the same time. The equation $V = IX_L$ is valid on the average but not at a particular instant of time.

AC Circuits: Capacitors

The current and voltage in a capacitor connected to an ac source are given by

$$I = I_0 \cos 2\pi ft \quad \text{and} \quad V = V_0 \sin 2\pi ft$$

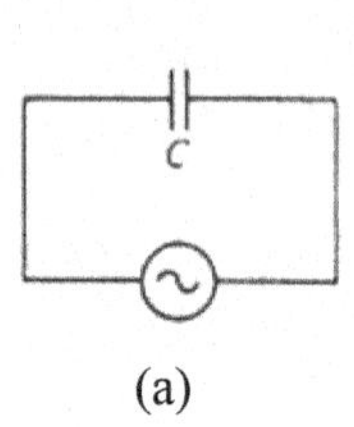

(a)

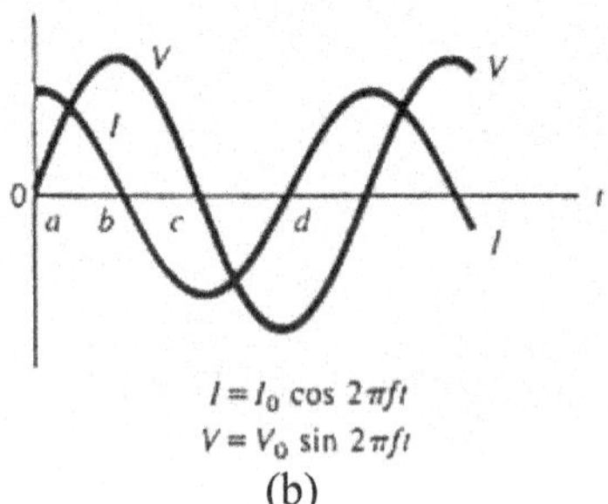

(b)

The current leads the voltage and is out of phase by $90°$; that is, the current reaches maximum and minimum values $\frac{1}{4}$ cycle before the voltage.

The average power dissipated in the capacitor is zero. Energy from the source is stored in the electric field between the plates of the capacitor. As the electric field decreases, the energy is transferred back to the source.

The magnitude of the current is related to the applied voltage at a given frequency by the equation

$$V = IX_C$$

where $X_C = \dfrac{1}{2\pi fC}$. X_C is the **capacitive reactance** and has units of ohms.

Capacitive reactance is inversely proportional to the frequency of the source and the capacitance of the capacitor. As in the case of the inductor, the current and voltage are not in phase, and while the equation $V = IX_C$ is valid on the average, it is not valid at a particular instant of time.

EXAMPLE PROBLEM 8. (Section 21-13) (*a*) Determine the reactance of a 100.0-mH inductor connected to a 120-Hz ac line. (*b*) The inductor is replaced with a 12.5-μF capacitor. Determine the reactance.

Part a. Step 1.

Determine the inductive reactance.

$$X_L = 2\pi fL$$

$$= 2\pi(120\text{ Hz})(100.0\times10^{-3}\text{ H})$$

$$X_L = 75.4\ \Omega$$

Part b. Step 1.

Determine the reactance of the capacitor.

$$X_C = \frac{1}{2\pi fC} = \frac{1}{2\pi(120\text{ Hz})(12.5\times10^{-6}\text{ F})}$$

$$X_C = 106\ \Omega$$

LRC Series Circuits

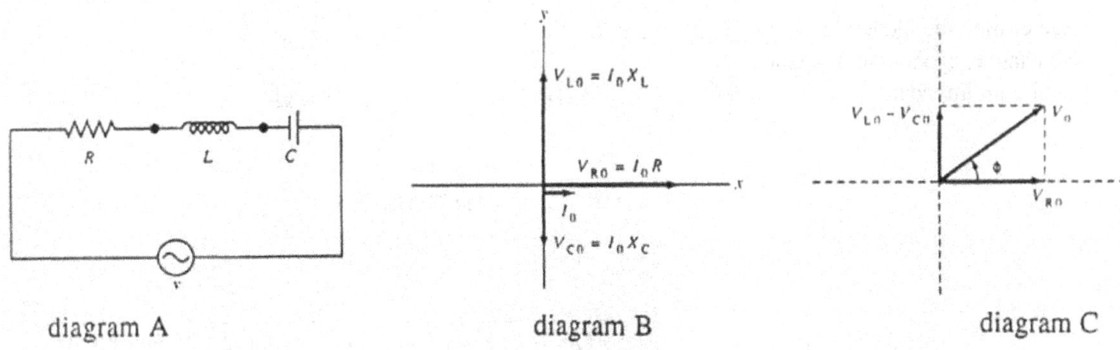

diagram A diagram B diagram C

An *LRC* circuit (see diagram A) can be analyzed by using a **phasor diagram**. In a phasor diagram, vector-like arrows are drawn in an *xy* coordinate system to represent V_{R0}, V_{L0}, and V_{C0}. As shown in diagram B, the magnitude of V_{R0} is represented by an arrow drawn along the $+x$ axis, V_{L0} along the $+y$ axis, and V_{C0} along the $-y$ axis.

When V_{L0} is greater than V_{C0}, diagram B reduces to diagram C. The resultant of $(V_{L0} - V_{C0})$ and V_{R0} is V_0. The peak voltage from the source (V_0) is out of phase with V_{R0} and the current by the angle ϕ, where ϕ is the phase angle. Using the Pythagorean theorem, it can be shown that

$$V = \sqrt{V_{R0}^2 + (V_{L0} - V_{C0})^2}$$

The **impedance** Z of an *LRC* circuit can be determined in a like manner.

$$Z = \sqrt{R^2 + (X_L - X_C)^2},$$

where Z is measured in ohms. The phase angle between the voltage and current can be determined from $\cos\phi = R/Z$, or

$$\tan\phi = (X_L - X_C)/R$$

The power dissipated by the impedance is given by

$$P = I_{rms}V_{rms}\cos\phi,$$

where $\cos\phi$ is the power factor. If the circuit contains only resistance, $\phi = 0°$, $\cos 0° = 1$, and

$$P = I_{rms}V_{rms}$$

If the circuit contains a pure capacitor and/or inductor, $\phi = 90°$ and $\cos 90° = 0$, and no power is dissipated; that is, $P = 0$.

EXAMPLE PROBLEM 9. (Section 21-13) An *LRC* circuit has a 200.0-mH inductor with $40.0\ \Omega$ of resistance connected in series with a $100\text{-}\mu\text{F}$ capacitor and a 60-Hz, 120-volt ac source. Determine the (*a*) total impedance, (*b*) phase angle, (*c*) rms current, and (*d*) power dissipated in the circuit.

Part a. Step 1.

Determine the inductive reactance, capacitive reactance, and total impedance.

$$X_L = 2\pi fL$$

$$X_L = 2\pi(60\text{ Hz})(200.0\times10^{-3}\text{ H}) = 75.4\ \Omega$$

$$X_C = \frac{1}{2\pi fC} = \frac{1}{2\pi(60\text{ Hz})(100\times10^{-6}\text{ F})}$$

$$X_C = 26.5\ \Omega$$

$$Z = \sqrt{R^2 + (X_L - X_C)^2} = \sqrt{(40.0\ \Omega)^2 + (75.4\ \Omega - 26.5\ \Omega)^2}$$

$$Z = 63.2\ \Omega$$

Part b. Step 1.

Determine the phase angle.

$$\tan\phi = \frac{X_L - X_C}{R}$$

$$\tan\phi = \frac{75.4\ \Omega - 26.5\ \Omega}{40.0\ \Omega} = 1.22$$

$$\phi = 50.7°$$

Part c. Step 1.

Determine the rms current.

$$I_{\text{rms}} = \frac{V_{\text{rms}}}{Z} = \frac{120\text{ V}}{63.2\ \Omega} = 1.90\text{ A}$$

Part d. Step 1.

Determine the power dissipated in the circuit.

$$P = IV\cos\phi$$

$$P = (1.90\text{ A})(120\text{ V})\cos 50.7° = 144\text{ W}$$

or

$$P = (I_{\text{rms}})^2 R = (1.90\text{ A})^2(40.0\ \Omega) = 144\text{ W}$$

Resonance in AC Circuits

Since $Z = \sqrt{R^2 + (X_L - X_C)^2}$, the impedance in an ac circuit is a minimum when $X_L - X_C$. At this point $I_{\text{rms}} = V_{\text{rms}}/R$ and I_{rms} is a maximum. Also, the energy transferred to the system is a maximum, and a condition known as **resonance** occurs.

For example, a resonant circuit in a radio is used to tune in a particular station. A range of frequencies reach the circuit; however, a significant current flows only at or near the **resonant frequency** (f_0). At resonance $X_L = X_C$; therefore,

$$2\pi f_0 L = \frac{1}{2\pi f_0 C}$$

and rearranging gives

$$f_0 = \frac{1}{2\pi}\sqrt{\frac{1}{LC}},$$

where f_0 is the resonant frequency of the circuit in hertz.

TEXTBOOK QUESTION 27. (Section 21-15) Describe how to make the impedance in an *LRC* circuit a minimum.

ANSWER: For an *LRC* circuit $Z = \sqrt{R^2 + (X_L - X_C)^2}$. The impedance is a minimum when $X_L = X_C$. At this point $Z = R$, the impedance is a minimum, and a condition known as resonance occurs. The resonant frequency (f_0) is inversely related to both *L* and *C*:

$$f_0 = \frac{1}{2\pi}\sqrt{\frac{1}{LC}}$$

EXAMPLE PROBLEM 10. (Sections 21-14 and 21-15) An *LRC* circuit has a 1.50-mH inductor with 100 Ω of resistance connected in series with a 100-μF capacitor and a 120-volt ac source. Determine the (*a*) resonant frequency and (*b*) peak current for the circuit.

Part a. Step 1.

Derive a formula for the resonant frequency.

The impedance reaches its minimum value at resonance. At resonance, $X_L = X_C$ and $Z = R$.

$$X_L = X_C$$

$$2\pi f_0 L = \frac{1}{2\pi f_0 C}$$

Rearranging gives

$$f_0^2 = \frac{1}{4\pi^2 LC}$$

$$= \frac{1}{4\pi^2 (1.50\times10^{-3}\text{ H})(100\times10^{-6}\text{ F})}$$

$$f_0^2 = 1.69\times10^5\text{ Hz}^2 \quad \text{and}$$

$$f_0 = 411\text{ Hz}$$

Part b. Step 1.

Determine the peak voltage.

$$V_{\text{rms}} = 120\text{ V}$$

The peak voltage $V_0 = \sqrt{2}\, V_{\text{rms}}$. Therefore,

$$V_0 = \sqrt{2}(120\text{ V}) = 170\text{ V}$$

Part b. Step 2.

Determine the peak current.

The peak current occurs at the resonant frequency when the impedance to current flow is at a minimum. At this frequency, $Z = R$, thus

$$I = \frac{V_0}{R} = \frac{170\text{ V}}{100\ \Omega}$$

$$= 1.70\text{ A}$$

PROBLEM SOLVING SKILLS

For a straight wire moving at speed v through a uniform magnetic field:

1. Use $\xi = B\ell v \sin\theta$ to determine the magnitude of the induced emf.
2. If the wire is part of a closed circuit, use Ohm's law to determine the current through the wire and the equation $F = I\ell B \sin\theta$ to determine the magnitude of the force that opposes the mechanical force acting on the wire.

For problems involving a changing magnetic flux through a loop of wire:

1. Determine the rate of change of flux $\frac{\Delta\Phi_B}{\Delta t}$.
2. Apply Faraday's law to determine the magnitude of the induced emf.
3. Apply Ohm's law to determine the magnitude of the induced current and Lenz's law to determine its direction.

For transformer problems:

1. Complete a data table listing the voltage, current, and number of turns of wire in both the primary and secondary coils.
2. Use the mathematical equation that relates the voltage and number of turns in the primary to the voltage and number of turns in the secondary.
3. An ideal transformer is 100% efficient. The power in the primary equals the power in the secondary. Use this concept to solve for the current in the secondary.

For problems involving ac circuits and impedance:

1. Use the formulas that relate the frequency to reactance to determine the inductive reactance and capacitive reactance.
2. Draw phasor diagrams and determine the voltage drop across each circuit element as well as the impedance in the circuit.
3. Use a phasor diagram to determine the phase angle. Determine the power factor.
4. If requested, determine the rms current and the average power dissipated.
5. If requested, determine the resonant frequency.

SOLUTIONS TO SELECTED TEXTBOOK PROBLEMS

TEXTBOOK PROBLEM 1. (Sections 21-1 to 21-4) The magnetic flux through a coil of wire containing two loops changes from -58 Wb to $+38$ Wb in 0.34 s. What is the emf induced in the coil?

Part a. Step 1.

Determine the change in flux through the loop.

$$\Delta\Phi_B = \Phi_f - \Phi_i$$

$$\Delta\Phi_B = (+38 \text{ Wb}) - (-58 \text{ Wb}) = +96 \text{ Wb}$$

Part a. Step 2.

Use Faraday's law to determine the magnitude of the induced emf.

$$\xi = -N\frac{\Delta\Phi}{\Delta t}$$

$$= -(2)\frac{+96\text{ Wb}}{0.34\text{ s}}$$

$$\xi = -565\text{ V}$$

TEXTBOOK PROBLEM 3. (Sections 21-1 to 21-4) The rectangular loop in Fig. 21-58 is being pushed to the right, where the magnetic field points inward. In what direction is the induced current? Explain your reasoning.

ANSWER: As shown in Fig. 21-58, the loop is pushed into the magnetic field. According to Lenz's law, the induced current in the loop will be in a direction to oppose the inwardly directed magnetic field. Therefore, the induced current in the loop will produce an outwardly directed magnetic field. Using the right-hand rule, it can be shown that the current in the loop will be in a counterclockwise direction.

TEXTBOOK PROBLEM 6. (Sections 21-1 to 21-4) A fixed 10.8-cm-diameter wire coil is perpendicular to a magnetic field 0.48 T pointing up. In 0.16 s, the field is changed to 0.25 T pointing down. What is the average induced emf in the coil?

Part a. Step 1.

Determine the area of the loop.

$$r = \left(\frac{10.8\text{ cm}}{2}\right)\left(\frac{1.0\text{ m}}{100\text{ cm}}\right) = 0.054\text{ m}$$

$$A = \pi r^2 = \pi(0.054\text{ m})^2 = 9.15\times10^{-3}\text{ m}^2$$

Part a. Step 2.

Determine the magnitude of the initial flux and final flux through the loop.

Let the initial upward flux through the loop be positive.

$$\Phi_i = B_i A = (+0.48\text{ T})(9.15\times10^{-3}\text{ m}^2) = +4.4\times10^{-3}\text{ Wb}$$

The magnetic field now points downward. Therefore, the final flux through the loop will be negative.

$$\Phi_f = B_f A = (-0.25\text{ T})(9.15\times10^{-3}\text{ m}^2) = -2.3\times10^{-3}\text{ Wb}$$

Part a. Step 3.

Use Faraday's law to determine the induced voltage in the loop.

$$\xi = -\frac{\Delta\Phi_B}{\Delta t} = -\frac{\Phi_f - \Phi_i}{\Delta t}$$

$$= -\frac{(-2.3\times10^{-3}\ \text{Wb} - +4.4\times10^{-3}\ \text{Wb})}{0.16\ \text{s}}$$

$$\xi = 4.2\times10^{-2}\ \text{V}$$

TEXTBOOK PROBLEM 17. (Section 21-3) In Fig. 21-11, the rod moves with a speed of 1.6 m/s on rails 30.0 cm apart. The rod has a resistance of 2.5 Ω. The magnetic field is 0.35 T, and the resistance of the U-shaped conductor is 21.0 Ω at a given instant. Calculate (*a*) the induced emf, (*b*) the current in the U-shaped conductor, and (*c*) the external force needed to keep the rod's velocity constant at that instant.

Part a. Step 1.

Determine the magnitude of the induced emf.

As shown in Fig. 21-11, every point in the wire is moving at the same speed. Therefore,

$$\xi = B\ell v \sin\theta$$

$$= (0.35\ \text{T})(0.300\ \text{m})(1.6\ \text{m/s})\sin 90^\circ$$

$$\xi = 0.168\ \text{V}$$

Part b. Step 1.

Determine the magnitude of the induced current.

$I = \xi/R$, where $R = 2.5\ \Omega + 21.0\ \Omega = 23.5\ \Omega$

$$I = \frac{0.168\ \text{V}}{23.5\ \Omega} = 7.15\times10^{-3}\ \text{A}$$

Part c. Step 1.

Use Lenz's law to determine the direction of the induced current.

Lenz's law predicts that the induced current must oppose the change that causes it. The current opposes the change by interacting with the external magnetic field to produce a force that opposes the rod's motion. Using the right-hand rule, the conventional current (the flow of positive charge) can be shown to be flowing in a clockwise direction through the U-shaped conductor while the force produced by the current is toward the left in the direction opposite to the rod's motion.

Part c. Step 2.

Determine the magnitude of the external force.

$$F = I\ell B \sin\theta$$

$$= (7.15\times10^{-3}\text{ A})(0.300\text{ m})(0.35\text{ T})\sin 90^\circ$$

$$F = 7.51\times10^{-4}\text{ N}$$

TEXTBOOK PROBLEM 22. (Section 21-5) A generator rotates at 85 Hz in a magnetic field of 0.030 T. It has 950 turns and produces an rms voltage of 150 V and an rms current of 70.0 A. (*a*) What is the peak current produced? (*b*) What is the area of each turn of the coil?

Part a. Step 1.

Use the formulas from Chapter 18 to determine the peak current and the peak voltage.

$$I_{\text{rms}} = 0.707 I_{\text{peak}}$$

$$I_{\text{peak}} = \frac{70.0\text{ A}}{0.707} = 99.0\text{ A}$$

$$V_{\text{rms}} = 0.707 V_{\text{peak}}$$

$$V_{\text{peak}} = \frac{150\text{ V}}{0.707} = 212\text{ V}$$

Part b. Step 1.

Use Eq. 21-5 to determine the area of each turn.

$$\xi = NB\omega A \sin\omega t$$

Note: At peak voltage, $\sin\omega t = 1.0$.

$$212\text{ V} = (950)(0.030\text{ T})(A)(2\pi)(85\text{ Hz})(1.0)$$

$$A = \frac{212\text{ V}}{(950)(0.030\text{ T})(2\pi)(85\text{ Hz})}$$

$$A = 1.39\times10^{-2}\text{m}^2$$

TEXTBOOK PROBLEM 31. (Section 21-7) A model-train transformer plugs into 120-V ac and 0.35 A while supplying 6.8 A to the train. (*a*) What voltage is present across the tracks? (*b*) Is the transformer step-up or step-down?

Part a. Step 1.

Complete a data table.	$V_P = 120\text{ V}$	$V_S = ?$
	$I_P = 0.35\text{ A}$	$I_S = 6.8\text{ A}$

Part a. Step 2.

Determine the current in the secondary.

Assuming that the transformer is 100% efficient, the power output in the secondary equals the power input in the primary.

$$P_{\text{input}} = P_{\text{output}}$$

$$I_P V_P = I_S V_S$$

$$(0.35\text{ A})(120\text{ V}) = (6.8\text{ A})V_S$$

$$V_S = 6.2\text{ V}$$

Part b. Step 1.

Is the transformer step-up or step-down?

Because V_P is greater than V_S, the transformer is a step-down transformer.

TEXTBOOK PROBLEM 40. (Section 21-7) What is the inductance of a coil if the coil produces an emf of 2.50 V when the current changes from -28.0 mA to $+31.0\text{ mA}$ in 14.0 ms?

Part a. Step 1.

Use Eq. 21-9 to solve for the inductance.

$$\xi = -L\frac{\Delta I}{\Delta t}$$

$$2.50\text{ V} = -L\left(\frac{(31.0\times10^{-3}\text{ A})-(-28.0\times10^{-3}\text{ A})}{14.0\times10^{-3}\text{ s}}\right)$$

$$2.50\text{ V} = -L(4.2\text{ A/s})$$

$$L = 0.593\text{ H}$$

TEXTBOOK PROBLEM 71. (Section 21-14) An *LRC* circuit has $L = 14.8 \text{ mH}$ and $R = 4.10\ \Omega$. (*a*) What value must *C* have to produce resonance at 3600 Hz? (*b*) What will be the maximum current if the peak external voltage is 150 V?

Part a. Step 1.

Derive a formula for the resonant frequency (f_0) and then determine the value of the capacitance.

For an *LRC* circuit, the impedance reaches its maximum value at resonance. At resonance, $X_L = X_C$ and $Z = R$.

$$X_L = X_C$$

$$2\pi f_0 L = \frac{1}{2\pi f_0 C}; \quad \text{rearranging gives}$$

$$C = \frac{1}{4\pi L f_0^2} = \frac{1}{4\pi^2 (14.8\times10^{-3}\text{ H})(3600\text{ Hz})^2}$$

$$C = 1.32\times10^{-7}\text{ F}$$

Part b. Step 1.

Use Ohm's law to determine the maximum current.

At resonance, the impedance equals the resistance *R*.

$$I = \frac{V}{Z} \quad \text{but} \quad Z = R$$

$$I = \frac{150\text{ V}}{4.10\ \Omega}$$

$$I = 36.6\text{ A}$$

TEXTBOOK PROBLEM 76. (Section 21-7) A high-intensity desk lamp is rated at 45 W but requires only 12 V. It contains a transformer that converts 120-V household voltage. (*a*) Is the transformer step-up or step-down? (*b*) What is the current in the secondary coil when the lamp is on? (*c*) What is the current in the primary coil? (*d*) What is the resistance of the bulb when on?

Part a. Step 1.

Determine whether the transformer is step-up or step-down.

The voltage in the secondary is less the voltage in the primary. Therefore, the transformer is step-down.

Part b. Step 1.

Determine the current in the secondary.

$P = I_S V_S$

$45\ \text{W} = I_S(12\ \text{V})$

$I_S = 3.75\ \text{A}$

Part c. Step 1.

Determine the current in the primary.

If the transformer is 100% efficient, then the power output in the secondary equals the power input in the primary:

$P_{\text{input}} = P_{\text{output}}$

$I_P V_P = I_S V_S$

$I_P(120\ \text{V}) = (3.75\ \text{A})(12\ \text{V})$

$I_P = 0.375\ \text{A}$

Part d. Step 1.

Use Ohm's law to determine the resistance of the bulb when it is on.

$V_S = I_S R$

$12\ \text{V} = (3.75\ \text{A})R$

$R = 3.2\ \Omega$

ELECTROMAGNETIC WAVES

OBJECTIVES

After studying the material of this chapter, the student should be able to:

- give a nonmathematical summary of Maxwell's equations.
- describe how electromagnetic waves are produced.
- draw a diagram representing the field strengths of an electromagnetic wave produced by a sinusoidally varying source of emf.
- calculate the velocity of electromagnetic waves in a vacuum if both the permittivity and permeability of free space are given.
- state the names given to the different segments of the electromagnetic spectrum.
- state the approximate range of wavelengths associated with each segment of the electromagnetic spectrum.
- state the equation that relates the speed of an electromagnetic wave to the frequency and wavelength and use this equation in problem solving.
- determine the peak magnitude of both the electric and magnetic field strength if the energy density of the electromagnetic wave is given.
- solve problems related to a wave's intensity at a particular point and calculate the peak values of both the electric and magnetic fields at this point.

KEY TERMS AND PHRASES

Maxwell's equations consist of four basic equations that describe all electric and magnetic phenomena.

electromagnetic waves are transverse waves that have both an electric and a magnetic component. Maxwell, based on his equations, concluded that a changing magnetic field (B) will produce a changing electric field (E) directed is at right angles to the magnetic field and that changing electric field will produce a changing magnetic field. The net result of the interaction of the changing E and B fields is the production of an electromagnetic wave that propagates away from the wave source at the speed of light.

electromagnetic spectrum includes radio waves, microwaves, infrared radiation, visible light, X-rays, and gamma rays. These EM waves differ in frequency (f) and wavelength (λ), but all travel at the speed of light.

intensity refers to the energy transported per unit time per unit area. The unit of intensity is watts per square meter, W/m^2.

SUMMARY OF MATHEMATICAL FORMULAS

speed of electromagnetic waves in a vacuum	$v = c = \dfrac{1}{\sqrt{\varepsilon_0 \mu_0}}$	The speed of EM waves in a vacuum is inversely related to the square root of the permittivity of free space (ε_0), where $\varepsilon_0 = 8.85 \times 10^{-12}\ \text{C}^2/\text{N}\cdot\text{m}^2$, and the permeability of free space (μ_0), where $\mu_0 = 4\pi \times 10^{-7}\ \text{T}\cdot\text{m/A}$.
speed of electromagnetic waves	$v = c = f\lambda$	The speed (v) of electromagnetic waves equals the product of the frequency (f) and the wavelength (λ). The speed of light in a vacuum (c) equals 3.00×10^8 m/s.
energy density	$u = \frac{1}{2}\varepsilon_0 E^2 + \frac{1}{2}\dfrac{B^2}{\mu_0}$ $u = \sqrt{\dfrac{\varepsilon_0}{\mu_0}} EB$	u refers to the total energy stored per unit volume at any particular instant. E and B represent the instantaneous magnitudes of the electric and magnetic fields at a particular point.
energy in an electromagnetic wave	$\bar{I} = \dfrac{E_0 B_0}{2\mu_0}$	The average energy transported per unit time per unit area perpendicular to the wave direction is referred to as the average intensity $(\bar{I})$. The unit of I is watts per square meter, W/m^2. E_0 represents the maximum value of the electric field vector, while B_0 represents the maximum value of the magnetic field vector.

CONCEPT SUMMARY

Maxwell's Equations

All electric and magnetic phenomena can be described by four basic equations known as **Maxwell's equations**. The following is a summary of each equation:

(1) Gauss's law for electricity is a generalized form of Coulomb's law and relates electric charge to the electric field that the charge produces.

(2) Gauss's law of magnetism describes magnetic fields and predicts that magnetic fields are continuous; that is, they have no beginning or ending point. The equation predicts that there are no magnetic monopoles.

(3) Faraday's law of induction describes the production of an electric field by a changing magnetic field.

(4) Maxwell's extension of Ampere's law describes the magnetic field produced by a changing electric field or by an electric current.

Production of Electromagnetic Waves

Maxwell concluded that a changing magnetic field (B) will produce a changing electric field (E), and the changing electric field will produce a changing magnetic field. The net result of the interaction of the changing E and B fields is the production of a wave, which has both an electric and a magnetic component and travels through empty space. This wave is referred to as an **electromagnetic (EM) wave**.

Electromagnetic waves are produced by accelerated electric charges. For example, accelerated electrons in atoms give off visible light while oscillating electric charges in an antenna are undergoing acceleration and produce radio waves. The following diagram represents the field strengths of an electromagnetic wave produced by a sinusoidally varying source of emf.

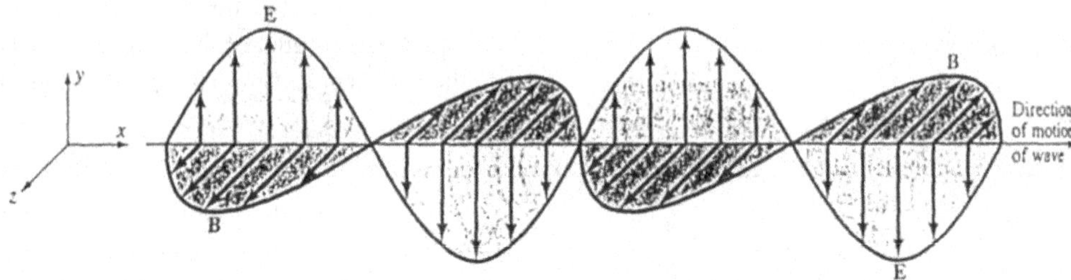

Note that electromagnetic waves are transverse waves with E and B at right angles to one another and perpendicular to the direction of travel.

The speed of EM waves in a vacuum is given by

$$v = c = \frac{1}{\sqrt{\varepsilon_0 \mu_0}},$$

where ε_0 is the permittivity of free space, $\varepsilon_0 = 8.85 \times 10^{-12}\ \text{C}^2/\text{N}\cdot\text{m}^2$, and μ_0 is the permeability of free space, $\mu_0 = 4\pi \times 10^{-7}\ \text{T}\cdot\text{m/A}$.

TEXTBOOK QUESTION 1. (Section 22-2) The electric field in an EM wave traveling north oscillates in an east–west plane. Describe the direction of the magnetic field vector in this wave.

ANSWER: The plane of the magnetic field vector is perpendicular to both the direction of travel of the wave and the plane of the electric field vector. Therefore, the magnetic field vector is in a vertical plane that is perpendicular to the surface of the Earth.

TEXTBOOK QUESTION 2. (Sections 22-3 and 11-8) Is sound an EM wave? If not, what kind of a wave is it?

ANSWER: Sound waves are not electromagnetic waves. Electromagnetic waves are transverse waves with an electric as well as a magnetic component that can travel through a vacuum. In Chapter 11, it was noted that sound waves are longitudinal waves composed of a series of compressions and expansions and require a medium such as air or water in which to travel.

EXAMPLE PROBLEM 1. (Sections 22-3 and 22-4) A communication satellite is placed in a circular orbit approximately 22,400 miles above the Earth. How long would it take a signal to travel from the Earth's surface to the satellite and return to the Earth? Note: The signal travels at 3.0×10^8 m/s.

Part a. Step 1.

Calculate the distance from the Earth to the satellite in meters.	$1.000 \text{ miles} = 1609 \text{ m}$ $d = (22{,}400 \text{ miles})\left(\frac{1609\,\text{m}}{1.000\,\text{mile}}\right)$ $d = 3.6\times10^7 \text{ m}$

Part a. Step 2.

Determine the time for signal to be received back on the Earth.	The round-trip distance that the signal must travel is $2(3.6\times10^7 \text{ m}) = 7.2\times10^7 \text{ m}$: $d = ct$ $7.2\times10^7 \text{ m} = (3.00\times10^8 \text{ m/s})t$ $t = 0.24 \text{ s}$

Electromagnetic Spectrum

Radio waves and visible light are only a small part of what is known as the **electromagnetic** (EM) **spectrum**. The spectrum includes radio waves, microwaves, infrared radiation, visible light, X-rays, and gamma rays. These EM waves differ in frequency (f) and wavelength (λ). Their velocity, frequency, and wavelength are related by the following:

$$v = c = f\lambda$$

where the speed of electromagnetic waves in a vacuum $(c) = 3.00\times10^8$ m/s. This equation represents the speed of all EM waves in free space.

The following table gives the approximate range of wavelengths for each portion of the EM spectrum:

Radio waves	10^4 m to 1 m	Ultraviolet	10^{-8} m to 10^{-9} m
Microwaves	1 m to 10^{-4} m	X-rays	10^{-9} m to 10^{-11} m
Infrared	10^{-4} m to 10^{-7} m	Gamma rays	less than 10^{-12} m
Visible light	4×10^{-7} m to 7×10^{-7} m		

TEXTBOOK QUESTION 5. (Section 22-3) Are the wavelengths of radio and television signals longer or shorter than those detectable by the human eye?

ANSWER: As Fig. 22-8 illustrates, the wavelengths of radio and television signals are longer than those of visible light.

TXTBOOK QUESTION 7. (Section 22-3) In the electromagnetic spectrum, what type of EM wave would have a wavelength of 10^3 km? 1 km? 1 m? 1 cm? 1 mm? 1 μm?

ANSWER: As shown in Fig. 22-8, 10^3 km is in the very long radio wave region of the EM spectrum, 1 km is in the radio wave region, 1 m is in the TV signal and microwave regions, 1 cm is in the microwave and satellite TV signal regions, 1 mm is in the microwave and infrared regions, and 1 μm is in the infrared region.

TEXTBOOK QUESTION 8. (Section 22-3) Can radio waves have the same frequencies as sound waves (20 Hz–20,000 Hz)?

ANSWER: Yes, a radio wave can have the same frequency as a sound wave. However, a radio wave is an electromagnetic transverse wave that travels at approximately 3.0×10^8 m/s, while a sound wave is a longitudinal wave that travels through air at approximately 340 m/s. An EM wave that has a frequency of 20,000 Hz would have a wavelength of approximately about 15 km, while the wavelength of a sound wave having the same frequency would be approximately 17 mm.

EXAMPLE PROBLEM 2. (Sections 22-3 and 22-4) Determine the frequency of green light of wavelength 5.56×10^7 meters.

Part a. Step 1.

Use $c = f\lambda$ to calculate the frequency of the green light.

The frequency, wavelength, and velocity of light are related by the equation $c = f\lambda$, where c refers to the speed of light in a vacuum.

$$f = \frac{c}{\lambda} = \frac{3.00\times10^8 \text{ m/s}}{5.56\times10^{-7} \text{ m}}$$

$$f = 5.40\times10^{14} \text{ Hz}$$

Energy Density of an EM Wave

Electromagnetic waves transport energy from one region of space to another. This energy is associated with the electric and magnetic fields of the wave. The energy per unit volume (energy density) stored in the electric field is given by

$$u = \tfrac{1}{2}\varepsilon_0 E^2,$$

where u is the energy density and has units of joules per cubic meter (J/m^3).

The energy stored per unit volume in the magnetic field is given by

$$u = \frac{1}{2}\frac{B^2}{\mu_0}$$

The total energy stored per unit volume at any particular instant is given by

$$u = \tfrac{1}{2}\varepsilon_0 E^2 + \frac{1}{2}\frac{B^2}{\mu_0}$$

E and B represent the instantaneous magnitudes of the electric and magnetic fields at a particular point. The total energy density can be stated in terms of only the electric field as $\mu = \varepsilon_0 E^2$, in terms of only the magnetic field as $u = B^2/\mu_0$, or in terms of both E and B as

$$u = \sqrt{\frac{\varepsilon_0}{\mu_0}}EB$$

EXAMPLE PROBLEM 3. (Section 22-5) The energy density of an EM wave is 6.00×10^{-5} J/m^3. Determine the peak magnitude of the (*a*) electric field strength and (*b*) magnetic field strength.

Part a. Step 1.

Determine the peak electric field strength.	The energy density is related to the electric field strength by the equation $u = \varepsilon_0 E^2$, but $\overline{E}^2 = \frac{E_0^2}{2}$. Therefore, $u = \tfrac{1}{2}\varepsilon_0 E_0^2$ Rearranging gives $E_0 = \sqrt{\frac{2\mu}{\varepsilon_0}} = \sqrt{\frac{(2)\left(6.00 \times 10^{-5}\,\frac{\text{J}}{\text{m}^3}\right)}{8.85 \times 10^{-12}\,\frac{\text{C}^2}{\text{N}\cdot\text{m}^2}}}$ $E_0 = 3.68\times10^3$ N/C

Part b. Step 1.

Determine the peak magnitude of the magnetic field strength.

The energy density can be expressed in terms of the magnetic field strength as $u = \frac{B^2}{\mu_0}$, but $\overline{B^2} = \frac{B_0^2}{2}$. Therefore,

$$u = \tfrac{1}{2}\frac{B_0^2}{\mu_0}$$

$$6.00 \times 10^{-5}\ \text{J/m}^3 = \frac{\tfrac{1}{2} B_0^2}{4\pi \times 10^{-7}\ \frac{\text{N} \cdot \text{s}^2}{\text{C}^2}}$$

Solving for B_0:

$$B_0 = 1.23 \times 10^{-5}\ \text{T}$$

Alternate solution: It can be shown that $c = \frac{E_0}{B_0}$. Therefore,

$$B_0 = \frac{E_0}{c} = \frac{3.68 \times 10^3\ \frac{\text{N}}{\text{C}}}{3.00 \times 10^8\ \text{m/s}}$$

$$B_0 = 1.23 \times 10^{-5}\ \text{T}$$

Energy in Electromagnetic Waves

The energy transported per unit time per unit area is the intensity *I*. The magnitude of the intensity is given by

$$I = \varepsilon_0 c E^2 \quad \text{or} \quad I = \frac{EB}{\mu_0}$$

The unit of intensity is watts per square meter, W/m^2. The average intensity transported per unit area per unit time is given by

$$\overline{I} = \tfrac{1}{2}\varepsilon_0 c E^2 = \tfrac{1}{2}\frac{c}{\mu_0} B_0^2 = \frac{E_0 B_0}{2\mu_0}$$

where E_0 and B_0 represent the maximum values of *E* and *B*.

EXAMPLE PROBLEM 4. (Section 22-5) A source of EM waves radiates uniformly in all directions. The amplitude of the electric field 1.50×10^4 m from the source is 200 V/m. Determine the average intensity $\overline{I}$.

Part a. Step 1.

Determine the time average value of the intensity.

The average intensity is given by

$$\bar{I} = \tfrac{1}{2}\varepsilon_0 c E_0^2$$

E_0 is the peak value (amplitude) of the electric field; thus

$$\bar{I} = \tfrac{1}{2}(8.85 \times 10^{-12}\ \text{C}^2/\text{N}\cdot\text{m}^2)(3.00 \times 10^8\ \text{m/s})(200\ \text{V/m})^2$$

$$\bar{I} = 53.1\ \text{W/m}^2$$

TEXTBOOK QUESTION 9. (Section 22-7) If a radio transmitter has a vertical antenna, should a receiver's antenna (rod type) be vertical or horizontal to obtain best reception?

ANSWER: As shown in Fig. 22-18a, the electric field produced by a vertical transmitter antenna is vertical. When the wave is received by the receiver antenna, the electrons in a vertical receiver antenna will begin to oscillate up and down. Therefore, the best reception would be obtained by a vertical receiver antenna.

PROBLEM SOLVING SKILLS

For problems involving the velocity, frequency, or wavelength of a periodic electromagnetic wave in a vacuum:

1. Express the wavelength of the wave in meters and the frequency of the wave in hertz.
2. Use the equation $c = f\lambda$ to solve the problem, where $c = 3.0 \times 10^8$ m/s.

For problems related to the energy in electromagnetic waves:

1. Note the instantaneous magnitude of the electric and/or magnetic field vector.
2. Solve for the energy density stored in the electric field using the equation $u = \tfrac{1}{2}\varepsilon_0 E^2$.

   Solve for the energy density stored in the magnitude field using the equation $u = \tfrac{1}{2}\dfrac{B^2}{\mu_0}$.

For problems related to the average intensity:

1. Determine the average intensity by using the equation $\bar{I} = \dfrac{E_0 B_0}{2\mu_0}$.

For problems related to the average power a specified distance from a source of electromagnetic waves which radiates uniformly in all directions:

1. Determine the surface area of an imaginary sphere through which the energy radiates.
2. Determine the average power using the equation $\bar{P} = \bar{I}A$.

SOLUTIONS TO SELECTED TEXTBOOK PROBLEMS

TEXTBOOK PROBLEM 6. (Section 22-2) If the magnetic field in a traveling EM wave has a peak magnitude of 10.5 nT, what is the peak magnitude of the electric field?

Part a. Step 1.

Use Eq. 22-2 to determine the peak electric field.

From Maxwell's equations,

$$c = \frac{E}{B}$$

$$E = (3.0 \times 10^8 \text{ m/s})(10.5 \times 10^{-9} \text{ T}) = 3.15 \text{ V/m}$$

TEXTBOOK PROBLEM 16. (Sections 22-3 and 22-4) How long would it take a message sent as radio waves from Earth to reach Mars when Mars is (*a*) nearest Earth, (*b*) farthest from Earth? Assume that Mars and Earth are in the same plane and that their orbits around the Sun are circles (Mars is $\approx 230 \times 10^6$ km from the Sun). [Hint: See Table 5-2.]

Part a. Step 1.

Use Table 5-2 to determine the shortest distance between Earth and Mars in meters.

From information provided in the textbook, the mean (average) distance from the Sun to the Earth is 149.6×10^6 km, while the mean distance from the Sun to Mars is 227.9×10^6 km. Assuming that Earth and Mars are on the same side of the Sun in their orbit, the distance the message would travel is

$$d = 227.9 \times 10^6 \text{ km} - 149.6 \times 10^6 \text{ km} = 7.83 \times 10^{10} \text{ m}$$

Part a. Step 2.

Determine the time for the message to travel this distance.

Radio waves travel at the speed of light, 3.0×10^8 m/s. The time required to travel this distance can be determined as follows:

$$t = \frac{d}{c} = \frac{7.83 \times 10^{10} \text{m}}{3.00 \times 10^8 \text{m/s}} = 261 \text{ s, or } 4.35 \text{ min}$$

Part b. Step 1.

Determine the farthest distance between Earth and Mars.

The farthest distance occurs when Earth and Mars are on the opposite sides of the Sun. The distance the message would travel is

$$d = 227.9 \times 10^6 \text{ km} + 149.6 \times 10^6 \text{ km} = 377.5 \times 10^6 \text{ km}$$

$$d = 3.775 \times 10^{11} \text{ m}$$

Part b. Step 2.

Determine the time for the message to travel this distance.

The time required to travel this distance can be determined as follows:

$$t = \frac{d}{c} = \frac{3.775 \times 10^{11} \text{ m}}{3.00 \times 10^8 \text{ m/s}} = 1260 \text{ s, or } 21.0 \text{ min}$$

TEXTBOOK PROBLEM 15. (Section 22-3) (*a*) What is the wavelength of a 22.75×10^9 Hz radar signal? (*b*) What is the frequency of an X-ray with wavelength 0.12 nm?

Part a. Step 1.

Determine the wavelength of the radar signal.

$$c = f\lambda$$

$$3.00 \times 10^8 \text{ m/s} = (22.75 \times 10^9 \text{ Hz})\lambda$$

$$\lambda = 0.013 \text{ m}$$

Part b. Step 1.

Determine the frequency of the X-ray.

$$c = f\lambda$$

$$3.00 \times 10^8 \text{ m/s} = f(0.12 \times 10^{-9} \text{ m})$$

$$f = 2.5 \times 10^{18} \text{ Hz}$$

TEXTBOOK PROBLEM 21. (Section 22-4) A student wants to scale down Michelson's light-speed experiment to a size that will fit in one room. An eight-sided mirror is available, and the stationary mirror can be mounted 12 m from the rotating mirror. If the arrangement is otherwise as shown in Fig. 22-10, at what the minimum rate must the mirror rotate?

Part a. Step 1.

Calculate the round-trip distance the light travels.

$d = 12\text{ m} + 12\text{ m} = 24\text{ m}$

Part a. Step 2.

Determine the time for light to travel 24 m.

$t = d/c$

$$t = \frac{24\text{ m}}{3.00 \times 10^8\text{ m/s}} = 8.0 \times 10^{-8}\text{ s}$$

Part a. Step 3.

Determine the minimum angular distance the apparatus must rotate for light to be reflected into the observer's eye.

In order for the light from the distant mirror to reflect into the observer's eye, the apparatus must rotate at least $\frac{1}{8}$ of a revolution during the time the light is traveling the 24 m.

$$\tfrac{1}{8}\text{ revolution}\left(\frac{2\pi\text{ rad}}{1\text{ rev}}\right) = 0.785\text{ rad}$$

Part a. Step 4.

Determine the minimum angular velocity of the apparatus in radians per second and revolutions per minute.

$$\omega = \frac{\Delta\theta}{\Delta t}$$

$$= \frac{0.785\text{ rad}}{8.0 \times 10^{-8}\text{ s}} = 9.8 \times 10^6\text{ rad/s}$$

$$= (9.8 \times 10^6\text{ rad/s})\left(\frac{1\text{ rev}}{2\pi\text{ rad}}\right)\left(\frac{60\text{ s}}{1\text{ min}}\right)$$

$$\omega = 9.4 \times 10^7\text{ rev/min}$$

TEXTBOOK PROBLEM 28. (Sections 11-9 and 22-5) A 15.8-mW laser puts out a narrow beam 2.40 mm in diameter. What are the rms values of E and B in the beam?

Part a. Step 1.

Given the diameter (d), determine the area of the beam.

$$A = \pi r^2 = \tfrac{1}{4}\pi d^2$$

$$= \tfrac{1}{4}\pi(2.40 \times 10^{-3}\text{ m})^2$$

$$A = 4.52 \times 10^{-6}\text{ m}^2$$

Part a. Step 2.

Determine E_{rms}.

From Sections 11-9 and 22-5, the intensity of energy a wave transports per unit time is the power (P) through an area (A):

$$I = \frac{P}{A} = \varepsilon_0 c E_{\text{rms}}^2$$

$$E_{\text{rms}}^2 = \frac{P}{Ac\varepsilon_0}$$

$$= \frac{15.8\times10^{-3}\ \text{W}}{(4.52\times10^{-6}\ \text{m}^2)(3.00\times10^{8}\ \text{m/s})(8.85\times10^{-12}\ \text{C}^2/\text{N}\cdot\text{m}^2}$$

$$E_{\text{rms}}^2 = 1.32\times10^{6}\ \text{V}^2/\text{m}^2$$

$$E_{\text{rms}} = 1.15\times10^{3}\ \text{V/m}$$

Part a. Step 3.

Determine B_{rms}.

Using Eq. 22-2 from the textbook gives

$$c = \frac{E_{\text{rms}}}{B_{\text{rms}}}$$

$$B_{\text{rms}} = \frac{E_{\text{rms}}}{c} = \frac{1.15\times10^{3}\ \text{V/m}}{3.00\times10^{8}\ \text{m/s}}$$

$$B_{\text{rms}} = 3.82\times10^{-6}\ \text{T}$$

TEXTBOOK PROBLEM 47. (Sections 22-3 and 22-4) The voice from an astronaut on the Moon (Fig. 22-22) was beamed to a listening crowd on Earth. If you were standing 28 m from the loudspeaker on Earth, what was the total time lag between when you heard the sound and when the sound entered a microphone on the Moon? Explain whether the microphone was inside the space helmet or outside, and why.

Part a. Step 1.

Determine the time for the radio wave to travel from the Moon to the Earth.

Radio waves travel at the speed of light, 3.0×10^{8} m/s. The distance between the Earth and Moon is approximately 3.48×10^{8} m.

$$t = \frac{d}{c}$$

$$t = \frac{3.84 \times 10^8 \text{ m}}{3.00 \times 10^8 \text{ m/s}} = 1.28 \text{ s}$$

Part a. Step 2.

Determine the time for the sound wave to travel 28 m.

At 20°C, the speed of sound is 343 m/s:

$$t = \frac{d}{v_{\text{sound}}}$$

$$t = \frac{28 \text{ m}}{343 \text{ m/s}} = 0.0816 \text{ s}$$

Part a. Step 3.

Determine the total time from the astronaut to the listener's ear.

$$t_{\text{total}} = 1.28 \text{ s} + 0.0816 \text{ s} = 1.36 \text{ s}$$

Sound does not travel through a vacuum. The Moon has no atmosphere; therefore, the microphone is inside the astronaut's helmet.

TEXTBOOK PROBLEM 40. (Section 22-7) A certain FM radio tuning circuit has a fixed capacitor $C = 810$ pF. Tuning is done by a variable inductance. What range of values must the inductance have to tune stations from 88 MHz to 108 MHz?

Part a. Step 1.

Derive a formula for resonant frequency.

Resonance in ac circuits was discussed in Section 21-14. At resonance, the inductive reactance (X_L) equals the capacitive reactance (X_C), and the impedance is at a minimum.

$$X_L = X_C$$

$$2\pi f_0 L = \frac{1}{2\pi f_0 C}$$

$$f_0^2 = \frac{1}{4\pi^2 LC}$$ and rearranging gives

$$L = \frac{1}{4\pi^2 f_0^2 C}$$

Part a. Step 2.

Determine the inductance to tune an 88-MHz station.

$$L = \frac{1}{4\pi^2 f_0^2 C}$$

$$L = \frac{1}{\left[4\pi^2 (88 \times 10^6 \text{ Hz})^2 (810 \times 10^{-12} \text{ F})\right]}$$

$$L = 4.04 \times 10^{-9} \text{ H} = 4.04 \text{ nH}$$

Part a. Step 3.

Determine the inductance to tune a 108-MHz station.

$$L = \frac{1}{4\pi^2 f_0^2 C}$$

$$= \frac{1}{\left[4\pi^2 (108 \times 10^6 \text{ Hz})^2 (810 \times 10^{-12} \text{ F})\right]}$$

$$L = 2.68 \times 10^{-9} \text{ H} = 2.68 \text{ nH}$$

Part a. Step 4.

State the range of required inductance.

The required inductance ranges from 2.68 nH to 4.04 nH.

TEXTBOOK PROBLEM 51. (Section 22-5) What are E_0 and B_0 at a point 2.50 m from a light source whose output is 18 W? Assume the bulb emits radiation of a single frequency uniformly in all directions.

Part a. Step 1.

Determine the energy radiated per unit area per unit time.

$$\overline{I} = \overline{P}/A, \quad \text{where } A = 4\pi r^2$$

$$= \frac{18.0 \text{ W}}{4\pi (2.50 \text{ m})^2}$$

$$\overline{I} = 0.23 \text{ W/m}^2$$

Part a. Step 2.

Determine the peak value (E_0) of the electric field vector.

$$\bar{I} = \tfrac{1}{2}\varepsilon_0 c E_0^2$$

$$0.23\ \text{W/m}^2 = \tfrac{1}{2}(8.85 \times 10^{-12}\ \text{C}^2/\text{N}\cdot\text{m}^2)(3.00 \times 10^8\ \text{m/s})E_0^2$$

$$E_0^2 = 1.73 \times 10^2\ \text{N}^2/\text{C}^2$$

$$E_0 = \sqrt{E_0^2} = \sqrt{173\ \text{N}^2/\text{C}^2}$$

$$E_0 = 13.2\ \text{N/C}$$

Part a. Step 3.

Determine the peak value (B_0) of the magnetic field vector.

$$B_0 = \frac{E_0}{c}$$

$$= \frac{13.2\ \text{N/C}}{3.0 \times 10^8\ \text{m/s}}$$

$$B_0 = 4.4 \times 10^{-8}\ \text{T}$$

TEXTBOOK PROBLEM 55. (Section 22-5) A radio station is allowed to broadcast at an average power not to exceed 25 kW. If an electric field amplitude of 0.020 V/m is considered to be acceptable for receiving the radio transmission, estimate how many kilometers away you might be able to detect this station.

Part a. Step 1.

The average intensity is given by Eq. 22-8.

$$\bar{I} = \tfrac{1}{2}\varepsilon_0 c E_0^2$$

$$= \tfrac{1}{2}(3.00 \times 10^8\ \text{m/s})(8.85 \times 10^{-12}\ \text{C}^2/\text{N}\cdot\text{m}^2)(0.020\ \text{V/m})^2$$

$$\bar{I} = 5.31 \times 10^{-7}\ \text{W/m}^2$$

Part a. Step 2.

Determine the radial distance from the station to the farthest point for reception.

$\overline{P} = \overline{I}A, \quad \text{where } A = 4\pi r^2$

$25 \times 10^3 \text{ W} = (5.31 \times 10^{-7} \text{ W/m}^2)(4\pi r^2)$

$r^2 = 3.7 \times 10^9 \text{ m}^2$

$r = 6.1 \times 10^4 \text{ m} \approx 61 \text{ km}$

23

LIGHT: GEOMETRIC OPTICS

OBJECTIVES

After studying the material of this chapter, the student should be able to:

- distinguish between specular reflection and diffuse reflection.
- draw a ray diagram and locate the position of the image produced by an object placed a specified distance from a plane mirror. State the characteristics of the image.
- distinguish between a convex and a concave mirror. Draw rays parallel to the principal axis and locate the position of the principal focal point of each type of spherical mirror.
- draw ray diagrams and locate the position of the image produced by an object placed a specified distance from a concave or convex mirror. State the characteristics of the image.
- use the mirror equations and the sign conventions to determine the position, magnification, and size of the image produced by an object placed a specified distance from a spherical mirror.
- state Snell's law and use this law to predict the path of a light ray as it travels from one medium into another. Explain what is meant by the index of refraction of a medium.
- explain what is meant by total internal reflection. Use Snell's law to determine the critical angle as light travels from a medium of higher index of refraction into a medium of lower index of refraction.
- distinguish between a convex and a concave lens. Draw rays parallel to the principal axis and locate the position of the principal focal points for each type of thin lens.
- draw ray diagrams and locate the position of the image produced by an object placed a specified distance from both types of thin lens. State the characteristics of the image.
- use the thin lens equation and the sign conventions to determine the position, magnification, and size of the image produced by an object placed a specified distance from a concave or convex lens.

KEY TERMS AND PHRASES

ray of light is a single beam of light that travels in a straight line until it strikes an obstacle. If the ray strikes a highly reflective surface, then **specular** (or mirror) **reflection** occurs. If the ray strikes a rough surface, then **diffuse reflection** results.

law of reflection state that (1) when a beam of light strikes a flat surface, the incident ray, the reflected ray, and the normal to the surface are all in the same plane, and (2) the angle of incidence equals the angle of reflection.

virtual image refers to an image that only appears to be behind the surface of a mirror (or lens). The image is not actually located at its apparent position and cannot be formed on a screen or film placed at that location.

real image of an object refers to an image that can be formed on a screen or film placed at the position where rays from a point on the object pass through a common point.

plane mirrors are flat mirrors that form an image that is erect or upright, virtual, and the same size as the object.

spherical mirrors are curved mirrors that are sections of a sphere. The surface of a **convex** mirror curves toward the observer, while the surface of a **concave** mirror curves away from the observer.

focal point of a spherical mirror is the point where rays parallel and very close to the principal axis all pass through (or appear to come from) after reflecting from the mirror.

focal point of a lens is the point where rays parallel and very close to the principal axis all pass through (or appear to come from) after refraction by the lens.

focal length of a mirror (or lens) is the distance from the center of the mirror (or lens) to the focal point.

magnification refers to the ratio of the height of the image to the height of the object.

refraction occurs when light changes direction as it passes from one transparent substance into another. The light changes direction at the interface between the two substances.

Snell's law describes the relationship between the incident ray and the refracted ray during refraction.

index of refraction equals the ratio of the speed of light in a vacuum to the speed of light in a transparent substance. The indices of refraction, along with the angle of incidence, determine the angle of bending of the light at the interface between two transparent substances.

total internal reflection occurs when light traveling through a medium of higher index of refraction is completely reflected at the interface with a medium of lower index of refraction.

critical angle is the minimum angle of incidence at which total internal reflection occurs.

converging lens causes light to converge as it passes through the lens. The surfaces of a converging lens curve outward toward the observer.

diverging lens causes light to diverge as it passes through the lens. The surfaces of a diverging lens curve away from the observer.

SUMMARY OF MATHEMATICAL FORMULAS

law of reflection	$\theta_i = \theta_r$	The angle of incidence (θ_i) equals the angle of reflection (θ_r).
mirror and lens equations	$\frac{1}{d_o} + \frac{1}{d_i} = \frac{1}{f}$	Equation relates the distance from the center of the mirror (or thin lens) to the object (d_o) and the center of the mirror (or thin lens) to the image (d_i) to the focal length (f) of the mirror (or thin lens).

	$f = \frac{1}{2}r$	The focal length (f) equals one-half the radius of curvature (r) of the mirror. The focal point is real for a concave mirror but virtual for a convex mirror.
	$m = \frac{h_i}{h_o} = -\frac{d_i}{d_o}$	The magnification (m) refers to the ratio of the size of the image (h_i) to the size of the object (h_o). The magnification equals the ratio of the image distance (d_i) to the object distance (d_o).
Snell's law, or law of refraction	$n_1 \sin \theta_1 = n_2 \sin \theta_2$	Snell's law states the relationship between the angle of incidence (θ_1) and the angle of refraction (θ_2). n_1 is the index of refraction of the incident medium, while n_2 is the index of refraction of the medium into which the light passes.
critical angle for total internal reflection	$\sin \theta_C = \frac{n_2}{n_1}$	At the critical angle (θ_C) and at all angles greater than this angle, light is totally reflected back into the medium of the incident ray.

CONCEPT SUMMARY

Laws of Reflection

A **ray** of light is a single beam of light that travels in a straight line until it strikes an obstacle. As shown in the diagram, if the ray strikes a highly reflective surface, then **specular** (or mirror) **reflection** occurs. If the ray strikes a rough surface, then **diffuse reflection** results. In specular reflection, it is found that (1) the incident ray (i), reflected ray (r), and the normal (N) to the surface are all in the same plane, and (2) the **angle of incidence** (θ_i) equals the **angle of reflection** (θ_r).

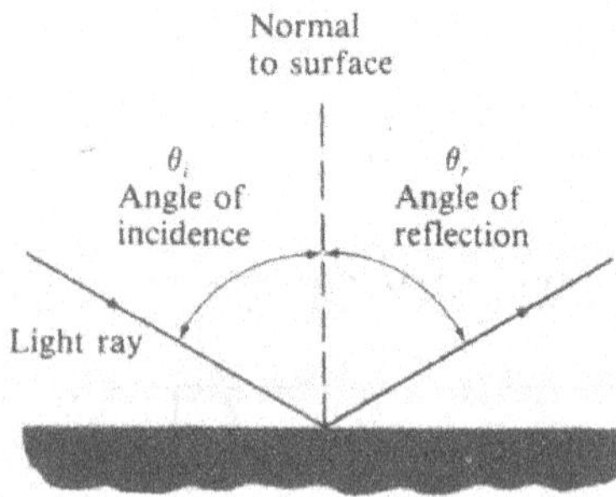

Plane Mirrors

By using the laws of specular reflection, it is possible to show that the **image** formed by an object placed in front of a **plane** (flat) **mirror** has the following characteristics: it is erect or upright, virtual, and the same size as the object. Also, the apparent distance from the mirror to the image is equal to the actual distance from the mirror to the object.

The word **virtual** refers to the fact that the image of the object appears to be behind the surface of the mirror, but the light does not pass through the mirror and the image is not actually located at its apparent position. It is not possible to form a virtual image on a screen or photographic film placed at the image position.

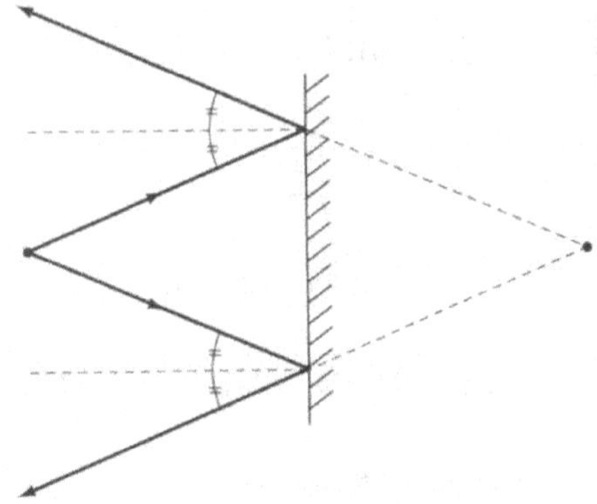

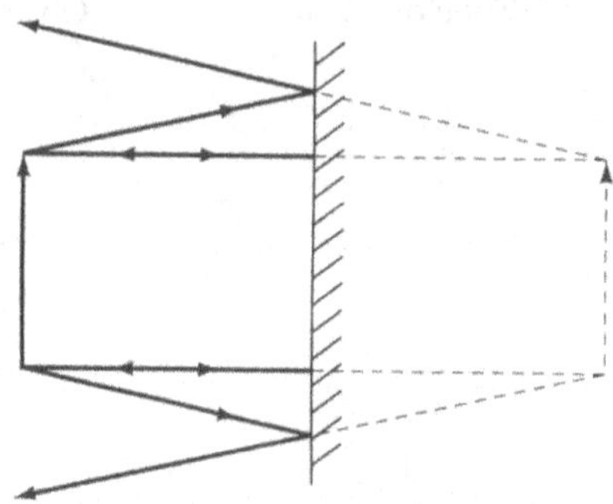

In order to locate the position of the image of any point on an object placed in front of a mirror, it is necessary to draw at least two rays of light that leave that point and reflect from the mirror. The above figures show the use of ray diagrams in locating the image produced by (1) a point and (2) an arrow placed in front of a plane mirror.

Spherical Mirrors

Spherical mirrors are curved mirrors that are sections of a sphere. **Convex** and **concave** mirrors are two types of spherical mirrors.

The principal focus or **focal point** of a spherical mirror is the point where rays parallel and very close to the **principal axis** all pass through (or appear to come from) after reflecting from the mirror. As shown in the following diagrams, the focal point (F) of the mirror is located halfway between the center of curvature and the center of the mirror. The **focal length** (f) equals one-half the radius of curvature of the mirror; that is, $f = \frac{1}{2}r$. The focal point is real for a concave mirror but virtual for a convex mirror.

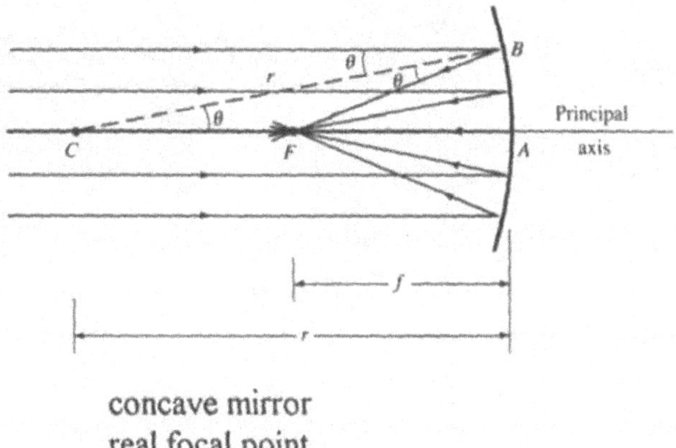

concave mirror
real focal point

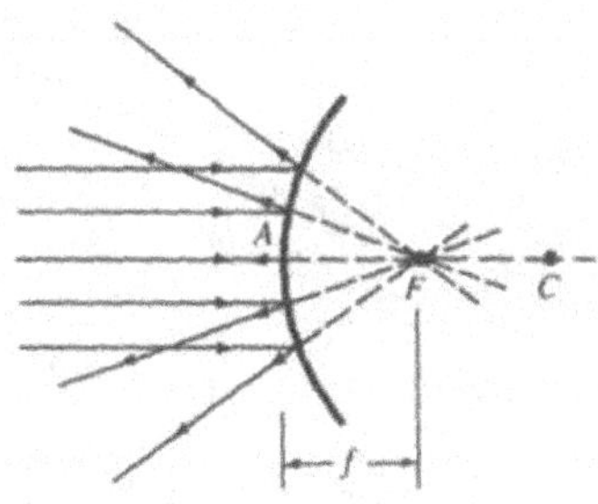

convex mirror
virtual focal point

Images Formed by Spherical Mirrors

The characteristics of the images formed by spherical mirrors depend on the distance from the mirror to the object (d_o), the focal length of the mirror (f), and whether the mirror is concave or convex. The following table summarizes the possibilities:

Type of mirror	Object distance as compared to the focal length	Characteristics of the image
concave	$d_o > 2f$	real, inverted, diminished
	$d_o = 2f$	real, inverted, same size
	$f < d_o < 2f$	real, inverted, magnified
	$d_o = f$	no image formed
	$d_o < f$	virtual, erect, magnified

Convex mirrors produce only virtual, erect, and diminished images.

The images formed by spherical mirrors may be located by using two rays. (1) A ray from the top of the object reflects from the center of the mirror. The principal axis is the normal to the mirror at this point. The angles of incidence and reflection are easily measured so that the reflected ray may be drawn. (2) A ray from the top of the object that is parallel to the principal axis will reflect through the focal point of the concave mirror and appear to be coming from the focal point of a convex mirror.

The image of the top of the object is located at the point where the rays cross or appear to cross. For a real image, the rays cross after reflecting from the mirror. For a virtual image, the rays appear to cross behind the mirror. The following diagrams represent each of the possibilities discussed in the table on the previous page. Note: d_o is the distance from the object to the mirror, while d_i is the distance from the image to the mirror. h_o represents the height of the object, while h_i represents the height of the image.

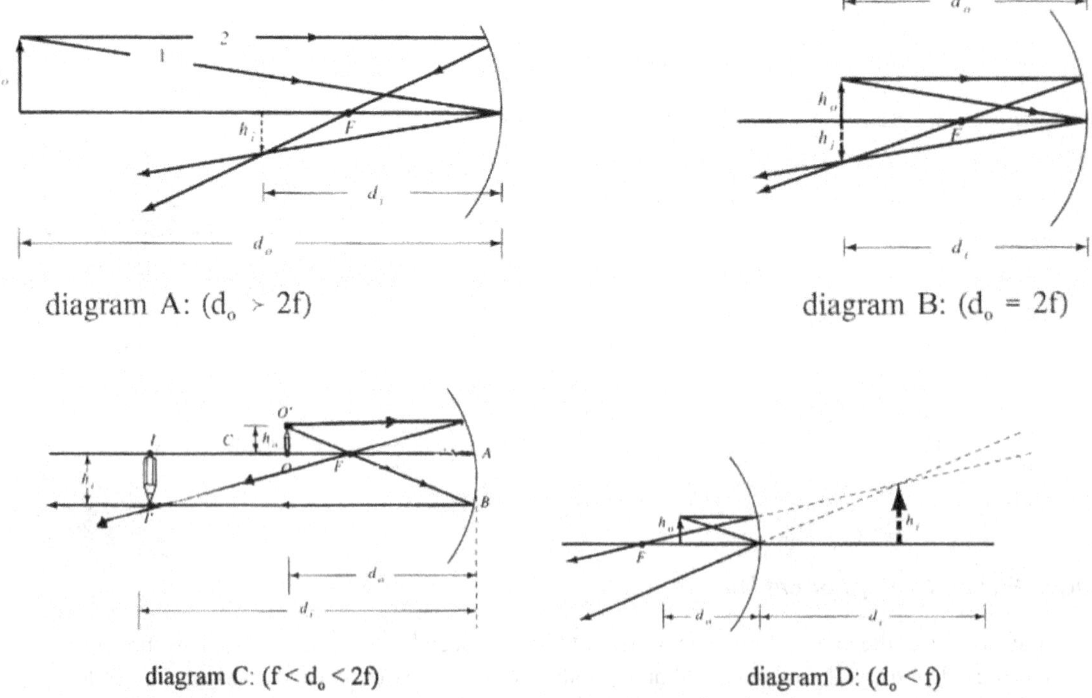

diagram A: (d_o > 2f)

diagram B: (d_o = 2f)

diagram C: (f < d_o < 2f)

diagram D: (d_o < f)

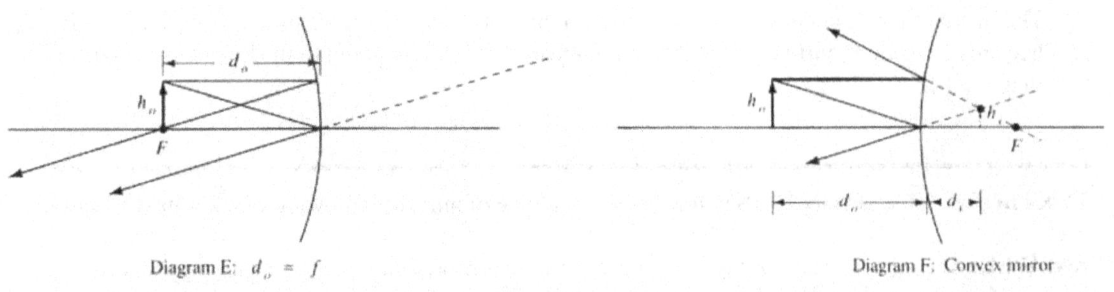

diagram E: ($d_o = f$)

diagram F: convex mirror

Mirror Equation

The mirror equation for both concave and convex spherical mirrors is as follows:

$$\frac{1}{d_o}+\frac{1}{d_i}=\frac{1}{f}$$

where $f=\frac{1}{2}r$ and r is the radius of curvature of the mirror.

Magnification and Size of Image

The magnification (m) refers to the ratio of the size of the image (h_i) to the size of the object (h_o). The magnification produced by a curved mirror is given by

$$m=\frac{h_i}{h_o}=-\frac{d_i}{d_o}$$

and

$$h_i = mh_o$$

Sign Conventions

The following sign conventions are given in the text to be used with the mirror equations. When the object, image, or focal point is on the reflecting side of the mirror (on the left side in all of the drawings), the corresponding distance is considered positive. If the image or focal point is behind the mirror (on the right), then the corresponding distance is considered to be negative. The object height and the image height are considered positive if they lie above the principal axis and negative if they lie below the principal axis. A negative sign is inserted in the magnification equation so that for an upright image the magnification is positive and for an inverted image it is negative.

TEXTBOOK QUESTION 4. (Section 23-3). An object is placed along the principal axis of a spherical mirror. The magnification of the object is –2.0. Is the image real or virtual, inverted or upright? Is the mirror concave or convex? On which side of the mirror is the image located.

ANSWER: Based on the sign convention, a magnification of –2.0 indicates that the image is real, inverted, and twice the height of the object. Diagram C (given earlier) shows an object placed before a concave mirror. The image produced on the reflecting side of the concave mirror matches the description of the image posed in the question.

The mirror cannot be convex, because an upright object place in front of a convex mirror will produce only an upright, virtual image that is diminished in size compared to the height of the object (see diagram F).

TEXTBOOK QUESTION 17. (Section 23-3) What type of mirror is shown in Fig. 23-50? Explain.

ANSWER: The mirror shown in Fig. 23-50 is a concave mirror. Only a object standing between the surface of the concave mirror and the focal point will produce a magnified, virtual image. Ray diagrams for this type of image are shown in Fig. 23-17 in the textbook and diagram D given earlier.

EXAMPLE PROBLEM 1. (Section 23-3) An object 1.0 cm high is placed 20 cm from a concave mirror of focal length 15 cm. (*a*) Draw a ray diagram and locate the position of the image formed. Draw in the image. (*b*) Mathematically determine the image distance from the mirror, magnification, and height of the image. (*c*) State the characteristics of the image.

Part a. Step 1.

Choose an appropriate scale factor and draw an accurate ray diagram.

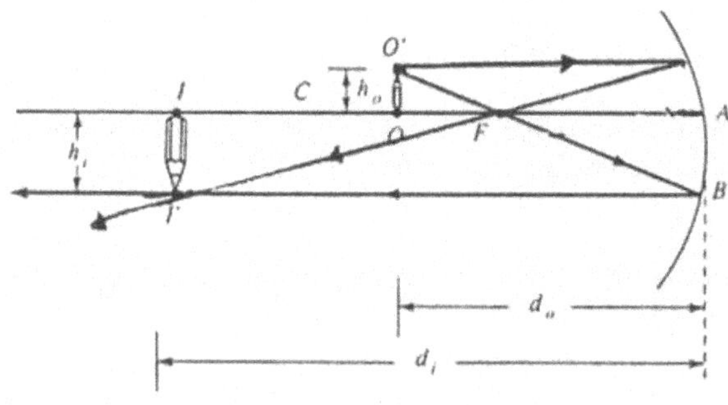

Part b. Step 1.

Mathematically determine the image distance from the mirror, magnification, and height of image.

$$\frac{1}{d_o}+\frac{1}{d_i}=\frac{1}{f}$$

$$\frac{1}{20.0\text{ cm}}+\frac{1}{d_i}=\frac{1}{15.0\text{ cm}}$$

$$\frac{1}{d_i}=\frac{20.0\text{ cm}-15.0\text{ cm}}{(15.0\text{ cm})(20.0\text{ cm})}=\frac{5.0\text{ cm}}{300\text{ cm}^2}$$

$$d_i=\frac{300\text{ cm}^2}{5.0\text{ cm}}=60.0\text{ cm}$$

$$m=-\frac{d_i}{d_o}=-\frac{60\text{ cm}}{30\text{ cm}}=-3.0$$

$$h_i=mh_o=(-3.0)(1.0\text{ cm})=-3.0\text{ cm}$$

Part c. Step 1.

State the characteristics of the image.

The image is real, inverted, and magnified.

EXAMPLE PROBLEM 2. (Section 23-3) An object 3.0 cm high is placed 5.0 cm from a concave mirror of focal length 10 cm. (*a*) Draw a ray diagram and locate the position of the image formed. Draw in the image. (*b*) Mathematically determine the image distance from the mirror, magnification, and height of the image. (*c*) State the characteristics of the image.

Part a. Step 1.

Choose an appropriate scale factor and draw an accurate ray diagram.

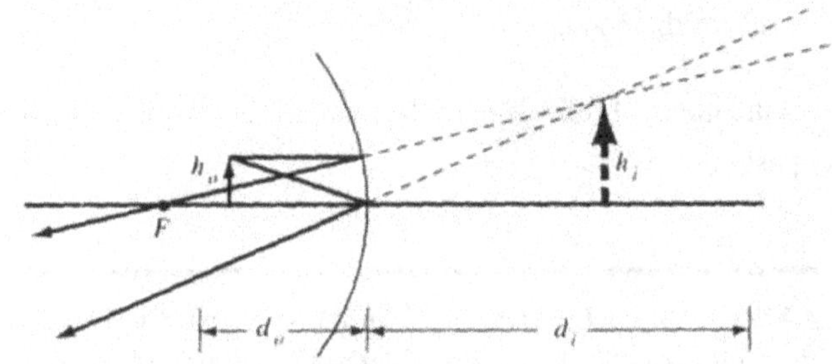

Part b. Step 1.

Mathematically determine the image distance from the mirror, magnification, and height of image.

$$\frac{1}{d_o}+\frac{1}{d_i}=\frac{1}{f}$$

$$\frac{1}{5.0\text{ cm}}+\frac{1}{d_i}=\frac{1}{10.0\,\text{cm}}$$

$$d_i=-10\text{ cm}$$

$$m=-\frac{d_i}{d_o}=-\left(\frac{-10\text{ cm}}{5.0\text{ cm}}\right)=+2.0$$

$$h_i=mh_o=(+2.0)(3.0\text{ cm})=+6.0\text{ cm}$$

Part c. Step 1.

State the characteristics of the image.

The positive sign for both the magnification and the image height indicates that the image is virtual and erect. The image is erect and magnified.

Refraction

As light passes from one medium into another, it changes direction at the interface between the two media. This change of direction is known as **refraction**. The relationship between the **angle of incidence** and the **angle of refraction** is given by **Snell's law**:

$$n_1 \sin \theta_1 = n_2 \sin \theta_2$$

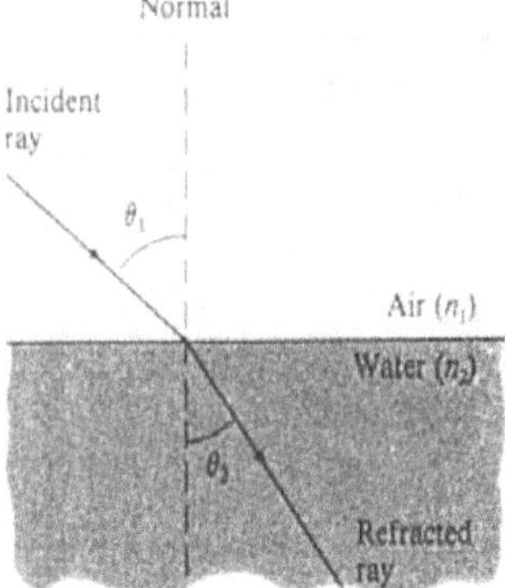

θ_1 is the angle of incidence and is measured from the normal to the surface to the line of the incident ray.

θ_2 is the angle of refraction and is measured from the normal to the surface to the line of the refracted ray.

n_1 is the index of refraction of the medium in which the light is initially traveling. The index of refraction is a dimensionless number which varies from a low of 1.00 for a vacuum to 2.42 for diamond.

n_2 is the index of refraction of the medium into which the light passes.

EXAMPLE PROBLEM 3. (Section 23-5) A ray of light strikes the surface of a flat glass plate at an incident angle of 30°. The index of refraction of air and glass is 1.00 and 1.50, respectively. Determine the (*a*) angle of reflection and (*b*) angle of refraction.

Part a. Step 1.

Draw an accurate diagram showing the incident ray, reflected ray, and approximate path of the refracted ray.

The reflected ray follows the laws of specular reflection; thus, the angle of reflection is 30°.

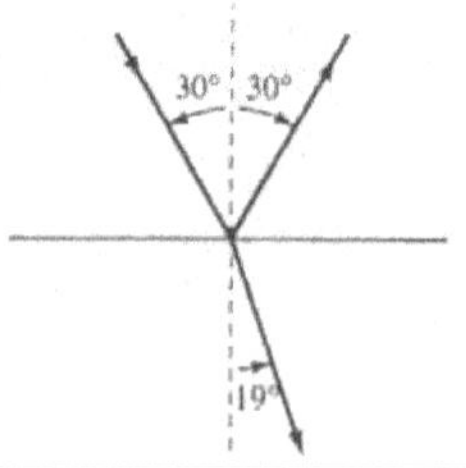

Part b. Step 1.

Use Snell's law to determine the angle of refraction.

$$n_1 \sin \theta_1 = n_2 \sin \theta_2$$

$$(1.00) \sin 30° = (1.50) \sin \theta_2$$

$$\sin \theta_2 = \frac{(1.00)(0.50)}{1.50} = 0.33$$

$$\theta_2 = 19°$$

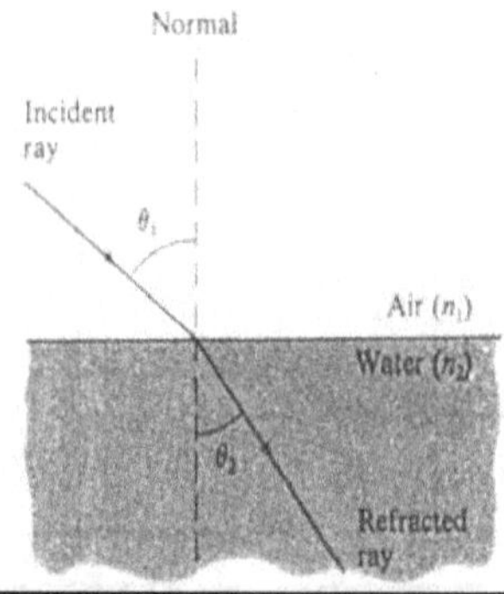

Total Internal Reflection

As light passes from a medium of higher index of refraction into a medium of lower index of refraction, an angle of incidence is reached at which the angle of refraction is 90°. This angle is known as the **critical angle** (θ_C), and at all angles greater than this angle, the light is totally reflected back into the medium of the incident ray. The equation for the critical angle can be determined by using Snell's law.

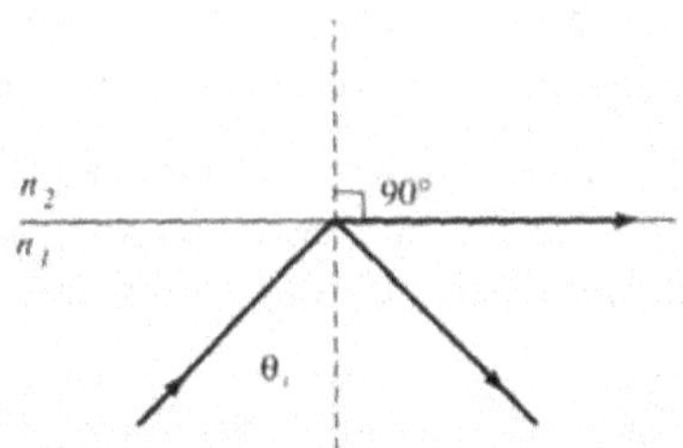

$$n_1 \sin \theta_C = n_2 \sin 90°, \text{ where } \sin \theta_C = \frac{n_2}{n_1}$$

EXAMPLE PROBLEM 4. (Section 23-6) Determine the critical angle for light passing from (*a*) diamond into air and (*b*) glass into water. The index of refraction for each substance is as follows: diamond 2.42, air 1.00, glass 1.50, water 1.33.

Part a. Step 1.

Use Snell's law to find the critical angle for the air–diamond interface.

At the critical angle, $\theta_{\text{air}} = 19°$. Using Snell's law,

$$n_{\text{diamond}} \sin \theta_{\text{diamond}} = n_{\text{air}} \sin \theta_{\text{air}}$$

$$(2.42) \sin \theta_C = (1.00) \sin 90°$$

$$\sin \theta_C = \frac{(1.00)(1.00)}{2.42} = 0.41$$

$$\theta_C = 24°$$

Part b. Step 1.

Use Snell's law to find the critical angle for the water–glass interface.

$$n_{\text{glass}} \sin \theta_{\text{glass}} = n_{\text{water}} \sin \theta_{\text{water}}$$

$$(1.50) \sin \theta_C = (1.33) \sin 90°$$

$$\sin \theta_C = \frac{(1.33)(1.00)}{1.50} = 0.89$$

$$\theta_C = 63°$$

Note: Total internal reflection occurs only when light is traveling from a medium of higher index of refraction into a medium of lower index of refraction.

Thin Lenses

Rays of light parallel to the principal axis of a **converging lens** converge at the focal point (F) of the lens after refraction. A converging lens, which is also known as a convex lens, has a real focal point.

Rays of light parallel to the principal axis of a **diverging lens** diverge after passing through the lens. If the refracted rays are traced on a straight line back through the lens, they appear to converge at a focal point. A diverging lens, which is also known as a concave lens, has a virtual focal point.

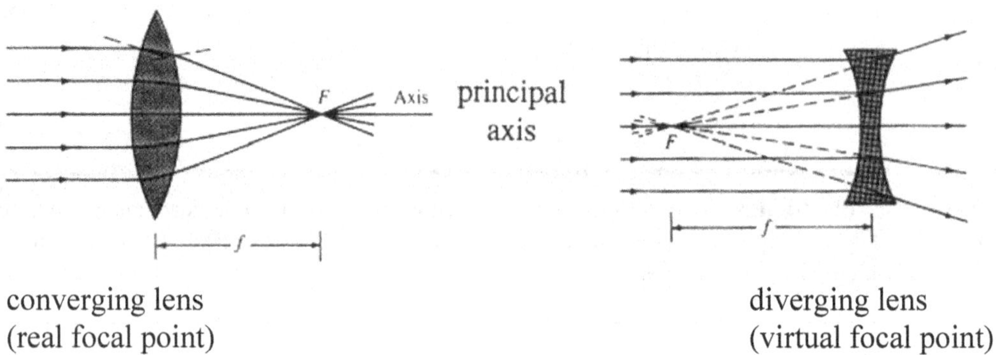

Images Formed by a Thin Lens

As in the case of images formed by spherical mirrors, the characteristics of images formed by thin lenses depend on the object distance (d_o), focal length of the lens (f), and whether or not the lens is convex or concave. The following table summarizes the possibilities:

Type of lens	Object distance as compared with the focal length	Characteristics of image
convex	$d_o > 2f$	real, inverted, diminished
	$d_o = 2f$	real, inverted, same size
	$f < d_o < 2f$	real, inverted, magnified
	$d_o = f$	no image formed
	$d_o < f$	virtual, erect, magnified

Concave lenses produce only virtual, erect, and diminished images.

The images formed by thin lenses may be located by using the following two rays. (1) A ray from the top of the object that strikes the center of the lens: At this point, the two sides of the lens are parallel. The ray emerges from the lens slightly displaced but traveling parallel to its original direction. (2) A ray from the top of the object that is parallel to the principal axis of a convex lens passes through the focal point of the side of the lens opposite the object: If the lens is concave, then the ray appears to have come from the virtual focal point located on the same side of the lens as the object.

For a real image, the image of any point on the object is located where the two rays from the point intersect after refraction. For a virtual image, it is necessary to trace the path of the refracted rays back through the lens. The image is located at the point where the rays appear to cross.

The following diagrams represent each of the situations described in the preceding table.

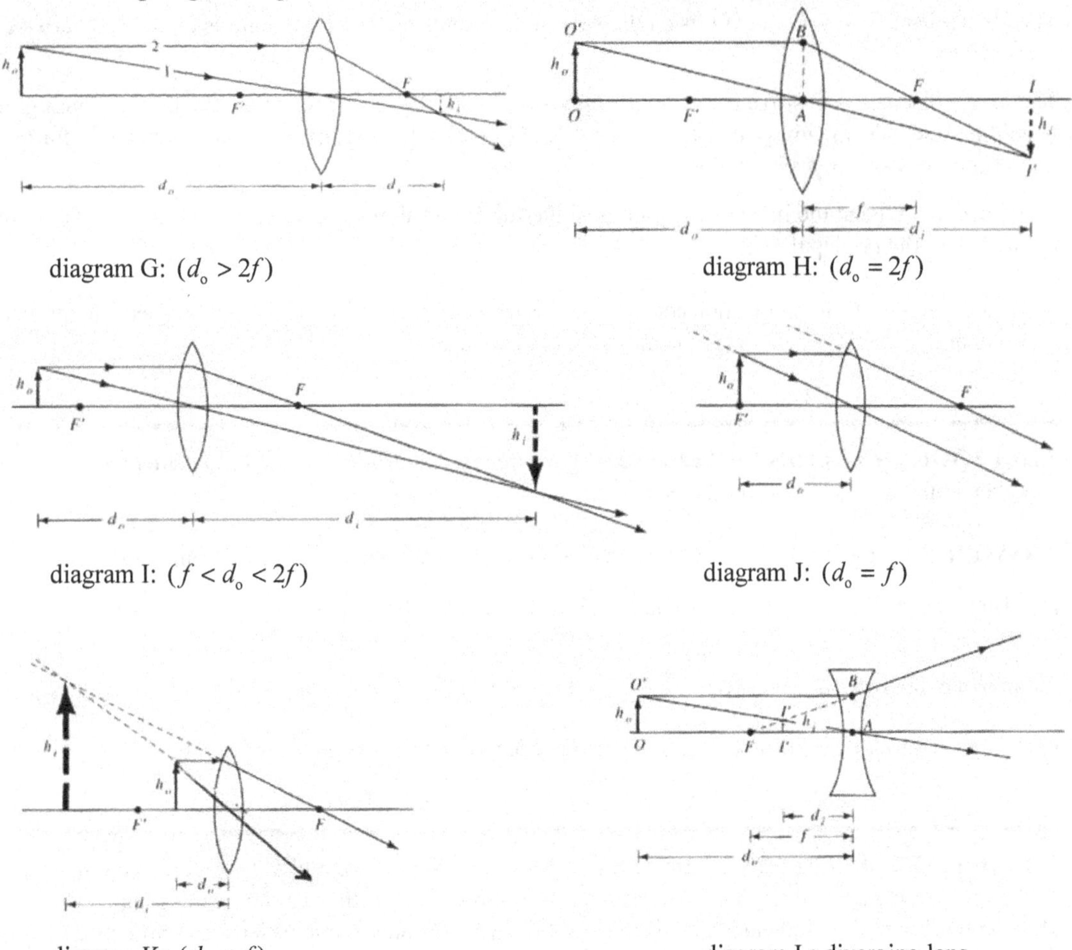

diagram G: ($d_o > 2f$)

diagram H: ($d_o = 2f$)

diagram I: ($f < d_o < 2f$)

diagram J: ($d_o = f$)

diagram K: ($d_o < f$)

diagram L: diverging lens

Thin Lens Equation

The thin lens equation for both converging and diverging lenses is

$$\frac{1}{d_o} + \frac{1}{d_i} = \frac{1}{f}$$

where d_o is the object distance from the lens, d_i is the image distance from the lens, and f is the focal length of the lens.

Magnification and Size

The magnification (m) of a lens is given by

$$m = \frac{h_i}{h_o} = -\frac{d_i}{d_o} \quad \text{and} \quad h_i = mh_o$$

where h_i is the image height and h_o is the object height.

Sign Conventions

The following sign conventions are given in the textbook to be used in connection with the lens equations:

1. The focal length is positive for a converging lens and negative for a diverging lens.
2. The object distance is positive if it is on the side of the lens from which the light is coming; otherwise, it is negative.
3. The image distance is positive if it is on the opposite side of the lens from where the light is coming; if it is on the same side, the image distance is negative. Equivalently, the image distance is positive for a real image and negative for a virtual image.
4. The object height and the image height are positive for points above the principal axis and negative for points below the principal axis.

The negative sign in the magnification equation has been inserted so that for an upright image the magnification (m) is positive and for an inverted image the magnification is negative.

TEXTBOOK QUESTION 19. (Section 23-7) Where must the film be placed if a camera lens is to make a sharp image of an object far away? Explain.

ANSWER: As shown in diagrams G, H, and I, as an object is moved farther away from a convex lens, the distance from the lens to the image decreases. Using the thin lens equation: $\frac{1}{d_o}+\frac{1}{d_i}=\frac{1}{f}$ However, for an object far away $d_o \approx \infty$. So, $\frac{1}{\infty}+\frac{1}{d_i}=\frac{1}{f}$. Therefore, $0+\frac{1}{d_i}=\frac{1}{f}$ and $d_i \approx f$. The film must be placed very close to the focal point of the lens in order for a clear image to be formed.

EXAMPLE PROBLEM 5. (Sections 23-7 and 23-8) An object 4.0 cm high is placed 15 cm from a convex lens of focal length 10 cm. (*a*) Draw a ray diagram and locate the position of the image formed. Draw in the image. (*b*) Mathematically determine the image distance from the lens, magnification, and height of the image. (*c*) State the characteristics of the image.

Part a. Step 1.

Choose an appropriate scale factor and draw an accurate diagram.

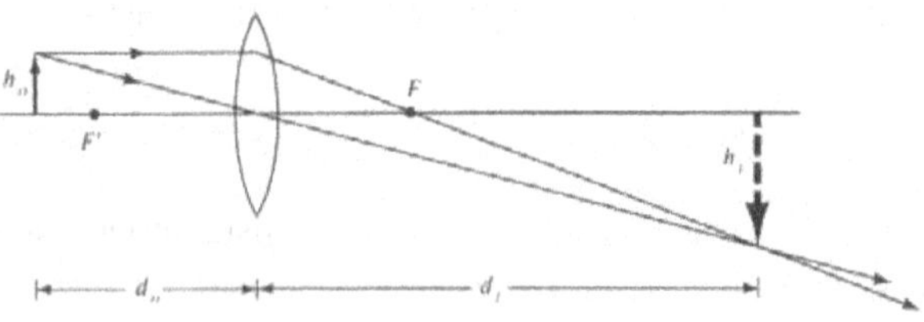

Part b. Step 1.

Mathematically determine the image distance from the mirror, magnification, and height of the image.

$$\frac{1}{d_o}+\frac{1}{d_i}=\frac{1}{f}$$

$$\frac{1}{15.0 \text{ cm}} + \frac{1}{d_i} = \frac{1}{10.0 \text{ cm}}$$

$$\frac{1}{d_i} = \frac{1}{10.0 \text{ cm}} - \frac{1}{15.0 \text{ cm}}$$

$$d_i = 30 \text{ cm}$$

$$m = -\frac{d_i}{d_o} = -\frac{(30 \text{ cm})}{(15 \text{ cm})} = -2.0$$

$$h_i = mh_o = (-2.0)(4.0 \text{ cm}) = -8.0 \text{ cm}$$

Part c. Step 1.

State the characteristics of the image.

The image is real, inverted, and magnified.

EXAMPLE PROBLEM 6. (Sections 23-7 and 23-8) An object 3.0 cm high is placed 5.0 cm from a convex lens of focal length 8.0 cm. (*a*) Draw a ray diagram and locate the position of the image formed. Draw in the image. (*b*) Mathematically determine the image distance from the lens, magnification, and height of the image. (*c*) State the characteristics of the image.

Part a. Step 1.

Choose an appropriate scale factor and draw an accurate ray diagram.

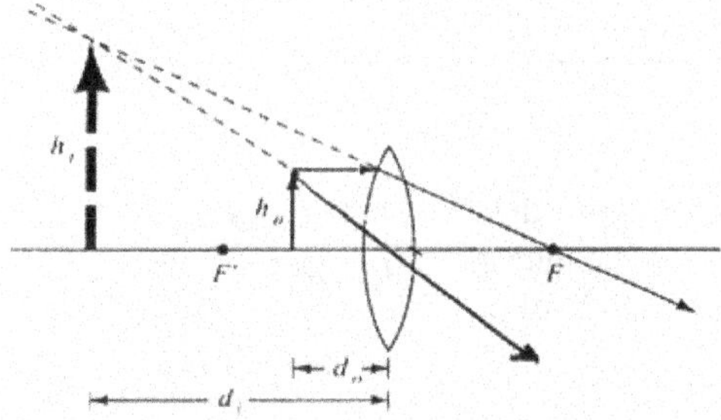

Part b. Step 1.

Mathematically determine the image distance from the lens, magnification, and height of the image.

$$\frac{1}{d_o} + \frac{1}{d_i} = \frac{1}{f}$$

$$\frac{1}{5.0 \text{ cm}} + \frac{1}{d_i} = \frac{1}{8.0 \text{ cm}}$$

$$d_i = -13 \text{ cm}$$

$$m = -\frac{d_i}{d_o} = -\frac{(13\text{ cm})}{(5.0\text{ cm})} = +2.6$$

$$h_i = mh_o = (+2.6)(3.0\text{ cm}) = +7.8\text{ cm}$$

Part c. Step 1.

State the characteristics of the image.

The positive sign for both the magnification and the image height indicates that the image is virtual and erect. The image is virtual, erect, and magnified.

EXAMPLE PROBLEM 7. (Sections 23-7 and 23-8) An object 5.0 cm high is placed 20 cm from a concave lens of focal length 10 cm. (*a*) Draw a ray diagram and locate the position of the image formed. Draw in the image. (*b*) Mathematically determine the image distance from the lens, magnification, and height of the image. (*c*) State the characteristics of the image.

Part a. Step 1.

Choose an appropriate scale factor and draw an accurate ray diagram.

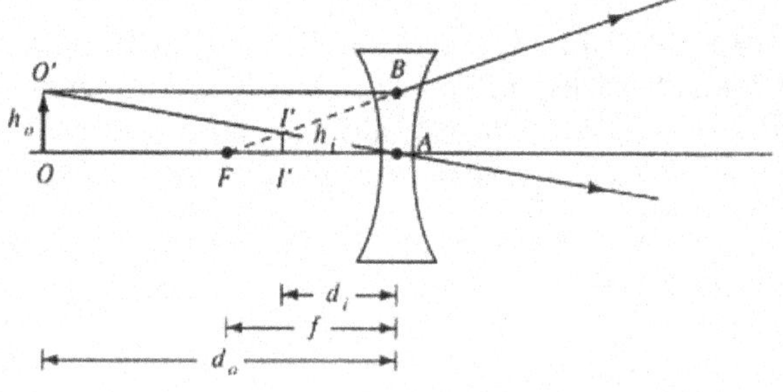

Part b. Step 1.

Mathematically determine the image distance from the mirror, magnification, and height of the image.

$$\frac{1}{d_o} + \frac{1}{d_i} = \frac{1}{f}$$

Hint: Based on the sign convention, the focal length of a concave lens is negative.

$$\frac{1}{20\text{ cm}} + \frac{1}{d_i} = \frac{1}{-10\text{ cm}}$$

$$d_i = -6.7\text{ cm}$$

$$m = -\frac{d_i}{d_o} = -\frac{(-6.7\text{ cm})}{(20\text{ cm})} = +0.33$$

$$h_i = mh_o = (0.33)(5.0\text{ cm}) = 1.7\text{ cm}$$

Part c. Step 1.

State the characteristics of the image.	The image is virtual, erect, and diminished.

PROBLEM SOLVING SKILLS

For problems involving image formation by a convex or a concave mirror:

1. Identify whether the mirror is concave or convex. The focal length is positive for a concave mirror and negative for a convex mirror.
2. Choose an appropriate scale factor to represent the focal length, object distance, and height of the object.
3. Use the two suggested rays to draw an accurate ray diagram. Draw in the image at the point where the two rays cross after reflection from the mirror.
4. Use the mirror equations and sign conventions to mathematically determine the image distance, image height, and magnification.
5. State the characteristics of the image: real or virtual; erect or inverted; and magnified, diminished, or same size as the object.

For problems involving refraction of light as it passes from one medium to another:

1. Draw an accurate diagram locating the incident ray and normal to the surface. Determine the angle of incidence.
2. Complete a data table using the information given in the problem.
3. Use Snell's law to solve the problem.
4. If the problem involves total internal reflection, then at the critical angle, the angle of refraction is 90°. Use Snell's law to determine the magnitude of the critical angle.

For problems involving image formation by a converging or a diverging lens:

1. Identify whether the lens is converging or diverging. The focal length is positive for a converging lens and negative for a diverging lens.
2. Lens problems involve refraction of light, while mirror problems involve reflection. Apply the steps listed above for mirror problems to solve problems involving lenses.

SOLUTIONS TO SELECTED TEXTBOOK PROBLEMS

TEXTBOOK PROBLEM 18. (Section 23-3) Some rearview mirrors produce images of cars to your rear that are smaller than they would be if the mirror were flat. Are the mirrors concave or convex? What is a mirror's radius of curvature if cars 16.0 m away appear 0.33 their normal size? Hint: See diagram F.

Part a. Step 1.

Determine the image distance from the mirror.

$$m = -\frac{d_i}{d_o}$$

$$+0.33 = -\frac{d_i}{16\text{ m}}$$

$$d_i = -5.28\text{ m}$$

Part a. Step 2.

Determine the focal length (f) of the mirror.

$$\frac{1}{d_o} + \frac{1}{d_i} = \frac{1}{f}$$

$$\frac{1}{16\text{ m}} + \frac{1}{-5.28\text{ m}} = \frac{1}{f}$$

$$f = -7.88\text{ m}$$

Part a. Step 3.

Determine the radius of curvature of the mirror.

$$r = 2f = 2(-7.88\text{ m})$$

$$r = -15.8\text{ m}$$

Part a. Step 4.

Determine whether the mirror is concave or convex.

The focal length and the radius of curvature of the mirror are negative. The mirror is convex because only a convex mirror has a negative focal length and radius of curvature.

TEXTBOOK PROBLEM 37. (Section 23-6) A beam of light is emitted in a pool of water from a depth of 82.0 cm. Where must it strike the air–water interface, relative to a spot directly above it, in order that the light does *not* exit the water?

Part a. Step 1.

Choose an appropriate scale factor and draw an accurate ray diagram.

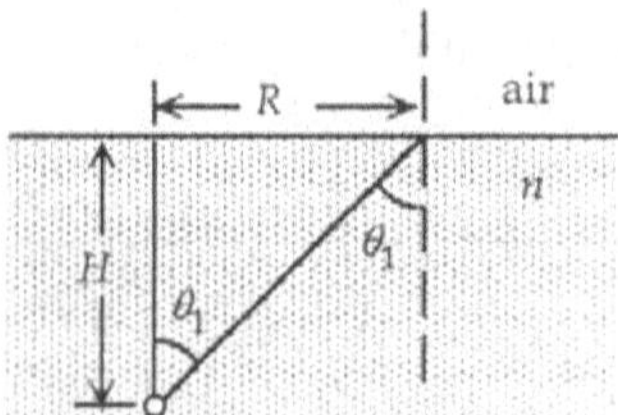

Part a. Step 2.

Use Snell's law to determine the critical angle for the water–air interface.

The beam is totally reflected back into the water. Therefore, $\theta_2 = 90°$:

$$n_{\text{water}} \sin \theta_1 = n_{\text{air}} \sin 90°$$

$$(1.33) \sin \theta_1 = (1.00)(1.00)$$

$$\theta_1 = 48.8°$$

Part a. Step 3.

Based on the diagram, determine the distance *R*.

$$\tan \theta_1 = \frac{R}{H}$$

$$\tan 48.8° = \frac{R}{82.0 \text{ cm}}$$

$$R = (1.14)(82.0 \text{ cm})$$

$$R = 93.5 \text{ cm}$$

TEXTBOOK PROBLEM 53. (Sections 23-7 and 23-8) How far from a converging lens with a focal length of 32 cm should an object be placed to produce a real image which is the same height as the object? *Hint*: See diagram H.

Part a. Step 1.

Determine the magnification of the image.

A converging lens produces an inverted real image of the object. The object and image are the same height; therefore, based on the sign convention in the textbook, $h_i = -h_o$.

$$m = \frac{h_i}{h_o} \quad \text{but} \quad h_i = -h_o$$

$$m = \frac{-h_o}{h_o} = -1$$

Part a. Step 2.

Determine d_o in terms of d_i.

$$m = -\frac{d_i}{d_o}$$

$$-1 = -\frac{d_i}{d_o}$$

$$d_o = d_i$$

Part a. Step 3.

Use the thin lens equation to determine the object distance.

$$\frac{1}{d_o} + \frac{1}{d_i} = \frac{1}{f} \quad \text{but} \quad d_o = d_i$$

$$\frac{2}{d_o} = \frac{1}{32 \text{ cm}}$$

$$d_o = 64 \text{ cm}$$

TEXTBOOK PROBLEM 55. (Section 23-5) A bright object and a viewing screen are separated by a distance of 86.0 cm. At what location(s) between the object and the screen should a lens of focal length 16.0 cm be placed in order to produce a sharp image on the screen? *Hint*: First draw a diagram.

Part a. Step 1.

Choose an appropriate scale factor and draw a ray diagram.

Ray diagrams G and I show the two possible distances where the object could be placed from the lens and produce a sharp image on the screen.

Part a. Step 2.

Use the thin lens equation to solve for the object distance (d_o).

$$\frac{1}{d_o} + \frac{1}{d_i} = \frac{1}{f} \quad \text{but}$$

$$d_o + d_i = 86.0 \text{ cm} \quad \text{and} \quad d_i = 86.00 \text{ cm} - d_o$$

$$\frac{1}{d_o} + \frac{1}{(86 \text{ cm} - d_o)} = \frac{1}{16.0 \text{ cm}}$$

Using algebra and rearranging gives

$$d_o^2 - (86.0 \text{ cm})d_o + 1376 \text{ cm}^2 = 0$$

Using the quadratic formula results in two possible solutions, either $d_o = 64.8$ cm or $d_o = 21.3$ cm.

TEXTBOOK PROBLEM 71. (Section 23-5) We wish to determine the depth of a swimming pool filled with water by measuring the width $(x = 6.50 \text{ m})$ and then noting that the bottom edge of the pool is just visible at an angle of 13.0° above the horizontal as shown in Fig. 23-62. Calculate the depth of the pool.

Part a. Step 1.

Draw a diagram based on Fig. 23-65 and determine the angle of refraction.

The angle of refraction (θ_2) is the angle measured from the vertical to the refracted ray as the ray travels through the air to the observer's eye.

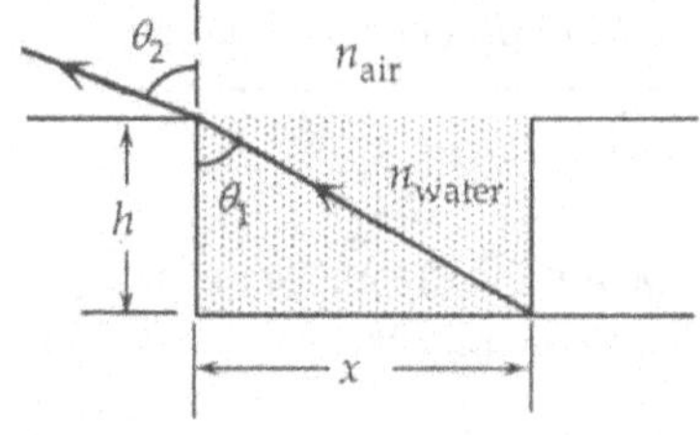

$\theta_2 = 90.0° - 13.0° = 77.0°$

Part a. Step 2.

Use Snell's law to determine the angle of incidence (θ_1).

$$n_1 \sin \theta_1 = n_2 \sin \theta_2$$

$$(1.33) \sin \theta_1 = (1.00) \sin 77.0°$$

$$\sin \theta_1 = \frac{(1.00)(0.974)}{1.33} = 0.732$$

$$\theta_1 = 47.1°$$

Part a. Step 3.

Based on the diagram shown in Step 1, use trigonometry to determine the depth h of the pool.

$$\tan \theta_1 = \frac{x}{h}$$

$$\tan 47.1° = \frac{6.50 \text{ m}}{h}$$

$$1.08 = \frac{6.50 \text{ m}}{h}$$

$$h = 6.04 \text{ m}$$

TEXTBOOK PROBLEM 79. (Section 23-3) An object is placed 18.0 cm from a certain mirror. The image is half the height of the object, inverted, and real. How far is the image from the mirror, and what is the radius of curvature of the mirror?

Part a. Step 1.

Determine the type of mirror used to produce the image.	Based on the description of the image, the mirror must be a concave mirror (see diagram A).

Part a. Step 2.

Determine the image distance from the mirror.

The object distance $d_o = 18.0 \text{ cm}$ and the magnification $m = -0.5$:

$$m = -\frac{d_i}{d_o}$$

$$-0.5 = -\frac{d_i}{18.0 \text{ cm}}$$

$$d_i = 9.0 \text{ cm}$$

Part a. Step 3.

Determine the focal length (f) of the mirror.

$$\frac{1}{d_o} + \frac{1}{d_i} = \frac{1}{f}$$

$$\frac{1}{18.0 \text{ cm}} + \frac{1}{9.0 \text{ cm}} = \frac{1}{f}$$

and solving for f gives

$$f = 6.0 \text{ cm}$$

Part a. Step 4.

Determine the radius of curvature (r) of the mirror.

$$f = \tfrac{1}{2}r$$

$$6.0 \text{ cm} = \tfrac{1}{2}r$$

$$r = 12 \text{ cm}$$

24

THE WAVE NATURE OF LIGHT

OBJECTIVES

After studying the material of this chapter, the student should be able to:

- use the wave model to explain reflection of light from mirrors and refraction of light as it passes from one medium into another.
- use the conditions for constructive and destructive interference of waves to explain the interference patterns observed in Young's double-slit experiment, single-slit diffraction, diffraction grating, and thin-film interference.
- solve problems involving a single slit, double slit, and diffraction grating for m, λ, d, D, and angular separation (θ) when the other quantities are given.
- solve problems involving thin-film interference for m, λ, n, or t when the other quantities are given.
- explain how the Michelson interferometer can be used to determine the wavelength of a monochromatic light source and solve problems to determine the wavelength of the light from the source.
- use the wave model to explain plane polarization of light and polarization by reflection.
- calculate the angle of maximum polarization for reflected light.

KEY TERMS AND PHRASES

interference patterns are produced when two (or more) coherent sources producing waves of the same frequency and amplitude superimpose.

coherent sources involve a fixed phase between the waves emitted by the sources.

crest is the highest point of that portion of a transverse wave above the equilibrium position.

trough is the lowest point of that portion of a transverse wave below the equilibrium position.

destructive interference occurs when the amplitude of the resultant of two interfering waves is smaller than the displacement of either wave.

constructive interference occurs when the amplitude of the resultant of two interfering waves is larger than the displacement of either wave.

diffraction occurs when waves bend and spread out as they pass an obstacle or narrow opening. The amount of diffraction depends on the wavelength of the waves and the size of the obstacle. In the case of a narrow opening, the amount of bending increases as the size of the opening decreases.

diffraction grating consists of a large number of closely spaced parallel slits that diffract light incident on the grating. The diffracted light will exhibit a pattern of constructive interference and destructive interference.

thin-film interference occurs when light waves are reflected at both the top and bottom surfaces of the film. The pattern of dark and bright lines observed results from light reflected from the top surface interfering with the light reflected from the bottom surface.

polarization is a property of light that indicates light is a transverse wave phenomenon. A transverse wave is a wave in which the particles of the medium move at right angles to the direction of motion of the wave. An electromagnetic wave in which the electric vector is vibrating in only one plane is said to be **plane-polarized.**

Brewster's angle is the angle of incidence at which maximum polarization of the reflected light occurs.

SUMMARY OF MATHEMATICAL FORMULAS

speed of light and the wavelength in a medium other than a vacuum	$v = \frac{c}{n}$	The speed of light (v) in a medium is the ratio of the speed of light in a vacuum (c) to the medium's index of refraction (n).
	$\lambda_n = \frac{\lambda_{\text{vacuum}}}{n}$	The wavelength (λ_n) of light in a medium equals the ratio of the wavelength in a vacuum to the index of refraction of the medium (n).
double-slit interference	$m\lambda = d \sin\theta$	The relationship between the wavelength (λ), the distance between the slits (d), and θ is the grating angle for constructive interference; m is the order of the interference fringe: $m = 0, 1, 2, 3,$ etc.
	$(m+\frac{1}{2})\lambda = d \sin\theta$	The relationship between the wavelength (λ), the distance between the slits (d), and θ is the grating angle for destructive interference.
single-slit diffraction bright fringes dark fringes	$(m+\frac{1}{2})\lambda = D \sin\theta$ $m\lambda = D \sin\theta$	The pattern for bright or dark fringes in a single-slit diffraction pattern depends on the wavelength (λ), the slit width (D), and the diffraction angle (θ), $m = 1, 2, 3,$ etc.
thin-film interference	$t = \frac{(2m+1)\,\lambda_n}{4}$ or $4t = (2m+1)\lambda_n$ where	If the indices of refraction of the media above and below the thin film are lower (or higher) than the index of refraction of the medium of the thin film, then maximum reflection of light of a particular wavelength occurs if the thickness (t) of the film is an odd-number multiple of quarter wavelengths. Note: $m = 0, 1, 2, 3,$ etc.

	$\lambda_n = \frac{\lambda}{n}$ $t = m\frac{\lambda_n}{2}$ or $2t = m\lambda_n$	Minimum reflection of the light occurs if the thickness of the film is a whole-number multiple of half wavelengths. Note: $m = 1, 2, 3,$ etc. Note: If the media above and below the thin film are such that a phase change occurs at both the top and bottom surfaces, then the equation for maximum reflection is given by $2t = m\lambda_n$ and for minimum reflection by $4t = (2m+1)\lambda_n$.
Michelson interferometer	$t = m\frac{\lambda}{2}$	The Michelson interferometer is a device that uses wave interference to determine the wavelength of light. By measuring the distance that one of the mirrors moves and the number of interference fringes (m) that pass the observer's field of view, the wave length (λ) of the light can be determined.
Brewster's angle	$\tan\theta_p = \frac{n_2}{n_1}$	The angle of incidence at which maximum polarization (θ_p) of the reflected light occurs is known as Brewster's angle. The tangent of Brewster's angle (θ_p) is related to the ratio of the index of refraction (n_2) of the substance from which the light is reflected to the index of refraction (n_1) of the substance in which the light is initially traveling.

CONCEPT SUMMARY

Reflection

In Chapter 11 it was observed that a **wave front** striking a straight barrier follows the law of reflection. This law was found to apply to light in Chapter 23.

Refraction

Water waves undergo **refraction** as they travel from deep to shallow water because the speed of the wave changes. Refraction of light occurs as light travels from one medium to another, and this was discussed in terms of Snell's law in Chapter 23. The wave theory proposed by Christian Huygens (1629–1695) predicts that the speed of light is less in water or glass than in air. Measurements of the speed of light in various materials agree with the wave theory. The speed of light in a medium (v) is inversely related to the medium's index of refraction (n); that is, $v = \frac{c}{n}$, where c is the speed of light in a vacuum and $c = 3.0 \times 10^8$ m/s.

For light traveling from medium 1 into medium 2,

$$\frac{n_2}{n_1} = \frac{v_1}{v_2} = \frac{\lambda_1}{\lambda_2}$$

where λ is the wavelength of the light in the medium.

EXAMPLE PROBLEM 1. (Section 24-2) Light of wavelength 546 nm passes from a vacuum into water of index of refraction 1.33. Determine the (*a*) speed of light in water and (*b*) wavelength of the light in the water.

Part a. Step 1.

Determine the speed of light in water.

The speed of light in a vacuum is 3.00×10^8 meters per second and the index of refraction of a vacuum is 1.00. Assume that medium 1 is the vacuum and medium 2 is the water:

$$\frac{n_2}{n_1} = \frac{v_1}{v_2}$$

$$\frac{1.33}{1.00} = \frac{3.00 \times 10^8 \text{ m/s}}{v_2}$$

$$v_2 = 2.26 \times 10^8 \text{ m/s}$$

Part b. Step 1.

Determine the wavelength of the light in water.

$1 \text{ nm} = 1 \times 10^{-9}$ m; however, it is appropriate to express the wavelength of light in nanometers (nm). There is no need to convert the wavelength to meters.

$$\frac{n_2}{n_1} = \frac{\lambda_1}{\lambda_2}$$

$$\frac{1.33}{1.00} = \frac{546 \text{ nm}}{\lambda_2}$$

$$\lambda_2 = 411 \text{ nm}$$

Interference

Two **coherent** sources producing waves of the same frequency and amplitude produce an interference pattern. The following diagram was used in Chapter 12 to demonstrate the pattern produced by sound waves. The sources of the waves are represented by S_1 and S_2, while A is a point of **constructive interference** and B is a point of **destructive interference**.

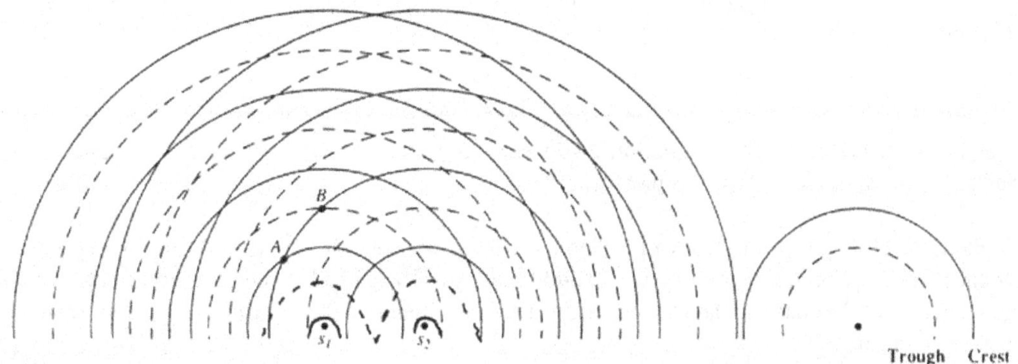

Young's Double-Slit Experiment

Young's double-slit experiment shows that mono-chromatic light passing through two openings also produces an interference pattern. The position of lines of constructive interference can be determined from the following equation:

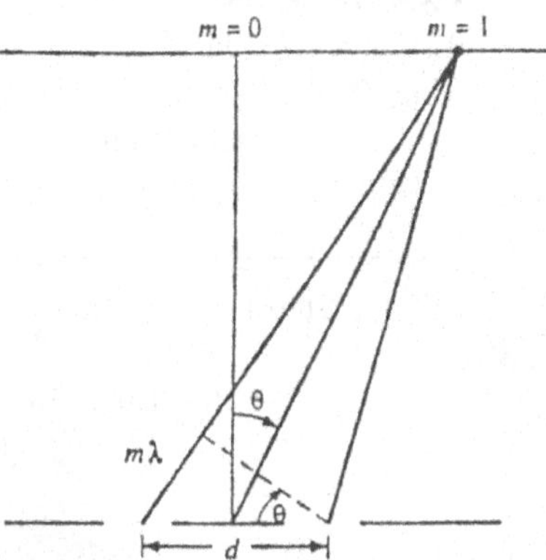

$$m\lambda = d \sin \theta$$

where d is the distance between the slits, and θ is the grating angle. As shown in the diagram, θ is the angle formed by a line drawn from the center point between the slits to the region of constructive interference and the normal drawn to the center of the line that connects the slits. m is the order of the interference fringe. Here m is a dimensionless number that takes on integer values starting with zero—that is, $m = 0, 1, 2, 3,$ etc.; λ is the wavelength of the monochromatic light incident on the double slit; and $m\lambda$ is the path difference between the sources and the region of constructive interference. The path difference is a whole number of wavelengths for constructive interference.

Regions of destructive interference alternate with regions of constructive interference in an interference pattern. The position of lines of destructive interference can be found from

$$(m + \tfrac{1}{2})\lambda = d \sin \theta$$

where m, λ, d, and θ are the same as described previously and $(m + \frac{1}{2})\lambda$ is the path difference between the sources to the points of destructive interference.

TEXTBOOK QUESTION 5. (Section 24-3) Monochromatic red light is incident on a double slit, and the interference pattern is viewed on a screen some distance away. Explain how the fringe pattern would change if the red light source is replaced by a blue light source.

ANSWER: The separation between the fringes in the interference pattern depends on the wavelength (λ). For constructive interference, $\sin \theta = m\frac{\lambda}{d}$, where $\sin \theta \approx \frac{\Delta X}{L}$, where ΔX is the separation between the bright fringes and L is the distance from the double slit to the screen. Rearranging the equation gives $\Delta X = \frac{m\lambda L}{d}$. The wavelength of blue light is shorter than the wavelength of red light. Therefore, if blue light replaces the red light, then the bright fringes will be closer together.

TEXTBOOK QUESTION 7. (Section 24-3) Why doesn't the light from the two headlights of a distant car produce an interference pattern?

ANSWER: The light from the two headlamps is not coherent. The light produced by one light is produced at random with respect to the second light. In order to produce an interference pattern, the light waves must be coherent, and the emitted waves must maintain the same phase relationship to one another at all times.

EXAMPLE PROBLEM 2. (Section 24-3) Monochromatic light is incident on two slits separated by 0.70 mm. The interference pattern is formed on a screen 1.0 m from the slits, and the second bright fringe is found to be 0.20 cm to the right of the central maximum. (*a*) Calculate the wavelength of the light in nanometers (nm) and (*b*) determine the maximum number of bright fringes that can be observed.

Part a. Step 1.	
Convert all of the data to SI units.	$0.70 \text{ mm} = 7.0 \times 10^{-4} \text{ m}$ $0.20 \text{ cm} = 2.0 \times 10^{-3} \text{ m}$
Part a. Step 2.	
Refer to the previous diagram and determine the angular deflection for $m = 2$.	$\tan\theta = \frac{\Delta X}{L} = \frac{2.0 \times 10^{-3} \text{ m}}{1.0 \text{ m}}$ $\tan\theta = 2.0 \times 10^{-3}$ The angle is small, and if the angle is expressed in radians, then $\sin\theta \approx \tan\theta \approx \theta$ and $\theta = 2.0 \times 10^{-3} \text{ rad}$
Part a. Step 3.	
Determine the wave-length of the light.	$m\lambda = d \sin\theta_1$ $2\lambda = (7.0 \times 10^{-4} \text{ m})(2.00 \times 10^{-3})$ $\lambda = 7.0 \times 10^{-7} \text{ m, or } 700 \text{ nm}$

Part b. Step 1.

Determine the maximum number of bright fringes that can be observed.

The maximum value that θ can have is 90°.

$$m\lambda = d \sin \theta$$

$$m(7.0 \times 10^{-7} \text{ m}) = (7.0 \times 10^{-4} \text{ m})(\sin 90°)$$

$$m = \frac{(7.0 \times 10^{-4} \text{ m})(1.00)}{7.0 \times 10^{-7} \text{ m}} = 1000$$

1000 fringes are formed on each side of the central maximum. Thus, counting the central fringe, the theoretical maximum number of lines that could be observed would be $1000 + 1000 + 1 = 2001$. However, it is necessary to round the number to two significant figures; thus, the answer is 2000.

EXAMPLE PROBLEM 3. (Section 24-3) Monochromatic light of wavelength 540 nm (5.40×10^{-7} m) is incident on two slits separated by 5.00×10^{-4} m. The interference pattern is formed on a screen 0.750 m from the slits. Determine the (*a*) angular deflection of the second-order dark fringe (dark line) and (*b*) distance from the central maximum to this fringe.

Part a. Step 1.

Determine the angular deflection of the second-order dark fringe.

$$(m + \tfrac{1}{2})\lambda = d \sin \theta_2$$

$$(2 + \tfrac{1}{2})(5.40 \times 10^{-7} \text{ m}) = (5.00 \times 10^{-4} \text{ m}) \sin \theta_2$$

$$\sin \theta_2 = 2.70 \times 10^{-3}$$

The angle is small; $\sin \theta_2 \approx \tan \theta_2$:

$$\theta_2 = 2.70 \times 10^{-3} \text{ radians}$$

Part b. Step 1.

Determine the distance from the central maximum to the second-order dark fringe.

$$\tan \theta_2 = \frac{\Delta X}{L}$$

$$\Delta X = L \tan \theta_3 = (0.750 \text{ m})(2.70 \times 10^{-3} \text{ rad})$$

$$\Delta X = 2.0 \times 10^{-3} \text{ m} = 2.0 \text{ mm}$$

Diffraction

The wave theory predicts and experiments confirm that waves will bend and spread out as they pass an obstacle or narrow opening. In the case of a narrow opening, the amount of bending increases as the size of the opening decreases.

Because the wavelength of light is much shorter than that of water waves or sound waves, the **diffraction** of light is only noticeable when the size of the opening is comparable to the wavelength of the light.

Single-Slit Diffraction

Due to the combined effects of diffraction and interference, monochromatic light passing through a single slit of width D produces an interference pattern of alternating bright and dark lines. The positions of lines of destructive interference can be determined by the following equation:

$$m\lambda = D \sin \theta$$

where D is the width of the opening and θ is the angle between the normal to the center of the slit and the center of the dark fringe. m is the order of the dark fringe; $m = 1, 2, 3,$ etc. For $m = 0$, the wave theory predicts that constructive rather than destructive interference will occur. Here λ is the wavelength of the light incident on the slit and $m\lambda$ is the path difference between the two edges of the slit and the position of the dark fringe.

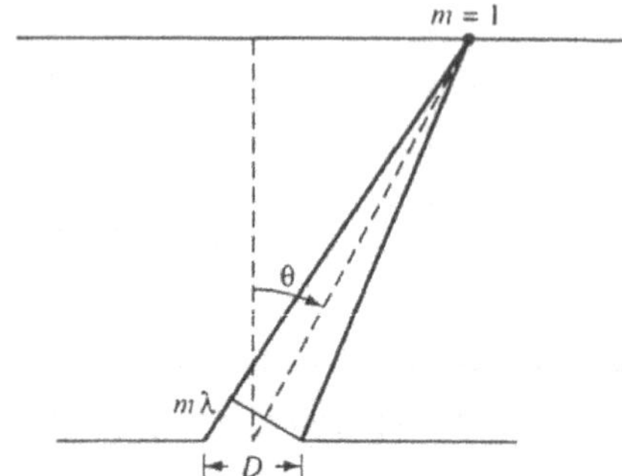

EXAMPLE PROBLEM 4. (Section 24-5) Light of wavelength 480 nm $(4.80 \times 10^{-7}\text{ m})$ is incident on a single slit of width 3.0×10^{-4} m. The diffraction pattern is formed on a screen 1.50 m from the slit. Calculate the distance from the center of the central maximum to the second-order dark fringe.

Part a. Step 1.

Refer to the previous diagram and determine the angular deflection (θ_2) of the second-order dark fringe.	$m\lambda = D \sin \theta_2$ $(2)(4.80 \times 10^{-7}\text{ m}) = (3.0 \times 10^{-4}\text{ m})(\sin \theta_2)$ $\sin \theta_2 = (9.60 \times 10^{-7}\text{ m})/(3.0 \times 10^{-4}\text{ m})$ $\sin \theta_2 = 3.2 \times 10^{-3}$ θ is small, so $\sin \theta \approx \tan \theta$, and $\theta_2 = 3.2 \times 10^{-3}$ radians

Part a. Step 2.

Determine the distance (ΔX) from the center of the central maximum to the second-order dark fringe.

$$\tan\theta_2 = \frac{\Delta X}{L}$$

$$\Delta X = L\tan\theta_2 = (1.50\text{ m})(3.2\times10^{-3}\text{ rad})$$

$$\Delta X = 4.80\times10^{-3}\text{ m, or } 4.80\text{ mm}$$

Diffraction Grating

A **diffraction grating** consists of a large number of closely spaced parallel slits that diffract light incident on the grating. The diffracted light will exhibit constructive interference at points given by the equation

$$m\lambda = d\sin\theta$$

where $m = 0, 1, 2,$ etc. is the order of the bright fringe, λ is the wavelength of the incident light, and d is the distance between the centers of the adjacent slits. The grating angle (θ) is measured from a line drawn normal to the center of the grating and a line drawn from the center of the grating to the position of the bright fringe.

The grating equation has the same form as the equation found in the Young double-slit experiment. In the double-slit experiment, the distance d was measured between the centers of the two slits. In the diffraction grating equation, d refers to the distance between the centers of adjacent slits. The amount of light passing through the diffraction grating is much greater than in the case of the double slit. As a result, the intensity of the bright lines is much greater.

A diffraction grating is particularly useful in separating the component wavelengths of the light incident on the grating. Because of this, it is frequently used in the analysis of spectra produced by various gases, such as mercury, hydrogen, and helium.

EXAMPLE PROBLEM 5. (Section 24-6) A laser beam is incident on a 15,000 groove per inch diffraction grating. The wavelength of the light produced by the laser is 628 nm (6.28×10^{-7} m), and the interference pattern is observed on a screen 2.00 m from the grating. Determine the (*a*) angular deflection of the first-order bright fringe, (*b*) angular deflection of the second-order bright fringe, (*c*) distance from the central maximum to the second-order bright fringe, and (*d*) maximum number of bright fringes that can be observed.

Part a. Step 1.

Determine the distance between adjacent grooves in the diffraction grating in meters.

There are 15,000 grooves per inch, so the distance between adjacent grooves is $\frac{1}{15,000}$ inch:

$$d = \left(\frac{1\text{ in.}}{15,000}\right)\left(\frac{2.54\text{ cm}}{1\text{ in.}}\right)\left(\frac{1\text{ m}}{100\text{ cm}}\right)$$

$$d = 1.69\times10^{-6}\text{ m}$$

Part a. Step 2.	
Determine the first-order angular deflection.	$m\lambda = d \sin \theta_1$ $(1)(6.28 \times 10^{-7}\text{ m}) = (1.69 \times 10^{-6}\text{ m}) \sin \theta_1$ $\sin \theta_1 = 0.372$ and $\theta_1 = 21.8°$
Part b. Step 1.	
Determine the second-order angular deflection.	$m\lambda = d \sin \theta_2$ $(2)(6.28 \times 10^{-7}\text{ m}) = (1.69 \times 10^{-6}\text{ m}) \sin \theta_2$ $\sin \theta_2 = 0.743$ and $\theta_2 = 48.0°$
Part c. Step 1.	
Determine the distance from the central maximum to the second-order bright fringe.	$\tan \theta_2 = \frac{\Delta X}{L}$ $\Delta X = L \tan \theta_2 = (2.00\text{ m})(\tan 48.0°)$ $\Delta X = 2.22\text{ m}$
Part d. Step 1.	
Determine the maximum number of bright fringes that can be observed.	$m\lambda = d \sin \theta$ The maximum possible value of θ is 90°; therefore, $m(6.28 \times 10^{-7}\text{ m}) = (1.69 \times 10^{-6}\text{ m})(\sin 90°)$ $m = 2.69$ Because m must be an integer, only two bright fringes can be observed on either side of the central bright fringe. The maximum number of fringes that can be observed is $2+2+1=5$.

Interference by Thin Films

In Chapter 11, it was noted that a transverse wave pulse traveling through a light section of rope will undergo partial transmission and partial reflection at a point where the pulse enters the heavier section. In the diagrams shown below, if the wave pulse is a crest, then the transmission will be a crest but the reflection will be a trough. A 180° phase change occurs for the reflected part of the wave. However, if the pulse is traveling in a heavy section and enters a light section, then no phase change occurs for either the transmitted or reflected wave pulse.

Light waves traveling from one medium to another undergo partial reflection and partial transmission at the interface of the two mediums. The medium with the lower index of refraction is analogous to the light section of the rope of Chapter 11. The medium with the higher index of refraction is analogous to the heavy section of rope.

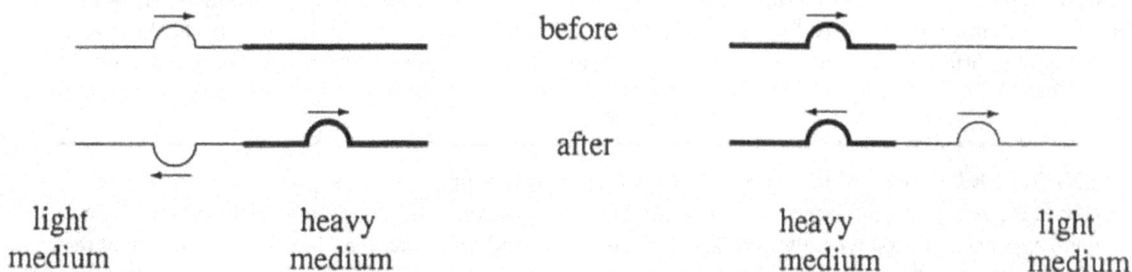

As shown in the following diagram, reflection and transmission of light waves in **thin films** occur at both the top and bottom surfaces of the film. The light reflected back to the observer is the result of light reflected from the top surface interfering with the reflection from the bottom surface.

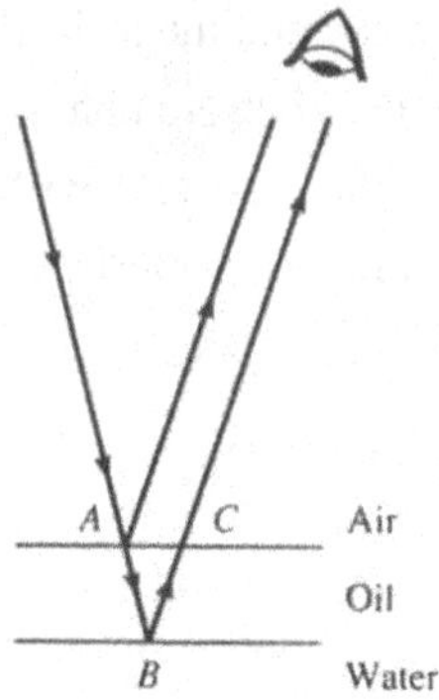

Light waves reflecting from both the top and bottom surfaces of a thin film.

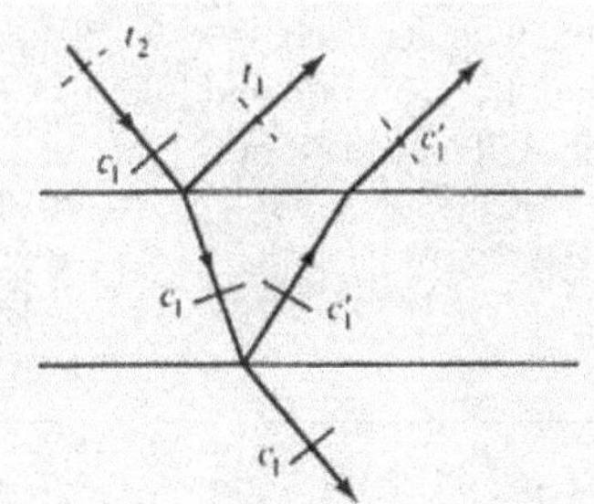

Crests and troughs are represented by a solid line for a crest and a dotted line for a trough. Waves are shown reflecting from both top and bottom surfaces of the thin film.

If the indices of refraction of the media above and below the thin film are lower (or higher) than the index of refraction of the medium of the thin film, then maximum reflection of light of a particular wavelength occurs if the thickness (t) of the film is an odd-number multiple of quarter wavelengths. The equation for thickness of the film in terms of wavelength is

$$t=\frac{(2m+1)\lambda_n}{4} \quad \text{or} \quad 4t=(2m+1)\lambda_n$$

where $\lambda_n=\dfrac{\lambda}{n}$ and $m=0, 1, 2,$ etc.

Minimum reflection of the light occurs if the thickness of the film is a whole-number multiple of half wavelengths. The equation for minimum reflection is given by

$$t=m\frac{\lambda_n}{2} \quad \text{or} \quad 2t=m\lambda_n$$

where $m=0, 1, 2,$ etc.

If the media above and below the thin film are such that a phase change occurs at both the top and bottom surfaces, then the equation for **maximum reflection** is given by $2t = m\lambda_n$ and for **minimum reflection** by $4t = (2m+1)\lambda_n$.

Since ordinary white light is made up of colors of varying wavelengths (400 nm–700 nm), there will be maximum reflection of certain colors but no reflection of other colors. This and the fact that the thickness of the thin film usually varies across the surface of the film result in the spectrum of colors seen reflected from soap bubbles and oil films.

EXAMPLE PROBLEM 6. (Section 24-8) A ring of wire is dipped into a soap solution and then held so that the soap film on the ring is vertical. As the soap film gradually drains toward the bottom, a dark band appears at the top with alternating bright and dark bands of light appearing along the length of the film. Determine the thickness of the film at the first three bright bands, as counted from the top, if the incident light has a wavelength of 670 nm and is directed perpendicular to the surface of the film. The index of refraction of the soap solution is 1.33.

Part a. Step 1.

Assume that a wave (c_1) strikes the top interface. Determine whether each subsequent transmission and reflection will be a crest or a trough.

Since $n_{\text{film}} > n_{\text{air}}$, a phase change occurs, and the reflection of c_1 at the top interface will be a trough (t_1). No phase change occurs for transmitted light, so the crest (c_1) will strike the lower surface as a crest. At the lower interface, the crest (c_1) will reflect as a crest (c_1'). This is because $n_{\text{film}} > n_{\text{air}}$. When c_1' strikes the top interface, it will be transmitted as a crest.

Part a. Step 2.

Draw a diagram using c for crest and t for trough for the situation described in Part a, Step 1.

air

soap film

air

Part a. Step 3.

Derive a formula for the film thickness that results in maximum reflection of the incident light.

If the thickness of the film is $\frac{1}{4}\lambda$, then c_1' will have traveled $\frac{1}{4}\lambda_n + \frac{1}{4}\lambda_n = \frac{1}{2}\lambda_n$ and c_1' will meet trough t_2, which is reflecting at the top interface as a crest c_2. The superposition of c_1' and c_2 will result in constructive interference and maximum reflection. Thus, maximum reflection occurs if the thickness of the film follows the equation

$$t = \frac{(2m+1)\lambda_n}{4} \quad \text{or} \quad 4t = (2m+1)\lambda_n,$$

where $\lambda_n = \frac{\lambda_{air}}{n_{film}} = \frac{670 \text{ nm}}{1.33} = 504 \text{ nm}$

Part a. Step 4.

Determine the thickness of the film at the position of the first three bright bands as counted from the top of the film.

The minimum thickness for maximum reflection occurs if $m = 0$. Since the film drains toward the bottom, the minimum thickness is located at the position of the first bright band observed near the top of the film.

$4t = (2m+1)\lambda_n$

If $m = 0$, then $4t = [2(0)+1](504 \text{ nm})$ and

$t = 126 \text{ nm} = 1.26 \times 10^{-7} \text{ m}$

Successive bright bands occur for $m = 1$, $m = 2$.

If $m = 1$, then $4t = [2(1)+1](504 \text{ nm})$ and

$t = 378 \text{ nm} = 3.78 \times 10^{-7} \text{ m}$

If $m = 2$, then $4t = [2(2)+1](504 \text{ nm})$ and

$t = 630 \text{ nm} = 6.30 \times 10^{-7} \text{ m}$

TEXTBOOK QUESTION 10. (Section 24-8) Some coated lenses appear greenish yellow when seen by reflected light. What reflected wavelengths do you suppose the coating is designed to eliminate completely?

ANSWER: Since the coated lens reflects greenish yellow, yellow and green are reflected and are not transmitted through the lens. While yellow and green are reflected, violet, blue, orange and red light are transmitted though the lens.

Michelson Interferometer

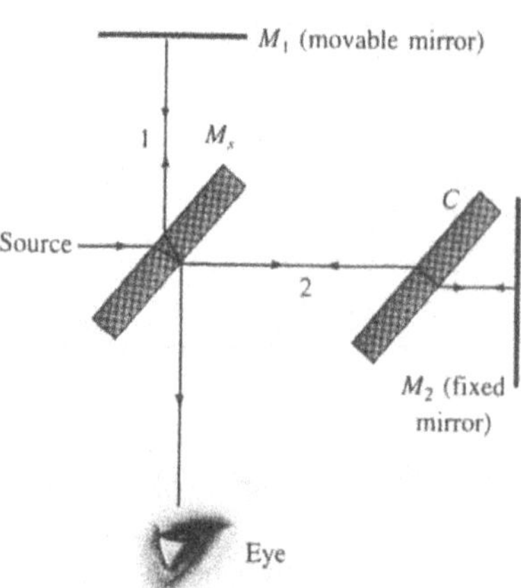

The **Michelson interferometer** is a device that uses wave interference to determine the wavelength of light. As shown in the diagram, the light from source strikes a half-silvered mirror (M_s). Part of the light is reflected to a movable mirror (M_1), while part of the light is transmitted and reflects from a fixed mirror (M_2). After reflection, the beams arrive at the observer's eye.

Because the speed of light is less in glass than in air, a glass plate called the compensator (C) is placed in the path of beam 2. Thus, both beams travel through the same thickness of glass before arriving at the observer's eye.

Assume that light reaching the observer's eye from mirrors M_1 and M_2 are in phase. Constructive interference occurs, and the observer sees light. If mirror M_1 is moved $\frac{1}{4}\lambda$ toward the observer, then the path difference changes by $\frac{1}{4}\lambda+\frac{1}{4}\lambda=\frac{1}{2}\lambda$ and destructive interference occurs. If the mirror is moved another $\frac{1}{4}\lambda$, then the total path difference has changed by $\frac{1}{2}\lambda+\frac{1}{2}\lambda=1\lambda$ from its original setting and constructive interference again occurs.

The distance (t) that the mirror M_1 moves can be determined by a device known as a micrometer. The wavelength of the monochromatic light from a source can be determined by counting the number of bright (or dark) fringes (m) that pass the field of view as M_1 is moved. The equation used is $t=m\dfrac{\lambda}{2}$.

EXAMPLE PROBLEM 7. (Section 24-9) The bright green line produced by the mercury spectrum is used as the source in an experiment involving the Michelson interferometer. When the movable mirror is moved 0.300 mm, 1500 bright fringes are counted. Determine the wavelength of the light.

Part a. Step 1.

Determine the wavelength of the light. Note: Refer to the preceding diagram.	Each time mirror M1 travels $\frac{1}{2}\lambda$, another bright fringe passes the field of view. Therefore, $t=m\dfrac{\lambda}{2}$ and rearranging gives $\lambda=2\dfrac{t}{m}=2\left(\dfrac{0.300\text{ mm}}{1500}\right)$ $\lambda=4.00\times10^{-7}\text{ m}=400\text{ nm}$

Polarization of Light

The interference phenomena previously studied in this chapter can be produced by longitudinal as well as transverse waves. **Polarization** is a property of light that indicates that light is a **transverse wave** phenomena.

An electromagnetic wave in which the electric vector is vibrating in only one plane is said to be **plane-polarized**. Ordinary light is not polarized, which means that its electric vector is vibrating in many planes at the same time.

There are several ways to polarize light. One method is to use a material such as a Polaroid sheet that removes all of the electric vectors except those in a particular plane. The Polaroid filter that causes the electric vector to vibrate in only one plane is known as the **polarizer**. The axis of a second Polaroid filter, known as the **analyzer**, can be crossed with the axis of the polarizer and block all light to the observer.

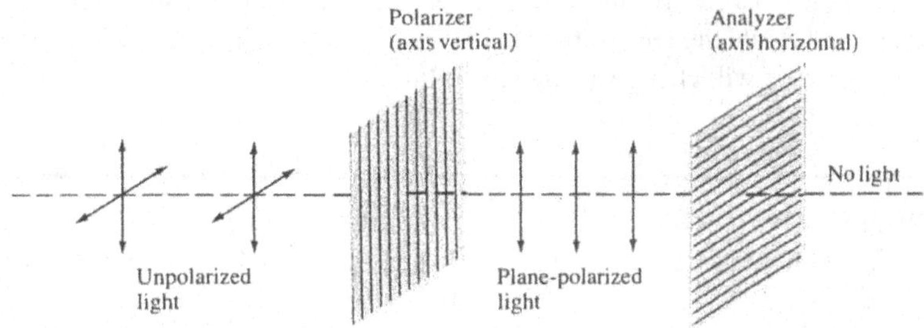

A second method to polarize unpolarized light is to reflect ordinary light from materials such as glass or water. The angle of incidence at which maximum polarization of the reflected light occurs is known as **Brewster's angle**. Brewster's angle θ_p can be found by using the equation

$$\tan\theta_p = \frac{n_2}{n_1}$$

where n_1 is the index of refraction of the medium in which the light is initially traveling and n_2 is the index of refraction of the medium from which the light is reflected.

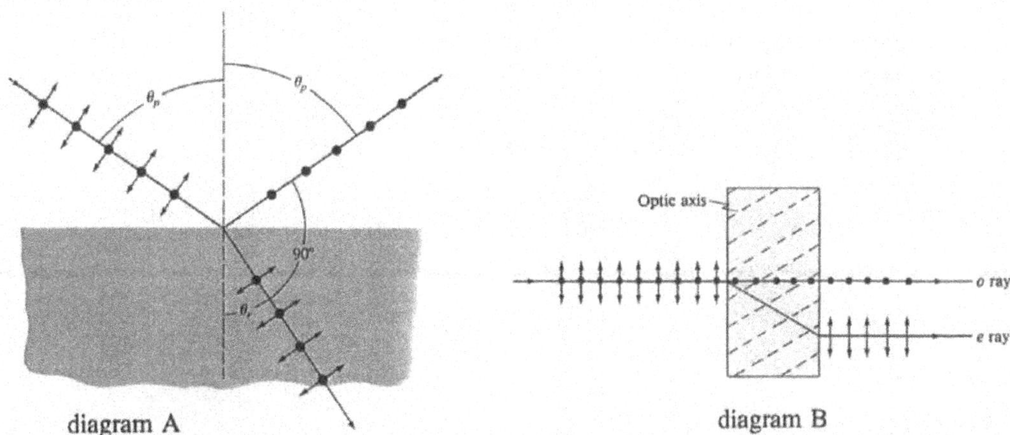

As shown in diagram A, at Brewster's angle the reflected light is plane-polarized parallel to the surface. The angle between the refracted ray and the reflected ray is 90°.

TEXTBOOK QUESTION 20. (Section 24-10) What does polarization tell us about the nature of light?

ANSWER: Polarization indicates that light must be a transverse wave rather than a longitudinal wave. As discussed in Chapter 12, sound waves are longitudinal waves that exhibit interference when emitted by two sources that are in phase. Sound waves are longitudinal waves; the molecules of the medium vibrate only along the direction of motion of the wave. There is no way that a longitudinal wave can be plane-polarized.

TEXTBOOK QUESTION 22. (Section 24-10) How can you tell if a pair of sunglasses is polarizing or not?

ANSWER: Look for glare reflecting from a waxed floor, a lake, or a nonmetallic reflecting surface. Make sure that the angle between the normal to the surface and your eye is 45° or more. Hold one lens in front of your eye and rotate the lens. If the lens is made of polarizing material, then the intensity of the light reaching your eye will change as the lens rotates.

EXAMPLE PROBLEM 8. (Section 24-10) Determine Brewster's angle for (*a*) a water–glass interface and (*b*) a water–diamond interface. Note: $n_{\text{diamond}} = 2.42,\ n_{\text{water}} = 1.33,$ and $n_{\text{glass}} = 1.50.$

Part a. Step 1.	
Determine Brewster's angle for an air–glass interface.	$\tan\theta_{\text{p}} = \dfrac{n_2}{n_1}$ $\tan\theta_{\text{p}} = \dfrac{1.50}{1.33} = 1.13$ $\theta_{\text{p}} = 48.4°$
Part b. Step 1.	
Determine Brewster's angle for a water–diamond interface.	$\tan\theta_{\text{p}} = \dfrac{2.42}{1.33}$ $\tan\theta_{\text{p}} = 1.82$ $\theta_{\text{p}} = 61.2°$

PROBLEM SOLVING SKILLS

For problems involving double-slit interference, single-slit diffraction, or a diffraction grating:

1. Determine whether the light is passing through a double slit, a single slit, or a diffraction grating.
2. Draw an accurate diagram labeling d or D, ΔX, L, the order of the image (m) and the path difference $m\lambda$.
3. Complete a data table based on the information given in the problem.
4. Note whether the problem involves constructive interference (bright fringes) or destructive interference (dark fringes).
5. Choose the appropriate formula and solve the problem.

For problems involving thin-film interference:

1. Assume a crest is incident on the top surface of the film.
2. Determine whether the subsequent reflections and transmissions from each interface will be a crest or trough.
3. Draw a diagram placing a t for trough and c for crest at each position where the light is reflected and transmitted.
4. Determine the minimum thickness of the film in terms of a fraction of a wavelength of the incident light required to produce (a) maximum reflection of the light and (b) minimum reflection of the light.
5. Based on Step 4, write a formula for the film thickness that will produce (a) maximum reflection of the light and (b) minimum reflection of the light.
6. If the wavelength of the light in air is given, then it is necessary to determine the wavelength of the light in the film.
7. Complete a data table, choose the appropriate formula, and solve the problem.

For problems involving the angle of maximum plane polarization of reflected light:

1. Determine the relative index of refraction of the two mediums—that is, the ratio of the index of refraction of the medium from which the light is reflected to the index of refraction of the medium in which the light is incident.
2. Use $\tan\theta_p = \dfrac{n_2}{n_1}$ to solve for Brewster's angle (θ_p).

SOLUTIONS TO SELECTED TEXTBOOK PROBLEMS

TEXTBOOK PROBLEM 2. (Section 24-3) The third-order bright fringe of 610-nm light is observed at an angle of 31° when the light falls on two narrow slits. How far apart are the slits?

Part a. Step 1.

Complete a data table. Note: Convert all data to meters.	$m = 3$	$\lambda = 610\text{ nm} = 6.10\times10^{-7}\text{ m}$
	$\theta_3 = 31°$	$d = ?$

Part a. Step 2.

Determine the distance (d) between the slits.

$m\lambda = d \sin \theta_1$

$(3)(6.10 \times 10^{-7} \text{ m}) = d(\sin 31°)$

$1.8 \times 10^{-6} \text{ m} = d(0.515)$

$d = 3.60 \times 10^{-6} \text{ m}$

TEXTBOOK PROBLEM 3. (Section 24-3) Monochromatic light falls on two very narrow slits 0.048 mm apart. Successive fringes on a screen 6.50 m away are 8.5 cm apart near the center of the pattern. Determine the wavelength and frequency of the light.

Part a. Step 1.

Complete a data table. Note: Convert all data to meters.

$d = 0.048 \text{ mm} = 4.8 \times 10^{-5} \text{ m}$ $\quad L = 6.50 \text{ m}$

$\Delta X = 8.5 \text{ cm} = 8.5 \times 10^{-2} \text{ m}$ $\quad \lambda = ?$ $\quad f = ?$

Part a. Step 2.

Determine the angular deflection for $m = 1$.

$$\tan \theta = \frac{\Delta X}{L} = \frac{8.5 \times 10^{-2} \text{ m}}{6.5 \text{ m}}$$

$\tan \theta = 1.3 \times 10^{-2}$

The angle is small, and if the angle is expressed in radians, then $\sin \theta \approx \tan \theta \approx \theta$ and

$\theta = 1.3 \times 10^{-2} \text{ rad}$

Part a. Step 3.

Determine the wavelength of the light.

$m\lambda = d \sin \theta_1$

$(1)(\lambda) = (4.8 \times 10^{-5} \text{ m})(1.3 \times 10^{-2})$

$\lambda = 6.3 \times 10^{-7} \text{ m}, \quad \text{or} \quad 630 \text{ nm}$

Part a. Step 4.

Determine the frequency of the light.

$c = f\lambda$

$$f = \frac{c}{\lambda} = \frac{3.0 \times 10^8 \text{ m/s}}{6.2 \times 10^{-7} \text{ m}}$$

$f = 4.8 \times 10^{14}$ Hz

TEXTBOOK PROBLEM 9. (Section 24-3) A parallel beam of light from a He–Ne laser, with a wavelength 633 nm, falls on two very narrow slits 0.068 mm apart. How far apart are the fringes in the center of the pattern on a screen 3.3 m away?

Part a. Step 1.

Complete a data table for the first-order fringe. Convert all data to meters.

$m = 1 \quad \lambda = 633 \text{ nm} = 6.33 \times 10^{-7}$ m

$d = 0.068 \text{ mm} = 6.8 \times 10^{-5}$ m $\qquad L = 3.3$ m

$\Delta X = ?$

Note: As shown in Fig. 24-9 of the textbook, the distance between adjacent fringes in the center of the pattern is the same.

Part a. Step 2.

Determine the angular deflection for the first-order light.

$m\lambda = d \sin \theta_1$

$(1)(6.33 \times 10^{-7} \text{ m}) = (6.8 \times 10^{-5} \text{ m})(\sin \theta_1)$

$\sin \theta_1 = 9.3 \times 10^{-3}$

Part a. Step 3.

Determine the distance (ΔX) from the central maximum to the first-order image.

The angle is small, and if the angle is expressed in radians, then $\sin \theta \approx \tan \theta \approx \theta$:

$$\tan \theta = \frac{\Delta X}{L}$$

$$9.3 \times 10^{-3} = \frac{\Delta X}{3.3 \text{ m}}$$

$\Delta X_1 = 3.07 \times 10^{-2}$ m

Part a. Step 4.

Repeat Steps 2 and 3 for the second-order image.

$m\lambda = d \sin \theta_2$

$(2)(6.33 \times 10^{-7}\text{ m}) = (6.8 \times 10^{-5}\text{ m})(\sin \theta_2)$

$\sin \theta_2 = 1.86 \times 10^{-2}$

Part a. Step 5.

Determine the distance (ΔX) from the central maximum to the second-order image.

The angle is small, and if the angle is expressed in radians, then $\sin \theta \approx \tan \theta \approx \theta$:

$$\tan \theta = \frac{\Delta X}{L}$$

$$1.86 \times 10^{-2} = \frac{\Delta X}{3.3\text{ m}}$$

$$\Delta X_2 = 6.14 \times 10^{-2}\text{ m}$$

Part a. Step 6.

Determine the distance between the first- and second-order fringes.

$\text{distance} = \Delta X_2 - \Delta X_1$

$\text{distance} = 6.14 \times 10^{-2}\text{ m} - 3.07 \times 10^{-2}\text{ m}$

$\text{distance} = 3.07 \times 10^{-2}\text{ m}$

TEXTBOOK PROBLEM 19. (Section 24-5) A light beam strikes a piece of glass at a $65.00°$ incident angle. The beam contains two wavelengths, 450.0 nm and 700.0 nm., for which the index of refraction of the glass is 1.4831 and 1.4754, respectively. What is the angle between the two refracted beams?

Part a. Step 1.

Use Snell's law to determine the angle of refraction of the 450.0-nm beam.

Note: Let n_2 represent the index of refraction of glass for the 450.0-nm beam.

$n_1 \sin \theta_1 = n_2 \sin \theta_2$

$(1.00) \sin 65.00° = (1.4831) \sin \theta_2$

$$\sin \theta_2 = \frac{(1.00)(0.906)}{1.4831}$$

$\sin \theta_2 = 0.611$

$\theta_2 = 37.66°$

Part a. Step 2.

Use Snell's law to determine the angle of refraction of the 700.0-nm beam.

Note: Let n_3 represent the index of refraction of glass for the 700.0-nm beam.

$n_1 \sin\theta_1 = n_3 \sin\theta_3$

$(1.00)\sin 65.00° = (1.4754)\sin\theta_3$

$\sin\theta_3 = \dfrac{(1.00)(0.906)}{1.4754}$

$\sin\theta_3 = 0.614$

$\theta_3 = 37.88°$

Part a. Step 3.

Determine the angle between the two beams.

The angle between the two beams is

$\theta_3 - \theta_2 = 37.66° - 37.88° = 0.224°$

TEXTBOOK PROBLEM 23. (Section 24-5) When blue light of wavelength 440 nm falls on a single slit, the first dark bands on either side of center are separated by 51.0°. Determine the width of the slit.

Part a. Step 1.

Draw an accurate diagram and complete a data table listing information both given and implied.

$m = 1$ $D = ?$

$\lambda = 440\text{ nm} = 4.4\times10^{-7}\text{ m}$

$\theta = \frac{1}{2}(51.0°) = 25.5°$

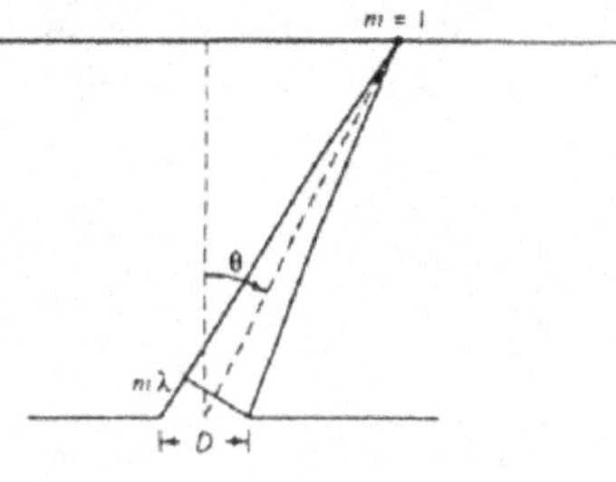

Part a. Step 2.

Determine the width of the slit.

$m\lambda = D\sin\theta$

$(1)(4.40\times10^{-7}\text{ m}) = D(\sin 25.5°)$, where $\sin 25.5° = 0.431$

$D = \dfrac{4.40\times10^{-7}\text{ m}}{0.431}$

$D = 1.02\times10^{-6}\text{ m}$

TEXTBOOK PROBLEM 42. (Sections 24-6 and 24-7) The first-order line of 589-nm light falling on a diffraction grating is observed at a 14.5° angle. How far apart are the slits? At what angle will the third order be observed?

Part a. Step 1.

Complete a data table for the first-order fringe. Convert all data to meters.

$m = 1$ $\lambda = 589 \text{ nm} = 5.89 \times 10^{-7} \text{ m}$

$d = ?$ $\theta = 14.5°$

Part a. Step 2.

Determine the distance between adjacent slits.

$$m\lambda = d \sin \theta_1$$

$$(1)(5.89 \times 10^{-7} \text{ m}) = d(\sin 14.5°)$$

$$d = \frac{(1)(5.89 \times 10^{-7} \text{ m})}{\sin 14.5°}$$

$$d = \frac{(5.89 \times 10^{-7} \text{ m})}{0.250}$$

$$d = 2.36 \times 10^{-6} \text{ m}$$

Part a. Step 3.

Determine the angle where the third order will be observed.

$$m\lambda = d \sin \theta_3$$

$$(3)(5.89 \times 10^{-7} \text{ m}) = (2.36 \times 10^{-6} \text{ m}) \sin \theta_3$$

$$\sin \theta_3 = 0.749$$

$$\theta_3 = 48.5°$$

TEXTBOOK PROBLEM 47. (Section 24-8) A lens appears greenish yellow ($\lambda = 570$ nm is strongest) when white light reflects from it. What minimum thickness of coating $(n = 1.25)$ do you think is used on such a glass lens $(n = 1.52)$?

Part a. Step 1.

Draw a diagram using *c* for crest and *t* for trough for the air–film–glass situation described in this Problem.

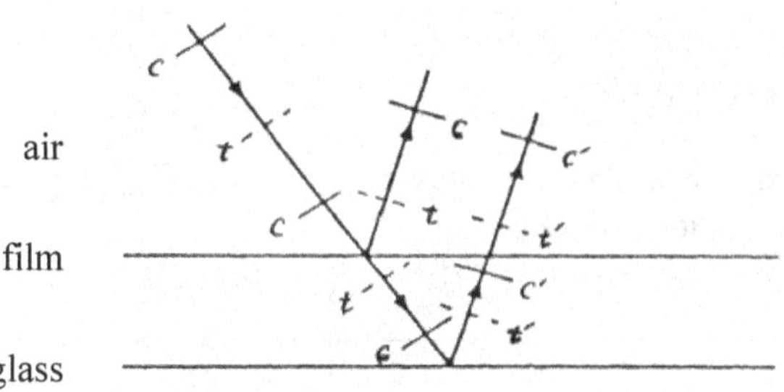

The diagram assumes that wave crest (c) strikes the top interface. Determine whether each subsequent transmission and reflection will be a crest or a trough.

Since $n_{\text{film}} > n_{\text{air}}$, a phase change occurs and the reflection of a c at the top interface will be a trough (t). No phase change occurs for transmitted light so the crest (c) will strike the lower interface as a crest. At the lower interface the crest (c) will reflect as a trough (t'). This is because $n_{\text{glass}} > n_{\text{film}}$. When t' reaches the top interface, it will be transmitted as a trough.

Part a. Step 2.

Derive a formula for film thickness that results in maximum reflection of the incident light.

If the thickness of the film is $\frac{1}{2}\lambda$, then the initial transmission will have traveled $\frac{1}{2}\lambda_n + \frac{1}{2}\lambda_n = 1\lambda_n$ before emerging from the top of the lens as a trough. The trough will meet the reflection of the next crest, which is reflecting at the top interface as a trough. The superposition of the two troughs will result in constructive interference and maximum reflection. Thus, maximum reflection occurs if the thickness of the film follows the equation

$$t = m\frac{\lambda_n}{2} \quad \text{or} \quad 2t = m\lambda_n,$$

where $\lambda_n = \dfrac{\lambda_{\text{air}}}{n_{\text{film}}} = \dfrac{570 \text{ nm}}{1.25} = 456 \text{ nm}$

Part a. Step 3.

Determine the minimum thickness of the film that reflects light of wavelength 570 nm.

If $m = 0$, then the thickness of the film is 0. The first nonzero thickness of the film occurs for $m = 1$.

$$t = m\frac{\lambda_n}{2}$$

$$t = (1)\left(\frac{456 \text{ nm}}{2}\right) = 228 \text{ nm} = 2.28 \times 10^{-7} \text{ m}$$

TEXTBOOK PROBLEM 57. (Section 24-9) What is the wavelength of the light entering an interferometer if 362 bright fringes are counted when the movable mirror moves 0.125 mm?

Part a. Step 1.

Complete a data table given the information provided.

$t = 0.125 \text{ mm} = 1.25 \times 10^{-4} \text{ m}$

$m = 362$

Part a. Step 2.

Determine the wavelength of the light.

The distance that mirror M_1 moves from one fringe to the next equals $\frac{1}{2}\lambda$. Therefore, the wavelength of the light can be found by using

$$t = m\frac{\lambda}{2}$$

$$1.25 \times 10^{-4} \text{ m} = (362)\frac{\lambda}{2}$$

$$\lambda = \frac{(2)(1.25 \times 10^{-4} \text{ m})}{362}$$

$$\lambda = 6.90 \times 10^{-7} \text{ m} = 690 \text{ nm}$$

25

OPTICAL INSTRUMENTS

OBJECTIVES

After studying the material of this chapter, the student should be able to:

- identify the major components of a camera and explain how these components combine to produce a clear image.
- identify the major components of the human eye and explain how an image is formed on the retina of the eye.
- explain the causes of nearsightedness and farsightedness, describe how these conditions may be corrected, and solve word problems related to corrective lenses for these conditions.
- explain how a magnifying glass can be used to produce an enlarged image, and solve word problems involving the magnifying glass.
- explain how two convex lenses can be arranged in order to form an astronomical telescope, and solve word problems related to this type of telescope.
- explain how two convex lenses can be arranged in order to form a compound microscope, and solve word problems related to this type of microscope.
- distinguish between spherical aberration and chromatic aberration and explain how each type of aberration can be corrected.
- describe the factors which affect resolution of an image and limit the magnification of a telescope or microscope.
- explain how the phenomena of X-ray diffraction can be used to determine either the distance between the atoms of a crystal or the wavelength of the incident X-rays. Use the Bragg equation to solve word problems involving X-ray diffraction.

KEY TERMS AND PHRASES

camera includes a convex lens, a light-tight box, a shutter, an iris diaphragm, and an electronic sensor (in a digital camera) or a piece of film or photographic plate (in a traditional camera). The image of the object is formed on the sensor or film.

human eye functions in a manner similar to a camera. Light passes through a transparent outer membrane called the **cornea**, a clear liquid called the aqueous humor, and the lens into a cavity that is filled with a second clear liquid called the vitreous humor. The image is formed on the back of the eye on the **retina** and transmitted to the brain via the optic nerve.

accommodation refers to the ability of the lens of the eye to change shape in order to bring objects into focus.

nearsightedness, or myopia, is a condition where a person can see nearby objects clearly, but objects at a distance are blurred.

farsightedness, or hyperopia, is a condition where a person can see distant objects clearly, but nearby objects are blurred.

diopter is a measure of the refractive power of a lens. The power in diopters is inversely related to the focal length expressed in meters.

magnifying glass is a converging lens that is used to produce a virtual, upright, and enlarged image of an object.

near point is the closest distance from the eye that an object can be placed and still be focused on the retina.

astronomical telescope, a refracting type, consists of a large-diameter convex lens with a long focal length (f_o) called the **objective lens** and a small-diameter convex lens with a short focal length (f_e) as the **eyepiece**. A magnified, inverted, virtual image of an object, such as a planet, is observed.

compound microscope consists of two convex lenses, an objective, and an eyepiece separated by a distance. The object is placed just beyond the focal point of the objective. The image observed is virtual, inverted, and greatly magnified.

spherical aberration occurs when rays of light from a point on the axis of the lens do not produce a point image after refraction. Instead, they produce a small circular patch of light.

chromatic aberration occurs because the index of refraction of transparent materials varies with wavelength. As a result, white light from a point source produces an image containing colors spread out over a small region.

achromatic doublet is a combination of two lenses used to correct chromatic aberration. One lens is a convex lens and the other a concave lens. The lenses have different indices of refraction and radii of curvature.

Rayleigh criterion states that two images are just resolvable when the center of the diffraction disk of one image is directly over the first minimum of the diffraction pattern of the other.

X-rays are electromagnetic waves of very short wavelength emitted when high energy electrons strike a metal target.

X-ray diffraction patterns are produced as a result of reflection of an X-ray beam from the surface of a crystal. The subsequent reflection of the X-rays from the planes of atoms results in a pattern of constructive and destructive interference.

SUMMARY OF MATHEMATICAL CONCEPTS

f-stop	$f\text{-stop} = \dfrac{f}{D}$	The *f*-stop refers to the adjustment to the size of the lens opening necessary to compensate for the outside brightness and the shutter speed. f is the focal length of the lens and D is the diameter of the opening.
angular magnification	$M = \dfrac{\theta'}{\theta}$	The angular magnification or magnifying power (M) is the ratio of the angle the object subtends with the eye (θ') when the magnifier is used to the angle the object subtends with the eye (θ) without the magnifier at a distance of 25 cm.

magnification of a magnifying glass with the image at the near point	$M = 1 + \frac{N}{f}$	The magnification depends on the near point (N) and the focal length of the lens. N is usually taken to be 25 cm.
magnification of a magnifying glass with the image at infinity	$M = \frac{N}{f}$	If the eye is relaxed, the image is at infinity and the object is then precisely at the focal point. The magnification (M) equals the ratio of the near point to the focal length.
magnification of a refracting telescope	$M = -\frac{f_o}{f_e}$	The magnification (M) equals the ratio of the focal length of the objective lens (f_o) to the focal length of the eyepiece (f_e). The negative sign indicates an inverted image.
magnification of a compound microscope	$M = M_e m_o$ $M \approx \frac{N\ell}{f_e f_o}$	The total magnification (M) of a compound microscope equals the product of the magnification of the eyepiece (M_e) and the magnification of the objective (m_o). The approximation is accurate when f_e and f_o are small compared to the distance between the lenses.
Rayleigh criterion	$\theta = 1.22\frac{\lambda}{D}$	The Rayleigh criterion is used to determine when two images can be resolved. The minimum angle (θ) for resolution depends on the wavelength (λ) of the light and the diameter of the objective lens (D).
resolving power of a microscope	$\text{RP} = s = f\theta = 1.22\frac{\lambda f}{D}$	For a microscope, the resolution is defined in terms of the resolving power (RP), where s is the minimum separation of two objects that can just be resolved and f is the focal length of the objective lens.
X-ray diffraction: Bragg equation	$m\lambda = 2d \sin \phi$	If an X-ray beam is incident on a crystal, then the subsequent reflection of the X-rays from the planes of atoms results in a diffraction pattern. Constructive interference occurs when $m\lambda = 2d \sin \phi$, where $m\lambda$ is a whole number of wavelengths and $m = 1, 2, 3,$ etc.; d is the distance between the layers of atoms; and ϕ is the angle of incidence of the X-ray beam with the surface of the crystal.

CONCEPT SUMMARY

The Camera

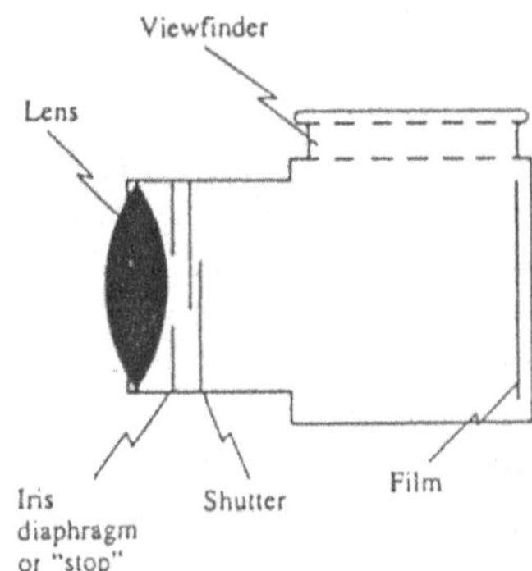

As shown in the diagram, a **camera** consists of a convex lens, a light-tight box, film or a photographic plate (in a traditional camera) or a sensor (in a digital camera), a shutter, and an iris diaphragm or stop.

Three main adjustments must be made in order to produce a clear image of the object on the film: shutter speed, *f*-stop, and distance from lens to sensor or film. The shutter speed determines the length of time that the shutter is open to allow light to pass through to the sensor or film. Shutter speed can vary from a second or more for a long exposure to 1/1000 second or less to capture the image of a moving object and avoid a blurred picture.

The amount of light reaching the film is controlled by the iris diaphragm or stop. The stop blocks light that has passed through the lens from reaching the sensor or film. The size of the opening is adjusted to compensate for the outside brightness as well as the shutter speed and is specified in terms of the *f*-stop. The *f*-stop is given by the equation

$$f\text{-stop} = \frac{f}{D}$$

where f is the focal length of the lens and D is the diameter of the opening.

Depending on the distance from the object to the lens, it may be necessary to adjust the distance between the lens and the sensor or film in order to produce a sharp image. This is accomplished by turning a ring on the lens that moves the lens toward or away from the sensor or film. On inexpensive cameras, the optics are arranged so that all objects beyond a certain distance, usually 6 feet, are in focus and no adjustment is necessary.

TEXTBOOK QUESTION 1. (Section 25-1) Why must a camera lens be moved farther from the sensor or film to focus on a closer object?

ANSWER: The relationship between the object distance, image distance, and focal length of the lens is given by the lens equation: $\frac{1}{d_o}+\frac{1}{d_i}=\frac{1}{f}$. The focal length of the lens is fixed, and the image distance is measured from the lens to the sensor or film where the image is formed. If the object is close to the lens, then the image distance must be increased so that a clear image will be formed on the sensor or film (see diagram C in Chapter 23 of this Study Guide). If the object is far away from the lens, then the image distance between the lens and sensor or film must be decreased in order to form a clear image (see diagram G in Chapter 23 of this Study Guide). Inexpensive cameras often are made so that the distance from the lens to the sensor or film is fixed. In this type of camera an object distance of 6 feet to infinity ensures that a clear image will be formed.

The Human Eye

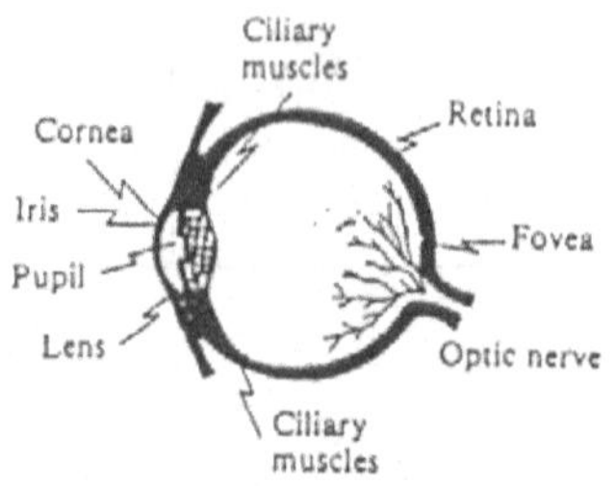

The basic structure of the human eye is shown in the diagram. Light passes through a transparent outer membrane called the **cornea**, a clear liquid called the aqueous humor, and the lens into a cavity that is filled with a second clear liquid called the vitreous humor. The image is formed on the back of the eye on the **retina** and transmitted to the brain via the optic nerve. The image formed on the retina is received by millions of light-sensitive receptors known as rods and cones. The rods are sensitive to the intensity level of light, while the cones distinguish color. Vision is most acute at the **fovea**, a region where the cones are most densely packed.

The amount of light passing through the eye is controlled by a diaphragm called the **iris**, which opens and closes automatically as the eye adjusts to light intensity. The opening in the center of the iris is called the **pupil** of the eye.

The lens of the eye changes shape and therefore its focal length in order to focus the image of an object on the retina. For distant objects, the ciliary muscles relax, the lens becomes thinner and the focal length greater, and the image is focused on the retina. For nearby objects, the muscles contract, causing the curvature of the lens to increase and decreasing the focal length. Again, the image comes to a focus on the retina. The ability of the eye to adjust in this manner is called **accommodation**.

Common defects of the eye include **nearsightedness**, or myopia, and **farsightedness**, or hyperopia. In nearsightedness, a person can see nearby objects clearly, but objects at a distance are blurred. This is usually because the eyeball is too long and the image comes to a focus in front of the retina. This condition can be corrected by a diverging lens that causes the rays to come to focus at the retina.

A farsighted person can see distant objects clearly, but a nearby object, for example, the print on this page, is blurred. This is usually caused by an eyeball that is too short. The rays from the object have not yet come to a focus when they strike the retina. A converging lens is used to converge the rays so that the image comes to a focus at the retina. When prescribing eyeglasses, the power (P) of a lens, expressed in diopters, is used in place of the focal length: $P = \frac{1}{f}$, where the focal length (f) is expressed in meters. P is positive for a converging lens and negative for a diverging lens.

TEXTBOOK QUESTION 8. (Section 25-2) Is the image formed on the retina of the human eye upright or inverted? Discuss the implications of this for our perception of objects.

ANSWER: As shown in Fig. 25-9 in the textbook, the lens of the eye is comparable to a convex lens. Also, as shown in Figs. 25-16a and 25-16b, the images of objects formed on the retina are real and inverted. Therefore, the image formed on the retina is similar to the image formed on the sensor or film in a camera. The implication is that the brain inverts the image formed on the retina so that we see the image upright.

TEXTBOOK QUESTION 4. (Section 25-2) Why are bifocals needed mainly by older persons and not generally by younger people?

ANSWER: Nearsightedness, or myopia, is usually caused by an eyeball that is too long. As a result, images of distant objects come to a focus in front of the retina. This problem is usually corrected by wearing glasses containing lenses that are slightly concave, that is, diverging lenses. If the focal length is correct, then the image comes to a focus on the retina.

With age, the eye loses the ability to focus on nearby objects as well as distant objects. This condition is known as presbyopia, literally, "old eyes." In older people this is usually caused by the inability of the lens of the eye to change shape sufficiently to shorten its focal length in order to read newsprint or a book. To correct this, the lower part of the eyeglasses is made slightly convex to accommodate to the changing eye. Since each lens then has an upper portion for distance viewing and a lower portion for close-up viewing, they are termed bifocals.

The Magnifying Glass

A **simple magnifier** or **magnifying glass** is a converging lens that is used to produce an enlarged image of an object on the retina of the eye. The diagram shown below shows the virtual image of an object produced when the lens is used as a magnifier compared to the object viewed by the unaided eye focused at its near point.

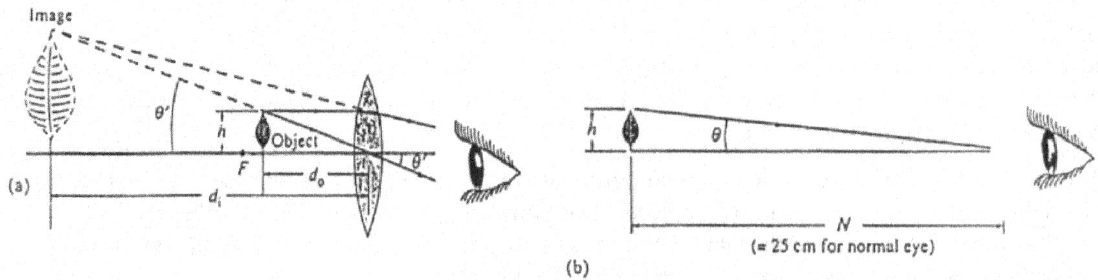

The **angular magnification** or **magnifying power** M is given by $M = \frac{\theta'}{\theta}$. θ' is the angle the object subtends with the eye when the magnifier is used, and θ is the angle the object subtends with the eye without the magnifier at a distance of 25 cm. The angular magnification can be written in terms of the focal length of the lens as follows:

$$M = 1 + \frac{N}{f} = 1 + \frac{25 \text{ cm}}{f}$$

N is the near point, which is the closest distance from the eye that an object can be placed and still be focused on the retina. N is usually taken to be 25 cm. If the eye is relaxed when using the magnifying glass, the image is then at infinity and the object is then precisely at the focal point. In this instance, $M = \frac{N}{f}$.

EXAMPLE PROBLEM 1. (Section 25-2) A student requires reading glasses of +0.75 diopters to read the print on this page when the page is held 25.0 cm from her eyes. Determine the minimum distance that she would have to hold a newspaper in order to read the print without her eyeglasses. Assume that the distance from the lens to the eye is negligible.

Part a. Step 1.

Determine the focal length of the lens.	The power of the lens is related to the focal length by the equation $P = \frac{1}{f}$. The power of the lens is +0.75 diopters, and the focal length of the lens is $f = \frac{1}{+75 \text{ diopters}} = +1.33 \text{ m} = 133 \text{ cm}$

Part a. Step 2

Use the lens equation to determine the image distance.

With her glasses on, the student holds the print 25.0 cm from her eyes and the image is formed at the near point. Thus, $d_o = 25.0$ cm and d_i is the distance from the lens to the near point of her vision:

$$\frac{1}{d_o} + \frac{1}{d_i} = \frac{1}{f}$$

$$\frac{1}{25 \text{ cm}} + \frac{1}{d_i} = \frac{1}{133 \text{ cm}}$$

$$d_i = -30.8 \text{ cm}$$

Without her glasses, the student would have to hold the page at least 30.8 cm from her eyes in order to read the print.

EXAMPLE PROBLEM 2. (Section 25-3) The maximum magnification produced by a particular converging lens—that is, a magnifying glass—is $3.0\times$ when the near point of a person's eye is 20 cm. Determine the (*a*) focal length of the lens and (*b*) magnification when that eye is relaxed.

Part a. Step 1.

Determine the focal length of the lens.

The maximum magnification is related to the focal length as is given by Eq. 25-2b.

$$M = 1 + \frac{N}{f}$$

$$3.0 = 1 + \frac{20 \text{ cm}}{f} \quad \text{and rearranging gives}$$

$$\frac{20 \text{ cm}}{f} = 3.0 - 1$$

$$\frac{20 \text{ cm}}{f} = 2.0$$

$$f = \frac{20 \text{ cm}}{2.0} = 10.0 \text{ cm}$$

Part b. Step1.

Use Eq. 25-2a to determine the magnification when the eye is relaxed.

When the eye is relaxed, the image is seen at infinity and the object is at the focal point; thus,

$$M = \frac{N}{f} = \frac{20 \text{ cm}}{10.0}$$

$$M = 2.0\times$$

Astronomical Telescope

A **refracting** type **astronomical telescope** consists of a large-diameter converging lens with a long focal length (f_o) called the **objective lens** and a small-diameter converging lens with a short focal length (f_e) as the **eyepiece**. As shown in the diagram, distant objects such as stars and planets can be considered to be an infinite distance from the telescope. As a result, the image (I_1) is located at the focal point of the objective. The eyepiece is positioned so that the image produced by the objective lens is at or just inside the focal point of the eyepiece. A magnified, inverted virtual image (I_2) of the object is observed.

For an object at infinity, the distance (ℓ) between the lenses equals $f_o + f_e$, and the magnifying power of the telescope is $M = \frac{\theta'}{\theta} = -\frac{f_o}{f_e}$, where the negative sign indicates an inverted image.

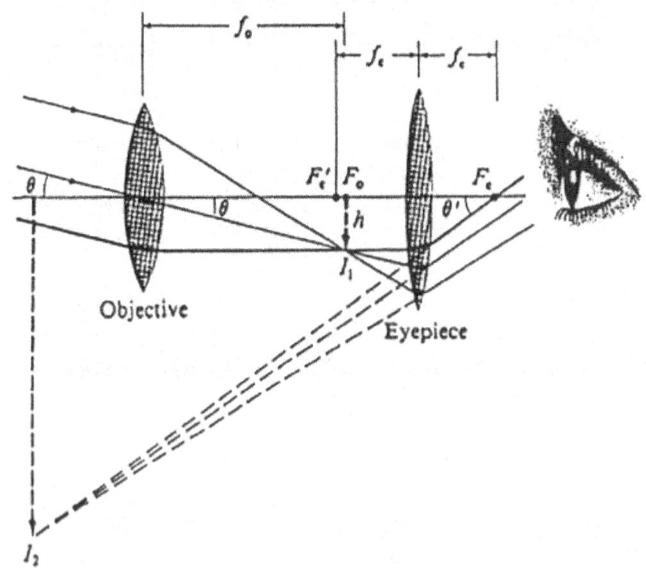

Terrestrial Telescope

A **terrestrial telescope** is designed so that an erect image is observed. The **Galilean** type uses a convex lens of long focal length as the objective and a concave lens of short focal length as the eyepiece. The eyepiece is a concave lens, its focal length is negative, and the angular magnification is positive. The angular magnification is given by $M = -\frac{f_o}{f_e}$.

EXAMPLE PROBLEM 3. (Section 25-4) An astronomical telescope consists of two convex lenses, an eyepiece of focal length 120 cm, and an objective of focal length 3.0 cm. Determine the magnification of the telescope if it is used to view a distant object.

Part a. Step 1.

Determine the magnification of the telescope. [Hint: Assume that the object is at infinity.]

The magnifying power can be found from Eq. 25-3 as follows:

$$M = -\frac{f_o}{f_e} = -\frac{120\text{ cm}}{3.0\text{ cm}}$$

$$M = -40\times$$

EXAMPLE PROBLEM 4. (Section 25-4) Two converging lenses are 20.0 cm apart. The focal lengths of the lenses are 5.0 cm and 10.0 cm, respectively. An object 3.0 cm high is placed 15.0 cm in front of the 5.0-cm lens. (*a*) Draw an accurate ray diagram and locate the final image formed by the combination. (*b*) Mathematically determine the position and size of the final image.

Part a. Step 1.

Draw an accurate diagram and locate the final image formed by the combination.

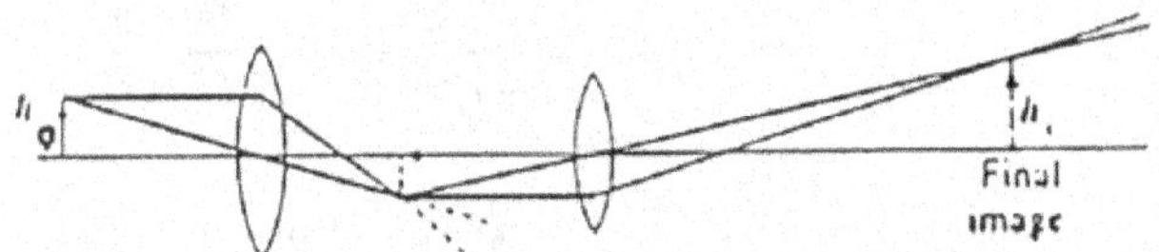

Part b. Step 1.

Determine the position and size of the image produced by the lens that is closest to the object.

$$\frac{1}{d_o} + \frac{1}{d_i} = \frac{1}{f}$$

$$\frac{1}{15.0\text{ cm}} + \frac{1}{d_i} = \frac{1}{5.0\text{ cm}}$$

$$-d_i = 7.50\text{ cm}$$

$$\frac{h_i}{h_o} = -\frac{d_i}{d_o}$$

$$\frac{h_i}{3.0\text{ cm}} = -\frac{7.50\text{ cm}}{15.0\text{ cm}}$$

$h_i = -1.5\text{ cm}$ The image is real, inverted, and diminished.

Part b. Step 2.

Determine the distance from the second lens to the image and also the height of the image.

State the characteristics the final image.

The image produced by the first lens now becomes the object for the second lens. The distance from the second lens to the image produced by the first lens is equal to the distance between the lenses minus the distance from the first lens to the image distance:

$$20.0\text{ cm} - 7.5\text{ cm} = 12.5\text{ cm}$$

$$\frac{1}{d_\text{o}} + \frac{1}{d_\text{i}} = \frac{1}{f}$$

$$\frac{1}{12.5\text{ cm}} + \frac{1}{d_\text{i}} = \frac{1}{10.0\text{ cm}}$$

$$d_\text{i} = 50.0\text{ cm}$$

The height of the final image is given by

$$\frac{h_\text{i}}{h_\text{o}} = -\frac{d_\text{i}}{d_\text{o}}$$

$$\frac{h_\text{i}}{-1.5\text{ cm}} = -\frac{50.0\text{ cm}}{12.5\text{ cm}}$$

$h_\text{i} = +6.0\text{ cm}$ The final image is real, upright, and magnified.

Compound Microscope

A **compound microscope** consists of two converging lenses, an objective and an eyepiece or ocular separated by a distance. The object is placed just beyond the focal point (F_o) of the objective. The image formed (I_1) is real, inverted, and enlarged. I_1 is formed inside the focal point (F_e) of the eyepiece, and the viewer observes an image (I_2) that is virtual, inverted, and greatly magnified.

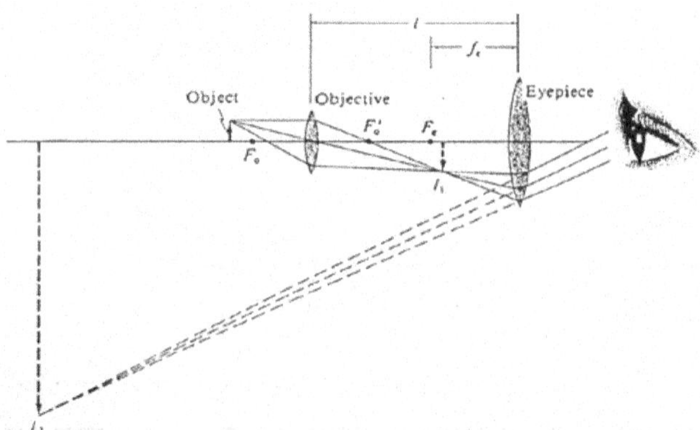

The magnification produced by the objective is given by

$$m_o = \frac{d_i}{d_o} = \frac{\ell - f_e}{d_o}$$

where d_i is the distance from the lens to the image, d_o is the distance from the lens to the object, and ℓ is the distance between the two lenses.

The magnification produced by the eyepiece is $M_e = \frac{N}{f_e}$, where N is the near point, which is usually taken to be 25 cm. The total magnification is given by

$$M = \frac{M_e}{m_o} = \left(\frac{N}{f_e}\right)\left(\frac{\ell - f_e}{d_o}\right) \approx \frac{N\ell}{f_e f_o}$$

The approximation is accurate when f_e and f_o are small compared with ℓ.

EXAMPLE PROBLEM 5. (Section 25-5) During a laboratory experiment, a student positions two lenses to form a compound microscope. The objective and eyepiece have focal lengths of 0.40 cm and 3.0 cm, respectively. The lenses are 4.9 cm apart, and an object is placed 0.50 cm in front of the objective lens. (*a*) Draw a ray diagram and locate the position of the final image. (*b*) Mathematically determine the position of the final image.

Part a. Step 1.

Draw an accurate diagram and locate the position of the final image.

The following diagram is not drawn to scale. This is necessary in order to show the position of the final image in the diagram.

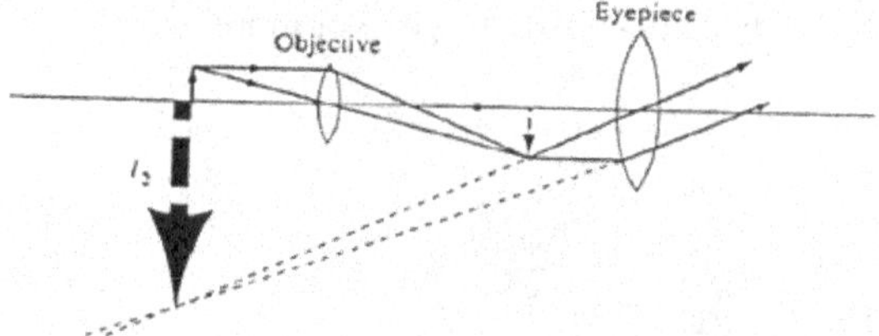

Part b. Step 1.

Mathematically determine the position of the image produced by the eyepiece.

Let f_o = focal length of the objective lens:

$$\frac{1}{d_o} + \frac{1}{d_i} = \frac{1}{f_o}$$

$$\frac{1}{0.50 \text{ cm}} + \frac{1}{d_i} = \frac{1}{0.40 \text{ cm}}$$

$$d_i = 2.0 \text{ cm}$$

Part b. Step 2.

Mathematically determine the position of the final image.

The image produced by the objective lens becomes the object for the eyepiece. The two lenses are 4.9 cm apart; thus the image produced is 2.9 cm from the eyepiece. Therefore,

$$\frac{1}{d_\text{o}}+\frac{1}{d_\text{i}}=\frac{1}{f_\text{o}}, \text{ where } f_\text{e} = \text{focal length of the eyepiece}$$

$$\frac{1}{2.9\text{ cm}}+\frac{1}{d_\text{i}}=\frac{1}{3.0\text{ cm}}$$

$$d_\text{i} = -87\text{ cm}$$

The final image is a virtual image formed 87 cm in front of the eyepiece.

Lens Aberrations

The ray diagrams drawn to locate images produced by thin lenses are only approximately correct. Only an "ideal" lens gives an undistorted image; the formation of images by real lenses is limited by what are referred to as **lens aberrations**.

In **spherical aberrations** (diagram a) rays of light from a point on the axis of the lens, which pass through different sections of the lens, do not produce a point image after refraction. Instead, they produce a small circular patch of light. The point at which the circle has its smallest diameter is referred to as the **circle of least confusion**. Spherical aberrations can be approximately corrected by the expensive method of grinding a nonspherical lens or, more frequently, by using a combination of two or more lenses.

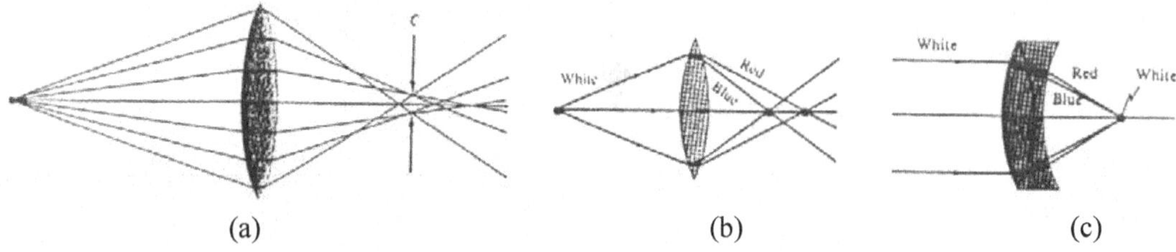

(a) (b) (c)

Chromatic aberration (diagram b) occurs because the index of refraction of transparent materials varies with wavelength. For example, blue light is bent more by glass than red light. As a result, white light from a point source produces an image containing colors spread out over a small region.

Chromatic aberration is approximately corrected through use of a combination of two lenses called an **achromatic doublet** (diagram c). One lens is a converging lens and the other a diverging lens. The lenses have different indices of refraction and radii of curvature.

TEXTBOOK QUESTION 13. (Sections 24-4 and 25-6) Explain why chromatic aberration occurs for thin lenses but not for mirrors.

ANSWER: As shown in Fig. 25-27, chromatic aberration occurs because the index of refraction of transparent materials varies with wavelength. For example, blue light is bent more by glass than red light. As a result, white light from a point source passing through a thin lens may produce an image containing colors spread out over a small region.

In one type of mirror, light reflects directly off a highly polished metallic surface. As a result, all of the light follows the laws of specular reflection, and no chromatic aberration results. Most mirrors, however, have a reflecting metallic surface that coats the back of a of thin glass plate. The sides of the glass plate are parallel. Although refraction and dispersion do occur to a minor degree in the thin glass, the light shows no chromatic aberration upon reflection from the mirror.

TEXTBOOK QUESTION 15. (Section 25-2) Which aberrations present in a simple lens are not present (or are greatly reduced) in the human eye?

ANSWER: The cornea of the human eye is less curved at the edges than at the center, and the lens is less dense at the edges than at the center. Both effects cause rays at the outer edges to be bent less strongly and thus help reduce spherical aberration. Also, the retina of the eye is curved, and this compensates for curvature of field that occurs due to the flat film used in cameras. In addition, wide-angle lenses commonly show distortion, but the small lens of the eye plus the curved retina compensate for this aberration.

There is significant absorption of the shorter wavelengths by the lens of the eye, and the retina is less sensitive to the blue and violet wavelengths. This is the region of the spectrum where chromatic aberration is the greatest, and as a result chromatic aberration is greatly reduced in the eye.

Limits of Resolution

The ability of a lens to produce distinct images of two point objects that are very close together is called the **resolution** of the lens. Two principal factors limit the resolution of a lens. One factor is lens aberration and the second is the wave nature of light.

The magnification produced by a microscope or telescope is limited by diffraction. Magnification beyond a certain point does not lead to an increase in sharpness or resolution of images. The **Rayleigh criterion** states that two images are just resolvable when the center of the diffraction disk of one image is directly over the first minimum of the diffraction pattern of the other. The first minimum is at an angle $\theta = 1.22\frac{\lambda}{D}$ from the central maximum, and therefore the objects are considered to be just resolvable at this angle. λ is the wavelength of the incident light and D is the diameter of the objective lens.

For a microscope, the resolution is defined in terms of the **resolving power (RP)**, where

$$\text{RP} = s = f\theta = 1.22\frac{\lambda f}{D}$$

where s is the minimum separation of two objects that can just be resolved and f is the focal length of the objective lens.

EXAMPLE PROBLEM 6. (Section 25-7) The objective lens of a compound microscope is 1.75 cm in diameter and has a 7.50-cm focal length. Light of wavelength 546 nm is used to illuminate objects to be viewed. (*a*) What is the angular separation of nearby objects when they are just resolvable? (*b*) What is the resolving power of this lens?

Part a. Step 1.

Find the angular separation of nearby objects when they are just resolvable.

The minimum angular separation is given by Eq. 25-7.

$$\theta = 1.22\frac{\lambda}{D} = (1.22)\frac{546\text{ nm}}{1.75\text{ cm}}$$

$$= (1.22)\frac{546 \times 10^{-9}\text{ m}}{1.75 \times 10^{-2}\text{ m}}$$

$$\theta = 3.81 \times 10^{-5}\text{ radians,}\quad \text{or}\quad 2.18 \times 10^{-3}\text{ degrees}$$

This is the limit on resolution for this lens due to diffraction.

Part b. Step 1.

Use Eq. 25-8 to solve for the resolving power of this lens.

$$\text{RP} = s = f\theta = 1.22\frac{\lambda f}{D}$$

$$\text{RP} = (1.22)\frac{(546\text{ nm})(7.50\text{ cm})}{1.75\text{ cm}}$$

$$\text{RP} = 2.85 \times 10^{-6}\text{ m,}\quad \text{or}\quad 2.85 \times 10^{-3}\text{ mm}$$

This distance is the minimum separation of two point objects that can just be resolved.

X-ray Diffraction

X-rays were discovered in 1895 by Wilhelm Roentgen. The nature of X-rays was not determined until 1913, when it was shown by Max von Laue that X-rays exhibit properties of electromagnetic waves of very short wavelength.

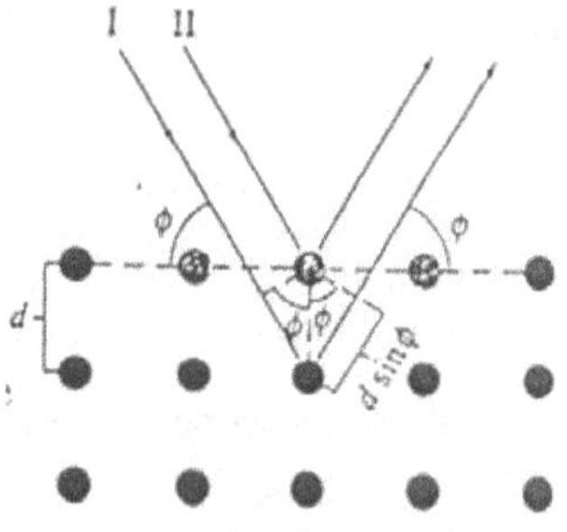

The wavelength of X-rays is comparable to the spacing of atoms in crystals such as sodium chloride (NaCl). If an X-ray beam is incident on the crystal, then the subsequent reflection of the X-rays from the planes of atoms results in a diffraction pattern. Based on the diagram, constructive interference will occur if the extra distance, known as the path difference, that ray I travels before rejoining ray II equals a whole number of wavelengths. The path difference can be shown to be equal to $2d\sin\phi$. Constructive interference occurs when $m\lambda = 2d\sin\phi$ (Bragg equation).

$m\lambda$ is a whole number of wavelengths and $m = 1, 2, 3,$ etc., d is the distance between the layers of atoms, and ϕ is the angle of incidence of the X-ray beam with the surface of the crystal.

EXAMPLE PROBLEM 7. (Section 25-11) X-rays having a wavelength of 0.250 nm are scattered by a crystal. Determine the angles for all diffraction maxima produced by one set of layers of ions spaced 0.412 nm apart.

Part a. Step 1.

Use the Bragg equation to determine the possible angles.

ϕ cannot exceed 90°; therefore, the possible angles are 90° or less:

$m\lambda = 2d \sin \phi_1$

If $m = 1$: $(1)(0.250 \text{ nm}) = 2(0.412 \text{ nm}) \sin \phi_1$

$\sin \phi_1 = 0.303$

$\phi_1 = 17.7°$

If $m = 2$: $(2)(0.250 \text{ nm}) = 2(0.412 \text{ nm}) \sin \phi_2$

$\sin \phi_2 = 0.607$

$\phi_2 = 37.4°$

If $m = 3$: $(3)(0.250 \text{ nm}) = 2(0.412 \text{ nm}) \sin \phi_3$

$\sin \phi_3 = 0.910$

$\phi_3 = 65.5°$

If $m = 4$: $(4)(0.250 \text{ nm}) = 2(0.412 \text{ nm}) \sin \phi_4$

$\sin \phi_4 = 1.21$

$\sin \phi_4 > 1$; the angle does not exist, and only three diffraction maxima are produced.

PROBLEM SOLVING SKILLS

For problems related to power of a lens:

1. Use the equation $P = \frac{1}{f}$ to determine either the power or the focal length of the lens.
2. Use the Gaussian form of the lens equation to determine the position of the image.
3. Use the equation $M = 1 + \frac{N}{f}$ to determine the magnification of a converging lens.

For problems related to the final image produced by two converging lenses:

1. Use the Gaussian form of the lens equations to determine the position and size of the image produced by the lens closest to the object.
2. Determine the distance from the second lens to the image produced by the first lens. This distance represents the object distance from the second lens.
3. Use the Gaussian form of the lens equation to determine the position and size of the image produced by the second lens.

To determine the limit of resolution of a lens for two point objects that are close together:

1. Apply the Rayleigh criterion to determine the minimum angle at which the objects are resolvable.
2. If the objects are being viewed through a microscope, use the formula for resolving power to determine the minimum separation of the objects that allows resolution.

For problems related to X-ray diffraction:

1. Apply the Bragg equation to determine either the distance between atoms in the crystal or the wavelength of the incident X-rays.

SOLUTIONS TO SELECTED TEXTBOOK PROBLEMS

TEXTBOOK PROBLEM 6. (Section 25-1) A nature photographer wishes to shoot a 34-m-tall tree from a distance of 65 m. What focal-length lens should be used if the image is to fill the 24-mm height of the sensor?

Part a. Step 1.

Complete a data table based on the information provided.

Note: The image is inverted; therefore, use the sign convention discussed in Chapter 23.

$h_o = 34\text{ m}$ $\quad d_o = 65\text{ m}$ $\quad f = ?$

$h_i = -24\text{ mm} = -0.024\text{ m}$

Part a. Step 2.

Determine the magnification of the image.

$$m = \frac{h_i}{h_o} = \frac{-0.024\text{ m}}{34\text{ m}}$$

$$m = -7.06 \times 10^{-4}$$

Part a. Step 3.

Determine the image distance d_i.

$$m = -\frac{d_i}{d_o}$$

$$-7.06 \times 10^{-4} = -\frac{d_i}{65\text{ m}}$$

$$d_i = 0.0459\text{ m}$$

Part a. Step 4.

Use the lens equation to determine the focal length (f) of the lens.

$$\frac{1}{d_o} + \frac{1}{d_i} = \frac{1}{f}$$

$$\frac{1}{65\text{ m}} + \frac{1}{0.0459\text{ m}} = \frac{1}{f}$$

Solving for f gives

$$f = 0.0459\text{ m} = 45.9\text{ mm}$$

TEXTBOOK PROBLEM 35. (Section 25-4) An astronomical telescope has its two lenses spaced 82.0 cm apart. If the objective lens has a focal length of 78.5 cm, what is the magnification of this telescope? Assume a relaxed eye.

Part a. Step 1.

Determine the focal length of the eyepiece.

As shown in Fig. 25-20, for an object at infinity, the distance between the two lenses (ℓ) equals the sum of the focal length of the objective lens (f_o) and the focal length of the eyepiece (f_e).

$$\ell = f_o + f_e$$

$$82.0\text{ cm} = 78.5\text{ cm} + f_e$$

$$f_e = 3.5\text{ cm}$$

Part a. Step 2.

Use Eq. 25-3 to determine the magnification.

$$M = -\frac{f_o}{f_e} = -\frac{78.5\text{ cm}}{3.5\text{ cm}}$$

$$M = -22.4\times$$

TEXTBOOK PROBLEM 38. (Section 25-4) The Moon's image appears to be magnified 150× by a reflecting astronomical telescope with an eyepiece having as focal length of 3.1 cm. What are the focal length and radius of curvature of the main (objective) mirror?

Part a. Step 1.

Use Eq. 25-3 to determine the focal length of the objective mirror.

$$M = -\frac{f_o}{f_e}$$

$$-150 = -\frac{f_o}{3.1\text{ cm}}$$

$$f_o = 465\text{ cm} = 4.65\text{ m}$$

Part a. Step 2.

Use Eq. 23-1 to determine the radius of curvature (r) of the objective mirror.

$$r = 2f_o = 2(4.65\text{ m})$$

$$r = 9.30\text{ m}$$

TEXTBOOK PROBLEM 45. (Section 25-5) A 17-cm-long microscope has an eyepiece with a focal length of 2.5 cm and an objective with a focal length of 0.33 cm. What is the approximate magnification?

Part a. Step 1.

Complete a data table based on the information provided.

$N = 25\text{ cm}$ $\quad \ell = 17\text{ cm}$

$f_e = 2.5\text{ cm}$ $\quad f_o = 0.33\text{ cm}$

Part a. Step 2.

Use Eq. 25-6b to solve for the approximate magnification.

$$M \approx \frac{N\ell}{f_e f_o}$$

$$M \approx \frac{(25\text{ cm})(17\text{ cm})}{(2.5\text{ cm})(0.33\text{ cm})}$$

$$M \approx 515\times$$

TEXTBOOK PROBLEM 60. (Section 25-11) X-rays of wavelength 0.138 nm fall on a crystal whose atoms, lying in planes, are spaced 0.285 nm apart. At what angle ϕ (relative to the surface, Fig. 25-38) must the X-rays be directed if the first diffraction maximum is to be observed?

Part a. Step 1.

Use the Bragg equation to determine the spacing.

$m\lambda = 2d \sin \phi_1$, where $m = 1$; so

$(1)(0.138 \text{ nm}) = 2(0.285 \text{ nm}) \sin \phi$

$\sin \phi = 0.242$

$\phi = 14.0°$

TEXTBOOK PROBLEM 70. (Section 25-1) A movie star catches a reporter shooting pictures of her at home. She claims the reporter was trespassing. To prove her point, she gives as evidence the film she seized. Her 1.65-m height is 8.25 mm high on the film, and the focal length of the camera lens is 220 mm. How far away from the subject was the reporter standing?

Part a. Step 1.

Complete a data table. Convert the data to meters.

$h_o = 1.65 \text{ m}$ and

$$h_i = (-8.25 \text{ mm})\left(\frac{1 \text{ m}}{1000 \text{ mm}}\right) = -0.00825 \text{ m} \text{ so}$$

$$f = (220 \text{ mm})\left(\frac{1 \text{ m}}{1000 \text{ mm}}\right) = 0.220 \text{ m}$$

Note: Using the sign convention from Chapter 23, the image height is negative because the image is inverted.

Part a. Step 2.

Determine the magnification of the image.

$$m = \frac{h_i}{h_o} = \frac{-0.00825 \text{ m}}{1.65 \text{ m}}$$

$m = -0.00500$

Note: Using the sign convention from Chapter 23, the magnification is negative because the image is inverted.

Part a. Step 3.

Use the lens equations from Chapter 23 to solve for the object distance.

$$m = -\frac{d_i}{d_o}$$

$$-0.00500 = -\frac{d_i}{d_o}$$

$$d_i = 0.00500 d_o$$

$$\frac{1}{d_o} + \frac{1}{d_i} = \frac{1}{f}$$

$$\frac{1}{d_o} + \frac{1}{0.00500 d_o} = \frac{1}{0.220 \text{ m}}$$ Solving for d_o gives

$$d_o = 44.2 \text{ m}$$

TEXTBOOK PROBLEM 72. (Section 25-3) A child has a near point of 15 cm. What is the maximum magnification the child can obtain using a 9.5-cm-focal-length magnifier? What magnification can a normal eye obtain with the same lens? Which person sees more detail?

Part a. Step 1.

Use Eq. 25-2b to determine the maximum magnification the child can obtain.

$$M = 1 + \frac{N}{f} = 1 + \frac{15 \text{ cm}}{9.5 \text{ cm}}$$

$$M = 2.6\times$$

Part a. Step 2.

Use Eq. 25-2b to determine the maximum magnification an adult can obtain.

$$M = 1 + \frac{N}{f},$$ where $N = 25$ cm for a normal eye

$$= 1 + \frac{25 \text{ cm}}{9.5 \text{ cm}}$$

$$M = 3.6\times$$

Part a. Step 3.

Which person sees more detail? The greater the magnification, the greater the detail observed. Therefore, the person with the normal eye sees more detail.

26

THE SPECIAL THEORY OF RELATIVITY

OBJECTIVES

After studying the material of this chapter, the student should be able to:

- state in the student's own words the postulates of the special theory of relativity.
- explain what is meant by a reference frame and distinguish between an inertial and a noninertial reference frame.
- explain what is meant by the principle of simultaneity and explain in the student's own words the thought experiment described in the text.
- explain what is meant by proper time and relativistic time and solve word problems involving time dilation.
- explain what is meant by proper length and relativistic length and solve word problems involving length contraction.
- use the principle of special relativity to determine the relative velocity of an object as measured by an observer moving with respect to the object.
- explain what is meant by rest energy and total energy and solve word problems involving Einstein's mass–energy equation.

KEY TERMS AND PHRASES

postulates of the special theory of relativity are (1) all inertial reference frames are equivalent, and (2) observers, regardless of their relative velocity or the velocity of the source, must measure the same value for the speed of light in a vacuum.

reference frame is a point in space with respect to which the motion of objects can be measured.

inertial reference frame is one that is either at rest or is moving in a straight line at a constant speed. In all inertial reference frames, the laws of physics are the same and hold in the same way.

principle of simultaneity Two events that are observed to occur simultaneously at different points in one reference frame will not occur simultaneously according to an observer in a second reference frame that is in motion relative to the first.

proper time is measured by an observer at rest with respect to the timing device.

time dilation refers to the difference in the measurement of time by an observer in motion relative to the timing device.

proper length is measured by an observer at rest with respect to the object measured.

length contraction refers to the decrease in measured length of an object when the object is in motion relative to an observer.

rest energy is the energy an object has due to its mass alone.

SUMMARY OF MATHEMATICAL FORMULAS

time dilation equation	$\Delta t = \dfrac{\Delta t_0}{\sqrt{1-v^2/c^2}}$	Δt_0 is the proper time measured by the observer at rest with the timing device, Δt is the relativistic time measured by the observer in motion relative to the timing device, v is the speed of the timing device relative to observer measuring Δt, and c is the speed of light in a vacuum.
length contraction equation	$\ell = \ell_0\sqrt{1-v^2/c^2}$	ℓ is the relativistic length measured by the observer in motion relative to the object, and ℓ_0 is the proper length measured by the observer at rest with respect to the object.
relativistic addition of velocities	$u = \dfrac{v+u'}{1+\dfrac{vu'}{c^2}}$	u is the speed of the object as measured by an observer in motion relative to the reference frame of the moving object, v is the speed of the moving reference frame, and u' is the speed of the object relative to an observer at rest in the moving reference frame. If u' is in the same direction as v, then u' is positive. If u' is opposite from v, then u' is negative.
total energy	$E = \dfrac{mc^2}{\sqrt{1-(v/c)^2}}$ If the object is at rest, then $E = mc^2$.	The total energy (E) equals the product of the mass and the speed of light squared. The total energy of an object is the sum of the object's mechanical energy and rest energy.
kinetic energy	$\text{KE} = E - E_0$	For a moving object, the kinetic energy equals the difference between an object's total energy (E) and its rest energy (E_0).

CONCEPT SUMMARY

Postulates of the Special Theory of Relativity

The **special theory of relativity** is based on two postulates formulated by Albert Einstein:

(1) All inertial frames of reference are equivalent.

(2) Observers, regardless of their relative velocity or the velocity of the source, must measure the same value for the speed of light in a vacuum.

TEXTBOOK QUESTION 3. (Section 26-1) A worker stands on top of a railroad car moving at constant velocity and throws a heavy ball straight up (from his point of view). Ignoring air resistance, explain whether the ball lands back in his hand or behind him.

ANSWER: A car that is at rest or moving at a constant speed is in an inertial reference frame. Therefore, if the car is not moving, then the ball would travel straight up and return to the worker's hand. However, if the car is moving at a constant speed, then the horizontal component of velocity of the ball is the same as the horizontal velocity of the car. The ball would travel upward but would travel the same horizontal distance as the car. This means that the ball would return to the worker's hand. See Fig. 26-2 of the textbook.

Inertial Reference Frames

A reference frame is a point in space with respect to which the motion of objects can be measured. An **inertial reference frame** is one that is either at rest or is moving in a straight line at a constant speed. In all inertial frames of reference, the laws of physics are the same and hold in the same way.

A noninertial reference frame is one that is undergoing accelerated motion. This means a frame in which the speed is changing, the direction of motion is changing, or both the speed and the direction of motion are changing.

As a consequence of the first postulate, it is possible to determine relative motion between objects traveling at constant velocity with respect to one another, but it is not possible to determine absolute motion.

Principle of Simultaneity

Two events that are observed to occur simultaneously at different points in one reference frame will not occur simultaneously according to an observer in a second reference frame that is in motion relative to the first.

The following **thought experiment** will be used to clarify this principle. Two railroad cars are directly opposite one another when lightning bolts strike points shown in diagram A. The cars are moving relative to one another with observer O_1 in car 1 assuming that he is at rest while car 2 is moving to the left at speed v. Likewise, observer O_2 in car 2 assumes that he is at rest and sees car 1 moving to the right at speed v.

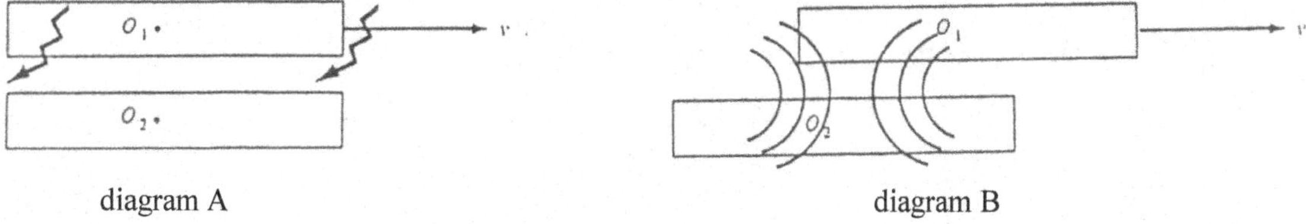

diagram A diagram B

As shown in diagram A, the lightning bolts strike the front and rear ends of each car simultaneously. As shown in diagram B, a moment later the light from the events simultaneously reaches O_2's position. However, in O_1's reference frame, the light from the front end has already reached O_1, while the light from the rear has not yet reached O_1. So from O_1's reference frame, the lightning strike at the front of the car must have occurred before the strike at the rear of the car.

Predictions from the Special Theory of Relativity

Time Dilation

Einstein's two postulates plus the principle of simultaneity lead to the prediction that time is measured differently in frames of reference moving relative to one another.

Observer A on a spaceship is moving toward the right at a speed v relative to an earthbound observer B. A pulse of laser light is fired vertically upward from the floor of the spacecraft just as the spacecraft is passing B's position.

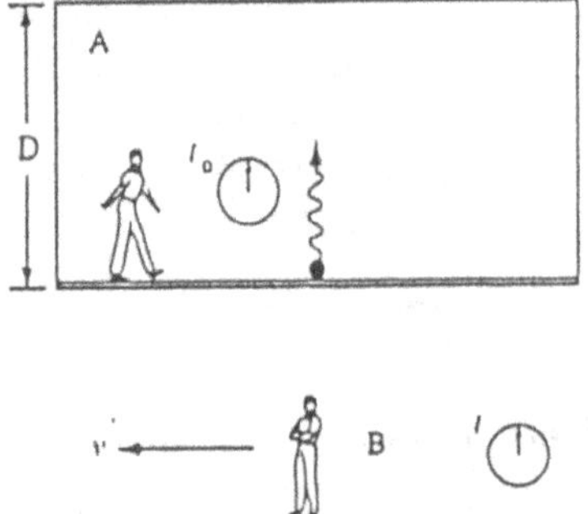

A is moving at a constant speed in a straight line, so he is in an inertial reference frame. From A's point of view, he is **not** moving; instead, B is moving toward the left at speed v. A sees the light travel vertically upward toward the ceiling a distance D at the speed of light (c). The time required for the light to travel from the floor to the ceiling, from A's viewpoint, can be determined from the equation

$$\text{time} = \text{distance/velocity}, \quad \text{or} \quad \Delta t_0 = D/c \quad \text{and} \quad D = c\,\Delta t_0$$

The time measured by observer A is the **proper time** Δt_0. D is the distance from the floor to the ceiling, and c is the speed of light in a vacuum.

As shown in diagram 1, from observer B's frame of reference he is at rest and it is observer A who is moving toward the right at speed v. B sees the beam of light travel from the floor to the ceiling, but A is traveling toward the right, and B observes that the beam does not travel vertically upward. B sees the beam travel at an angle to the vertical. According to the second postulate, both A and B must measure the speed of light to be c. Thus, B sees the beam travel at speed c at an angle to the vertical.

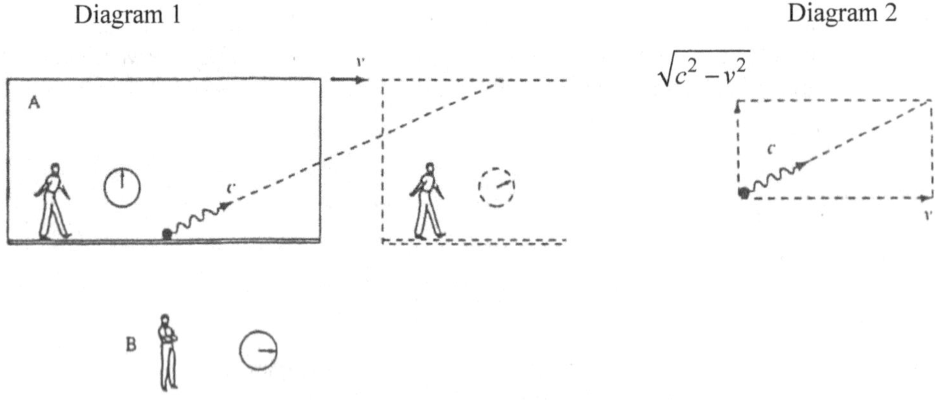

As shown in diagram 2, observer B uses the Pythagorean theorem and determines that the component of the light's speed in the vertical direction is $\sqrt{c^2 - v^2}$. The time required for the light to travel from the floor to the ceiling, the vertical component of motion, is given by

$$\Delta t = \frac{D}{\sqrt{c^2 - v^2}} \quad \text{and} \quad D = \sqrt{c^2 - v^2}\,\Delta t$$

where Δt is the relativistic time.

Both observers measure D to be the same because there is no relative motion between their frames of reference in the vertical direction. Therefore, $c\,\Delta t_{o} = \sqrt{c^2 - v^2}\;\Delta t$. Rearranging gives

$$\Delta t = \frac{\Delta t_0}{\sqrt{1 - v^2/c^2}}$$

This equation is known as the time dilation equation.

TEXTBOOK QUESTION 8. (Section 26-4) A young-looking woman astronaut has just arrived home from a long trip. She rushes up to an old gray-haired man and in the ensuing conversation refers to him as her son. How might this be possible?

ANSWER: The answer to this question is based on the time dilation formula, Eq. 26-1a. Clocks moving relative to an observer are measured by that observer to run more slowly (as compared with clocks at rest). Therefore, if the woman astronaut traveled at an extremely high speed relative to her earthbound son, then her clock would run slow as compared to her son's clock.

EXAMPLE PROBLEM 1. (Sections 26-4 and 26-5) Observer A in the diagram shown at the beginning of the "Time Dilation" section of this Study Guide measures the passage of 10.0 years in the spacecraft. Determine the amount of time that has passed according to observer B's clock if the spacecraft is moving at (*a*) 0.60c and (*b*) 0.98*c*.

Part a. Step 1.

Use the time dilation equation to determine the amount of time that passes on B's clock if $v = 0.6c$.

Observer A measures the proper time (Δt_0), while B measures the relativistic time (Δt) and $v = 0.80c$. Then

$$\sqrt{1 - v^2/c^2} = \sqrt{1 - (0.60c/c^2)} = \sqrt{1 - 0.36} = 0.80 \quad \text{and}$$

$$\Delta t = \frac{\Delta t_0}{\sqrt{1 - v^2/c^2}}$$

$$\Delta t = \frac{10 \text{ years}}{0.80} = 12.5 \text{ years}$$

Part b. Step 1.

Use the time dilation equation to determine the amount of time that passes on B's clock if $v = 0.98c$.

If $v = 0.98c$, then

$$\sqrt{1 - v^2/c^2} = \sqrt{1 - \frac{0.96c^2}{c^2}}$$
$$= \sqrt{1 - 0.96} = \sqrt{0.04} = 0.20$$

$$\Delta t = \Delta t_0 \sqrt{1 - \frac{v^2}{c^2}}$$

$$\Delta t = \frac{10.0 \text{ years}}{0.20} = 50.0 \text{ years}$$

Length Contraction

Not only would observer B measure time differently than observer A, but B would also measure lengths differently. Objects in the spacecraft are at rest relative to observer A. Observer A measures the **proper length** (ℓ_0) of objects in the spacecraft.

The spacecraft shown in diagram 1 of the "Time Dilation" section of this Study Guide is moving in the x direction relative to observer B. Observer B will see the length of objects in the spacecraft shortened in the x direction but not in the y and z directions. The length of an object (ℓ) as measured by observer B will differ from the length measured by observer A according to the following equation:

$$\ell = \ell_0 \sqrt{1 - v^2/c^2}$$

where ℓ is the contracted length and is measured by the observer in motion relative to the object. The proper length ℓ_0 is the length measured by the observer at rest relative to the object.

TEXTBOOK QUESTION 9. (Sections 26-2 to 26-5) If you were traveling away from Earth at speed $0.6c$, would you notice a change in your heartbeat? Would your mass, height, or waistline change? What would observers on Earth using telescopes say about you?

ANSWER: You are at rest with respect to objects in the spacecraft, and of course you are at rest with respect to your own body. Because of this, you would not notice any change in your heartbeat, mass, height, or waistline. However, you are in motion relative to an earthbound observer and

$$\sqrt{1 - v^2/c^2} = \sqrt{1 - \frac{(0.6c)^2}{c^2}} = \sqrt{1 - 0.36} = 0.80$$

Using the relativity equations, it can be shown that according to the earthbound observer, your heartbeat and waistline are reduced by the factor 0.80 while your mass appears to have increased. Your height does not change because there is no relative motion between the earthbound observer and your height.

EXAMPLE PROBLEM 2. (Sections 26-4 and 26-5) Observer A in the diagram shown at the beginning of the "Time Dilation" section of this Study Guide measures the length of a piece of wood to be 1.50 meter and its mass to be 3.00 kg. The spacecraft is moving at $0.90c$ relative to observer B. Determine the length of the wood.

Part a. Step 1.

Determine which observer measures the proper length and which observer measures the contracted length.	Observer A is at rest relative to the piece of wood; therefore, observer A measures the proper length and proper mass. $\ell_0 = 1.50\ \text{m}$ and $m_0 = 3.00\ \text{kg}$ The piece of wood is in motion relative to observer B, so B measures the contracted length (ℓ).

Part a. Step 2.

Determine the contracted length of the wood.	In Example Problem 1, it was determined that if $v = 0.98c$, then $\sqrt{1-v^2/c^2} = 0.20$. $\ell = \sqrt{1-v^2/c^2}\ \ell_0$ $\ell = (0.20)(1.50\ \text{m}) = 0.30\ \text{m}$

E = mc²; Mass and Energy

When mechanical work is done on an object, the object's energy must change. However, according to the special theory of relativity, as the object's speed increases, $\text{KE} = \frac{1}{2}mv^2$ no longer applies. The total energy that the object possesses is given by

$$E = \frac{mc^2}{\sqrt{1-(v/c)^2}}$$

If the object is at rest, then $E = mc^2$.

The total energy of an object can be shown to be equal to the sum of the object's mechanical energy and **rest energy**. The object's rest energy is the energy that is the product of the mass (m) and the speed of light squared (c^2). For a moving object, the total energy can be expressed as follows:

$$\text{total energy} = \text{KE} + \text{rest energy}$$

The object's kinetic energy can be determined by rearranging the above formula:

$$\text{KE} = E - E_0$$

According to the theory, even an object at rest has energy, and it is possible to convert mass into energy and vice versa.

Although the relation $E = mc^2$ applies to all processes involving energy transfer, it is usually detected only in certain nuclear processes, for example, nuclear fission and fusion. However, any situation where an object's energy changes should result in a change in its mass. For example, lifting a book and placing it on a table should result in an increase in the book's mass, and heating a rod in a fire should result in an increase in its mass.

TEXTBOOK QUESTION 19. (Section 26-9) It is not correct to say that "matter can neither be created nor destroyed." What must we say instead?

ANSWER: The formula $E = mc^2$ relates mass and energy and can be used to determine how much energy can be obtained from mass and vice versa. As stated in the textbook: "Mass and energy are interconvertible. The law of conservation of energy must include mass as a form of energy."

TEXTBOOK QUESTION 18. (Section 26-9) If mass is a form of energy, does this mean that a spring has more mass when compressed than when relaxed? Explain.

ANSWER: According to the special theory of relativity, the total energy that the spring possesses is given by $E = mc^2$. When mechanical work is done on a spring, the spring's energy increases. Since mass is a form of energy, as the spring is compressed the increased energy is found in the increased mass of the spring.

EXAMPLE PROBLEM 3. (Section 26-7) The mass of an alpha particle is 6.6×10^{-27} kg. Determine the alpha particle's relativistic momentum when it is traveling at 0.98*c*.

Part a. Step 1.

Use Eq. 26-4 to determine the relativistic momentum.

$$p = \frac{mv}{\sqrt{1-(v/c)^2}} = \left(\frac{6.6 \times 10^{-27}\ \text{kg}}{\sqrt{1-(0.98\ c/c)^2}}\right)(0.98)(3.0 \times 10^8\ \text{m/s})$$

$$p = 9.7 \times 10^{-18}\ \text{kg} \cdot \text{m/s}$$

Relativistic Addition of Velocities

Suppose observer A in the spacecraft shown earlier in diagram 1 throws an object in the positive x direction with a speed of u'. Observer B will not measure the relative speed of the object to be u where $u = v + u'$, because the addition of the two velocities might give a value larger than c. Instead, the speed of the object as measured by B is given by

$$u = \frac{v + u'}{1 + \frac{vu'}{c^2}}$$

where u is the speed of the object as measured by observer B, v is the speed of observer A relative to observer B, and u' is the speed of the object relative to observer A. If u' is in the same direction as v, then u' is positive. If u' is opposite from v, then u' is negative.

EXAMPLE PROBLEM 4. (Section 26-10) The spacecraft shown earlier in diagram 1 is traveling at $0.80c$ relative to observer B. Observer A throws an object in the same direction that the spacecraft is moving. The object's velocity relative to observer A is $0.60c$. Determine the velocity of the object as measured by observer B.

Part a. Step 1.

Use the equation for the relativistic addition of velocities.

$$u = \frac{v+u'}{1+\frac{vu'}{c^2}} = \frac{0.80c+0.60c}{1+\frac{(0.80c)(0.60c)}{c^2}}$$

$$u = \frac{1.40c}{1+\frac{0.48c^2}{c^2}} = \frac{1.40c}{1+0.48} = 0.946\text{c}$$

PROBLEM SOLVING SKILLS

For problems related to time dilation and length contraction:

1. Determine which observer is at rest relative to the objects to be measured. This observer measures the proper time, proper length, and proper mass. The observer who is moving relative to the objects measures the dilated time and the contracted length.
2. Use the equations for time dilation and length contraction to solve the problem.

For problems related to the relativistic addition of velocities:

1. Determine the speed of the object as measured by one of the observers.
2. Determine the relative speed of the two observers.
3. Use the equation for relativistic addition of velocities to determine the speed of the object relative to the second observer.

For problems related to the total energy of an object:

1. Determine the object's mechanical energy and rest energy.
2. The object's total energy equals the sum of the mechanical energy and rest energy.

SOLUTIONS TO SELECTED TEXTBOOK PROBLEMS

TEXTBOOK PROBLEM 5. (Sections 26-4 and 26-5) In an Earth reference frame, a star is 49 light-years away. How fast would you have to travel so that to you the distance would be only 35 light-years?

Part a. Step 1.

Use the length contraction equation to determine the spacecraft's speed.

The observer on the spacecraft measures the contracted length $\ell = 35$ light-years. An earthbound observer measures the distance to be $\ell_0 = 49$ light-years:

$$\ell = \ell_0\sqrt{1-\frac{v^2}{c^2}}$$

$$35 \text{ light-years} = (49 \text{ light-years})\sqrt{1-\frac{v^2}{c^2}}$$

$$0.714 = \sqrt{1-\frac{v^2}{c^2}}$$

$$0.510 = 1-(v/c)^2$$

$$-0.490 = -(v/c)^2$$

$$0.700 = v/c$$

$$v = 0.700c$$

TEXTBOOK PROBLEM 11. (Sections 26-4 and 26-5) A fictional news report stated that starship *Enterprise* had just returned from a 5-year voyage while traveling at 0.70*c*. (*a*) If the report meant 5.0 years of *Earth time*, how much time elapsed on the ship? (*b*) If the report meant 5.0 years of *ship time*, how much time passed on Earth?

Part a. Step 1.

Use the time dilation equation to measure the amount of time that passed on the ship's clock.

Earth observers measured 5.0 years, while the crew of the *Enterprise* measured a shorter time:

$$\Delta t = \frac{\Delta t_0}{\sqrt{1-\frac{v^2}{c^2}}}$$

$$5.0 \text{ years} = \frac{\Delta t_0}{\sqrt{1-\frac{(0.70c)^2}{c^2}}} = \frac{\Delta t_0}{\sqrt{0.51}}$$

$$\Delta t_0 = (5.0 \text{ years})(0.714) = 3.57 \text{ years}$$

Part b. Step 1.

Use the time dilation equation to measure the amount of time that passed on an Earth observer's clock.

The crew of the *Enterprise* measured 5.0 years, while an Earth observer measured a longer time:

$$\Delta t = \frac{\Delta t_0}{\sqrt{1-\frac{v^2}{c^2}}}$$

$$\Delta t = \frac{5.0\text{ years}}{\sqrt{1-\frac{(0.70c)^2}{c^2}}}$$

$$\Delta t = \frac{5.0\text{ years}}{\sqrt{0.51}} = 7.0\text{ years}$$

TEXTBOOK PROBLEM 16. (Section 26-7) What is the momentum of a proton traveling at $v = 0.68c$?

Part a. Step 1.

Use Eq. 26-4 of the textbook to determine the momentum.

$$p = \frac{mv}{\sqrt{1-(v/c)^2}} = \left(\frac{1.67\times10^{-27}\text{ kg}}{\sqrt{1-(0.68c/c)^2}}\right)(0.68)(3.0\times10^8\text{ m/s})$$

$$p = 4.65\times10^{-19}\text{ kg}\cdot\text{m/s}$$

TEXTBOOK PROBLEM 30. (Section 26-9) At what speed will an object's kinetic energy be 33% of its rest energy?

Part a. Step 1.

Use Eq. 26-5a of the textbook to determine the speed of the object.

total energy = kinetic energy + rest energy

$$\text{KE} = \left(\frac{m}{\sqrt{1-(v/c)^2}}\right)c^2 - mc^2 \quad \text{but} \quad \text{KE} = 0.33\,mc^2$$

$$0.33mc^2 = \left(\frac{m}{\sqrt{1-(v/c)^2}}\right)c^2 - mc^2$$

Note: mc^2 cancels from each term.

$$1.33 = \left(\frac{1}{\sqrt{1-(v/c)^2}}\right)$$

$$\sqrt{1-\frac{v^2}{c^2}} = \frac{1}{1.33} = 0.752$$

$$1-(v/c)^2 = 0.565$$

$$-(v/c)^2 = 0.565 - 1$$

$$(v/c)^2 = 0.435$$

$$v = 0.66c$$

TEXTBOOK PROBLEM 47. (Section 26-10) A person on a rocket traveling at $0.40c$ (with respect to the Earth) observes a meteor come from behind and pass her at a speed she measures as $0.40c$. How fast is the meteor moving with respect to the Earth?

Part a. Step 1.

Use the equation for the relativistic addition of velocities.

Let u represent the speed of the meteor as measured by an observer on Earth:

$$u = \frac{v+u'}{1+\frac{vu'}{c^2}} = \frac{0.40c + 0.40c}{1+\frac{(0.40c)(0.40c)}{c^2}}$$

$$u = \frac{0.80c}{1+0.16}$$

$$u = 0.69c$$

TEXTBOOK PROBLEM 75. (Sections 26-4 and 26-5) An astronaut on a spaceship traveling at $0.75c$ relative to Earth measures his ship to be 23 m long. On the ship, he eats his lunch in 28 min. (*a*) What length is the spaceship according to observers on Earth? (*b*) How long does the astronaut's lunch take to eat according to observers on Earth?

Part a. Step 1.

Use Eq. 26-3a to determine the length as measured by an observer on Earth.

The astronaut measures the proper length $(\ell_0 = 23 \text{ m})$, while the observer on Earth measures the contracted length ℓ:

$$\ell = \ell_0\sqrt{1-\frac{v^2}{c^2}}$$

$$\ell = (23 \text{ m})\sqrt{1-\frac{(0.750c)^2}{c^2}}$$

$$\ell = (23 \text{ m})(0.661) = 15.2 \text{ m}$$

Part b. Step 1.

Use Eq. 26-1a to determine the time as measured by an observer on Earth.

The astronaut measures the proper time $(t_0 = 28 \text{ min})$, while the observer on Earth measures the dilated time t:

$$\Delta t_0 = \Delta t\sqrt{1-\frac{v^2}{c^2}}$$

$$28 \text{ min} = \Delta t\sqrt{1-\frac{(0.750c)^2}{c^2}}$$

$$28 \text{ min} = \Delta t\sqrt{1-0.56}$$

$$28 \text{ min} = \Delta t(0.661)$$

$$\Delta t = \frac{28 \text{ min}}{0.661} = 42.3 \text{ min}$$

TEXTBOOK PROBLEM 73. (Section 26-9) Suppose a 14,500 kg spaceship left Earth at a speed of $0.90c$. What is the spaceship's kinetic energy? Compare this with the total U.S. annual energy consumption (about 10^{20} J).

Part a. Step 1.

Use Eq. 26-5a to determine the spaceship's kinetic energy.

$$\text{total energy} = \text{kinetic energy} + \text{rest energy}$$

$$\text{kinetic energy} = \text{total energy} - \text{rest energy}$$

$$\text{KE} = \frac{m}{\sqrt{1-\frac{v^2}{c^2}}}c^2 - mc^2$$

$$= \frac{m}{\sqrt{1-\frac{(0.90c)^2}{c^2}}}c^2 - mc^2 \quad \text{but} \quad \sqrt{1-\frac{(0.90c)^2}{c^2}} = 0.436$$

$$= \frac{m}{0.436}c^2 - mc^2$$

$$= 2.29mc^2 - mc^2 = 1.29mc^2$$

$$\text{KE} = 1.29mc^2 = (1.29)(14{,}500 \text{ kg})(3.0 \times 10^8 \text{ m/s})^2$$

$$\text{KE} = 1.68 \times 10^{21} \text{ J}$$

Part a. Step 2.

Compare the spaceship's KE to the annual energy consumption of the U.S.

$$\frac{\text{energy of spaceship}}{\text{U.S. output}} = \frac{1.68 \times 10^{21} \text{ J}}{10^{20} \text{ J}} \approx 16.8$$

The spaceship's kinetic energy is approximately 16.8 times greater than the total U.S. annual energy consumption.

27

EARLY QUANTUM THEORY AND MODELS OF THE ATOM

OBJECTIVES

After studying the material of this chapter, the student should be able to:

- describe the method used by J. J. Thomson to determine the ratio of the charge on an electron to its mass.
- use Wien's law to determine the peak wavelength emitted by a blackbody at a given temperature.
- describe Planck's quantum hypothesis and calculate the energy of a photon at a given frequency or wavelength.
- state the experimental results of the photoelectric effect and use the photon theory to explain these results.
- use the photon theory to determine the maximum kinetic energy of photons emitted from the surface of a metal or the threshold wavelength for the metal.
- use the photon theory and Compton's hypothesis to calculate the wavelength of a photon after it has been scattered as a result of a collision with an electron.
- use $E = mc^2$ to determine the minimum energy required for pair production.
- explain the significance of the principle of complementarity.
- use de Broglie's hypothesis to determine the wavelength of moving particle.
- describe the apparatus used in the Rutherford scattering experiment and describe the experimental results.
- describe Rutherford's model of the atom and list two problems with the model.
- determine the wavelength of a photon emitted as an electron drops from a higher energy level to a lower energy level. Determine the frequency and energy of this photon.
- use Bohr's postulates to explain the emission spectra produced by the hydrogen atom.
- determine the Bohr radius and angular momentum of an electron in a given energy level.

KEY TERMS AND PHRASES

cathode rays were shown by J. J. Thomson's *e/m* experiment to consist of charged particles now known as **electrons**. As a result of this experiment, the ratio of the charge on an electron (e) to its mass (m) was determined.

Planck's quantum hypothesis predicts that the molecules in a heated object can vibrate only with discrete amounts of energy. Thus the energy of the vibrating atom is quantized.

oil-drop experiment, done by Robert Millikan, determined that the charge on a microscopically small oil drop is always a small whole-number multiple of 1.6×10^{-19} C. This value equals the charge on the electron. Once the value of the charge on an electron was determined, the accepted value for the mass of the electron was determined to be 9.1×10^{-31} kg.

blackbody radiation refers to the intensity of spectral radiation emitted by a "perfectly" radiating object.

photoelectric effect indicates that light has characteristics of particles. Light particles are called **photons.**

work function (W_0) is the minimum energy required to break the electron free from the attractive forces that hold the electron to the surface of a metal.

pair production occurs when a high-energy photon known as a gamma ray traveling near the nucleus of an atom disappears and an electron and a positron may appear in its place.

wave–particle duality refers to the phenomenon in which both particles, such as electrons and protons, and light exhibit both the properties of waves and the properties of particles.

Compton effect shows that the interaction of a photon with an electron can be viewed as a two-particle collision.

de Broglie wavelength is the wavelength of a particle of mass m traveling at speed v. The wavelength is given by $\lambda = \frac{h}{mv}$, where λ is the wavelength of the particle.

emission spectra are produced by a high voltage placed across the electrodes of a tube containing a gas under low pressure. The light produced can be separated into its component colors by a diffraction grating. Such analysis reveals a spectrum of discrete lines and not a continuous spectrum.

ionization energy refers to the energy required to remove an electron from an atom.

SUMMARY OF MATHEMATICAL FORMULAS

Wien's law	$\lambda_p T = 2.90 \times 10^{-3}$ m·K	The relationship between absolute temperature (T) and the peak wavelength (λ_p) in blackbody radiation
photon energy	$E = hf$	The energy (E) of a photon is related to the frequency (f) of the light.
photoelectric effect	$\text{KE}_{\text{max}} = hf - W_0$	The maximum kinetic energy (KE_{max}) of the emitted photoelectrons equals the difference between the energy of the incident photon (hf) and the work function (W_0) of the metal surface.
Compton effect	$\lambda' - \lambda = \left(\frac{h}{m_e c}\right)(1 - \cos\phi)$	A collision between a photon and an electron results in a change of wavelength for the photon.
de Broglie wavelength	$\lambda = \frac{h}{mv}$	λ represents the wavelength of a particle of mass m traveling at speed v.

Balmer equation	$\frac{1}{\lambda}=R\left(\frac{1}{2^2}-\frac{1}{n^2}\right)$	An electron dropping from a higher energy level to a lower energy level emits a photon of wavelength λ, where $R=1.097\times 10^7\ \text{m}^{-1}$ is the Rydberg constant and $n=3, 4,$ etc.
ionization energy for an electron from a hydrogen-like atom	$E=(-13.6\ \text{eV})\frac{Z^2}{n^2}$	Ionization energy (E) of an electron located in the nth level of the hydrogen-like atom
Bohr radius	$r=n^2(0.53\times 10^{-10}\ \text{m})$	Radius (r) of the orbit of an electron in the hydrogen atom, where $n=1, 2,$ etc.
Bohr's quantum condition (angular momentum of an electron)	$L=mvr_n=\frac{nh}{2\pi}$	The angular momentum (L) of an electron orbiting the hydrogen nucleus

CONCEPT SUMMARY

The Electron

In 1897, J. J. Thomson performed the *e/m* experiment, which showed that **cathode rays** consist of charged particles now known as **electrons**. As a result of his experiment, he was able to determine that the ratio of the charge on an electron (e) to its mass (m) is

$$\frac{e}{m}=\frac{v}{Br}$$

where v is the electron's velocity, B is the magnetic field strength that is directed perpendicular to the electron's path, and r is the radius of the circular path in which the particle travels.

B and r are readily measured, while the velocity of the particle can be determined by using a device known as a velocity selector. In a velocity selector (see Section 20-11), an electric field is arranged so that the electric force $F=qE$ balances the force exerted on the charged particle by the magnetic field $F=qvB\sin 90°$. As a result, $qE=qvB$ and $v=E/B$. The particle passes through the selector undeflected.

Since $\frac{e}{m}=\frac{v}{Br}$ and $v=\frac{E}{B}, \frac{e}{m}=\frac{E}{B^2r}$. Substituting experimental values for E, B, and r, Thomson obtained a value for the electron's charge to mass ratio close to the modern value of 1.76×10^{11} C/kg.

In 1909, R. A. Millikan was able to determine the charge on an electron in an experiment referred to as the **oil-drop experiment.** Millikan found that the charge on a microscopically small oil drop is always a small whole-number multiple of 1.6×10^{-19} C. Once the charge on the electron became known, it was possible to determine its mass. The accepted value for the mass of the electron is 9.1×10^{-31} kg.

Planck's Quantum Hypothesis

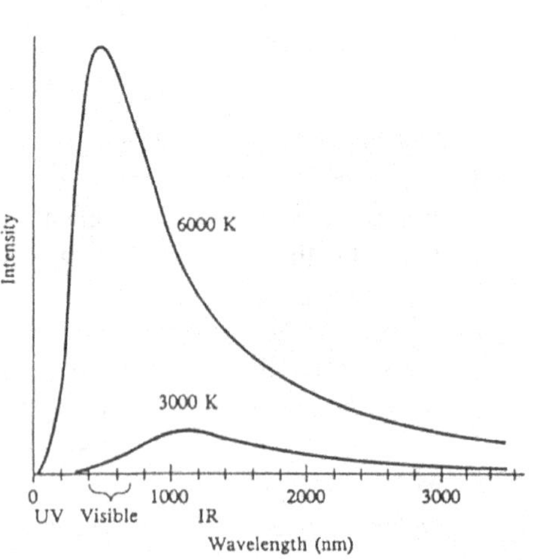

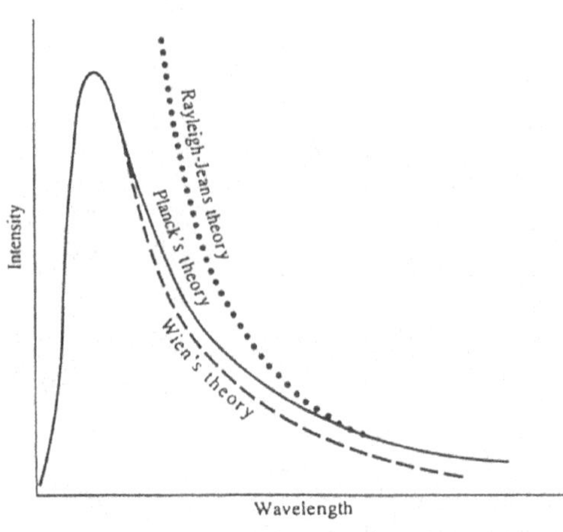

The intensity of spectral radiation emitted by a "perfectly" radiating object, known as **blackbody radiation**, when graphed as a function of wavelength, produces curves as shown in the diagram above.

The wavelength at which peak intensity occurs decreases as the Kelvin temperature of the object increases. The relationship between the peak wavelength and the absolute temperature is given by **Wien's law,**

$$\lambda_{\mathrm{p}} T = 2.90 \times 10^{-3}\ \mathrm{m \cdot K}$$

where λ_{p} is the peak wavelength, the wavelength of the emitted light at the point where the intensity of the emitted light is a maximum. T is the temperature of the blackbody in degrees Kelvin.

Max Planck produced a theory that agreed with the experimental data by assuming that the molecules in the heated object can vibrate only with discrete amounts of energy. Thus the energy of the vibrating atom is quantized.

The energy is related to the natural frequency of vibration of the molecules of the radiating object by the formula

$$E = nhf$$

where h is **Planck's constant,** $h = 6.626 \times 10^{-34}\ \mathrm{J \cdot s}$. E is the energy in joules, n is a whole number, $n = 1, 2, 3$, etc., and f is the object's natural frequency of vibration in hertz.

TEXTBOOK QUESTION 2. (Section 27-2) If energy is radiated by all objects, why can we not see them in the dark? (See also Section 14-8.)

ANSWER: The human eye is sensitive only to visible light, that is, violet (400 nm) through red (700 nm). The energy emitted by objects such as this book or your body is far below that required to produce visible light. For example, let us assume that the skin temperature of your body is approximately 34°C (307 K). Using Wien's law to determine the peak wavelength of light, $\lambda_{\mathrm{p}}(307\ \mathrm{K}) = 2.90 \times 10^{-3}\ \mathrm{m \cdot K}$, gives a peak wavelength of 9.45×10^{-6} m, or 9450 nm. This radiation is in the infrared portion of the electromagnetic spectrum.

EXAMPLE PROBLEM 1. (Section 27-2) (*a*) Use Wien's law to determine the peak wavelength of light emitted by a blackbody whose temperature is 5000 K. (*b*) Is this wavelength in the visible portion of the electromagnetic spectrum?

Part a. Step 1.

Determine the peak wavelength.	$\lambda_{\text{p}}T = 2.90 \times 10^{-3}\ \text{m}\cdot\text{K}$ $\lambda_{\text{p}}(5000\ \text{K}) = 2.90 \times 10^{-3}\ \text{m}\cdot\text{K}$ $\lambda_{\text{p}} = 5.8 \times 10^{-7}\ \text{m} = 580\ \text{nm}$

Part b. Step 1.

Is this wavelength in the visible spectrum?	The visible spectrum extends from 400 to 700 nm. The peak wave-length is 580 nm; therefore, it is in the visible range. 580 nm is located in the yellow portion of the electromagnetic spectrum.

TEXTBOOK QUESTION 3. (Section 27-2) What can be said about the relative temperatures of whitish-yellow, reddish, and bluish stars? Explain.

ANSWER: Wien's law states that $\lambda_{\text{p}}T = 2.90 \times 10^{-3}\ \text{m}\cdot\text{K},$ where λ_{p} is the peak wavelength and T is the temperature in degrees Kelvin. Based on this law, the higher the temperature of the star, the shorter the peak wavelength of the light. From Fig. 24-12, blue stars have the shortest wavelength and are the hottest. Whitish-yellow stars are hot, while red stars are the coolest.

TEXTBOOK QUESTION 7. (Section 27-3) UV light causes sunburn, whereas visible light does not. Suggest a reason.

ANSWER: Based on Planck's quantum hypothesis $(E = hf)$, the energy of a photon is directly proportional to its frequency. Figure 24-12 shows that a UV photon has a higher frequency than visible light and therefore higher energy. Therefore, the higher energy UV photons are able to cause sunburn, while the lower energy photons of visible do not.

Photoelectric Effect

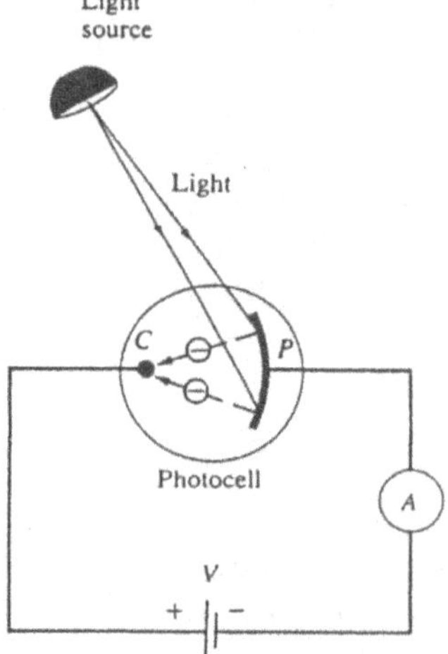

When light is incident on a metal surface, electrons are emitted. In the photocell shown in the diagram, the incident light causes the electrons to be emitted from the plate (*P*), and the difference in potential causes the electrons to travel to the collector (*C*).

The experimental results indicate that (1) the number of electrons emitted per second (electron current) increases as the light intensity increases, (2) the maximum kinetic energy of the electrons is not affected by the intensity of the light, and (3) below a certain frequency, called the threshold frequency, no electrons are ejected no matter how intense the light beam.

Wave theory agrees with the first result but cannot explain the second and third experimental results. In 1905, Albert Einstein proposed an extension of the quantum theory to explain the results of the **photoelectric effect**. His theory says that light is emitted in particles that are now called **photons**. Each photon has energy that is related to the frequency of the light according to the formula

$E = hf$ The experimental results of the photoelectric effect can be explained by the photon theory:

(1) An electron in the metal making up the plate of the photocell absorbs the energy of the incident photon. If the photon energy is large enough, then the electron escapes from the metal. As the intensity of the light beam increases, the number of photons increases and the number of electrons emitted each second increases. However, the intensity of the light does not affect the energy of the incident photons.

(2) The maximum kinetic energy of the emitted electrons is related to the energy of the incident photon by the equation $hf = \text{KE}_{\text{max}} + W_0$, where hf is the energy of the incident photon in joules and KE_{max} is the kinetic energy of the most energetic electrons emitted from the metal in joules. The **work function** (W_0) is the minimum energy required to break the electron free from the attractive forces that hold the electron to the metal.

(3) If the energy of the incident photon is below that of the work function, then no electrons are emitted. The minimum frequency required to eject an electron is called the threshold frequency or "cutoff" frequency.

EXAMPLE PROBLEM 2. (Section 27-3) The threshold wavelength for incident photons to eject electrons from the surface of gold is 230 nm. If light of wavelength 100 nm shines on the metal, determine the (*a*) work function of the metal and (*b*) energy of the most energetic electrons.

Part a. Step 1.

Determine the work function of the metal.	At the threshold wavelength, the energy absorbed by the electron is used to break free from the attractive forces that hold it to the metal. As a result, it has no kinetic energy when it finally breaks free. Therefore,

$$E = \text{KE}_{\text{max}} + W_0 \quad \text{but} \quad \text{KE}_{\text{max}} = 0$$

$$hf = 0 + W_0, \qquad \text{where} \quad hf = \frac{hc}{\lambda}$$

$$W_0 = \frac{(6.63 \times 10^{-34} \text{ J}\cdot\text{s})(3 \times 10^8 \text{ m/s})}{2.30 \times 10^{-7} \text{ m}}$$

$$W_0 = 8.64 \times 10^{-19} \text{ J, or } 5.40 \text{ eV}$$

Alternate solution:

$$W_0 = hf = \frac{hc}{\lambda} = \frac{1240 \text{ eV}\cdot\text{nm}}{\lambda}$$

$$W_0 = \frac{1240 \text{ eV}\cdot\text{nm}}{230 \text{ nm}} = 5.40 \text{ eV}$$

Part b. Step 1.

Determine the energy of the most energetic electrons.

$$E = \text{KE}_{\text{max}} + W_0,$$

where $E = hf = \dfrac{1240 \text{ eV}\cdot\text{nm}}{100 \text{ nm}} = 12.4 \text{ eV}$

$$12.4 \text{ eV} = \text{KE}_{\text{max}} + 5.40 \text{ eV}$$

$$\text{KE}_{\text{max}} = 7.00 \text{ eV}$$

Compton Effect

In 1922, Arthur Compton used the photon theory to explain why X-rays scattered from certain materials have different wavelengths than the incident X-rays. According to Compton, an incident photon transferred some of its energy to an electron in the material and was scattered with lower energy and therefore longer wavelength.

According to the theory, the incident photon carries both energy ($E = hf$) and momentum ($p = \frac{E}{c} = \frac{hf}{c} = \frac{h}{\lambda}$). The photon interaction with the electron can be considered to be a two-particle collision, as shown in the diagram.

By applying the laws of conservation of energy and momentum to the collision, the theory correctly predicts a wavelength change given by the following equation:

$$\lambda' - \lambda = \frac{h}{m_e c}(1 - \cos\phi)$$

λ is the wavelength of the incident light, λ' is the wavelength of the scattered photon, m_e is the mass of the recoil electron, and ϕ is the angle between the direction of the incident photon and the scattered photon.

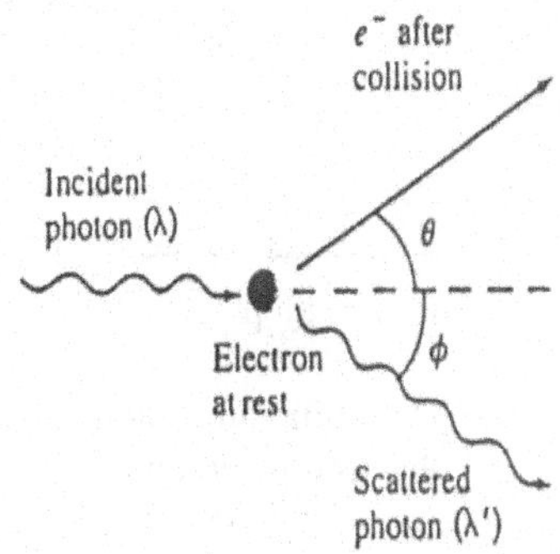

TEXTBOOK QUESTION 12. (Section 27-5) If an X-ray photon is scattered by an electron, does the photon's wavelength change? If so, does it increase or decrease? Explain.

ANSWER: The incident photon transfers some of its energy to the electron. The result is a scattered photon that has less energy and a different wavelength than the incident photon. Since $E = hf = \frac{hc}{\lambda}$, if the energy is less, the wavelength must increase.

EXAMPLE PROBLEM 3. (Section 27-5) An X-ray of wavelength 0.1500 nm is scattered by an electron. In the resulting collision, the scattered photon is reflected directly backward, while the electron travels in the direction of the incident photon. Determine the (*a*) wavelength of the scattered photon and (*b*) energy of the recoil electron.

λ → electron — before collision

← λ' ○ → v — after collision

Part a. Step 1.

Determine the wavelength of the scattered photon.

The collision is head on, and the angle between the incident photon and the scattered photon is 180°. Therefore,

$$\lambda' - \lambda = \left(\frac{h}{m_e c}\right)(1 - \cos\phi), \quad \text{where} \quad \cos\phi = \cos 180° = -1$$

$$m_e c = (9.1 \times 10^{-31}\ \text{kg})(3.0 \times 10^{8}\ \text{m/s})$$

$$m_e c = 2.7 \times 10^{-22}\ \text{kg} \cdot \text{m/s}$$

$$\lambda' - 0.1500\ \text{nm} = \frac{(6.63 \times 10^{-34}\ \text{J} \cdot \text{s})(1\ -\ -1)}{2.7 \times 10^{-22}\ \text{m}}$$

$$\lambda' - 0.1500\ \text{nm} = 4.86 \times 10^{-12}\ \text{m} \quad \text{but}$$

$$4.86 \times 10^{-12}\ \text{m} = 4.86 \times 10^{-3}\ \text{nm}, \quad \text{so}$$

$$\lambda' = 0.1500\ \text{nm} + 4.86 \times 10^{-3}\ \text{nm} = 0.155\ \text{nm}$$

Part b. Step 1.

Determine the kinetic energy of the recoil electron.

The kinetic energy of the recoil electron can be determined by using the law of conservation of energy. The energy of the incident photon equals the sum of the energy of the scattered photon and the kinetic energy of the recoil electron.

$$\frac{1240\ \text{eV} \cdot \text{nm}}{\lambda} = \frac{1240\ \text{eV} \cdot \text{nm}}{\lambda'} + \text{KE}_{\text{electron}}$$

$$\frac{1240\ \text{eV}\cdot\text{nm}}{0.150\ \text{nm}}=\frac{1240\ \text{eV}\cdot\text{nm}}{0.155\ \text{nm}}+\text{KE}_{\text{electron}}$$

$$8270\ \text{eV}=8000\ \text{eV}+\text{KE}_{\text{electron}}$$

$$\text{KE}_{\text{electron}}=270\ \text{eV}$$

TEXTBOOK QUESTION 14. (Sections 27-3 and 27-5) Why do we say that light has wave properties? Why do we say that light has particle properties?

ANSWER: Experiments described in Chapter 24 such as those involving interference, refraction, dispersion, and polarization all indicate that light is a wave. Experiments described in Chapter 27 such as the photoelectric effect and the Compton effect indicate that light is has particle properties.

Pair Production

The equation $E=mc^2$ implies that it is possible to convert mass into energy and vice versa. One example of the conversion of energy to mass is **pair production.** In this process, a high-energy photon known as a gamma ray traveling near the nucleus of an atom may disappear, and an electron and a positron may appear in its place. The electron and the positron have the same mass and carry the same magnitude of electric charge; however, while the electron is negatively charged, the positron carries a positive charge. Thus, in addition to the laws of conservation of energy and momentum, the law of conservation of electric charge also holds true.

The minimum energy of a gamma ray required for the pair production of electron and positron can be shown to be about 1.02 MeV. If the energy of the gamma ray is above this amount, then excess energy is shared equally between the particles in the form of kinetic energy.

Wave–Particle Duality; the Principle of Complementarity

Experiments such as Young's interference experiment and those on polarization and single-slit diffraction indicate that light is a wave. The photoelectric effect and the Compton effect indicate that light is a particle. Light is a phenomenon that exhibits both the properties of waves and the properties of particles. This is known as **wave–particle duality.**

Niels Bohr proposed the **principle of complementarity**, which says that for any particular experiment involving light, we must use either the wave theory or the particle theory, but not both. The two aspects of light complement one another.

Wave Nature of Matter

Just as light exhibits properties of both particles and waves, particles such as electrons, protons, and neutrons also exhibit wave properties. Thus the wave–particle theory extends to matter as well as light. In 1923, Louis de Broglie suggested that the wavelength of a particle of mass m traveling at speed v is given by

$$\lambda=\frac{h}{mv}$$

where λ is the **de Broglie wavelength** of the particle.

In 1927, two Americans, Davisson and Germer, produced diffraction patterns by scattering electron beams from the surface of a metal crystal. The calculated wavelength of the electron waves agreed with de Broglie's prediction. It was later shown that protons and neutrons as well as other particles exhibit wave properties as well as particle properties.

Rutherford's Model of the Atom

In 1909, Ernest Rutherford suggested to two of his students, Hans Geiger and Ernest Marsden, that they bombard a piece of thin gold foil with high-energy, positively charged alpha particles.

Alpha particles are emitted spontaneously from a radioactive source. Because of their high energy, Rutherford expected that the alpha particles would pass through the gold foil without being significantly deflected from their original direction. However, as shown in the diagram, Geiger and Marsden found that while most of the alpha particles did pass through undeflected, some were deflected by 30° or more, and a few were deflected by 90° or more.

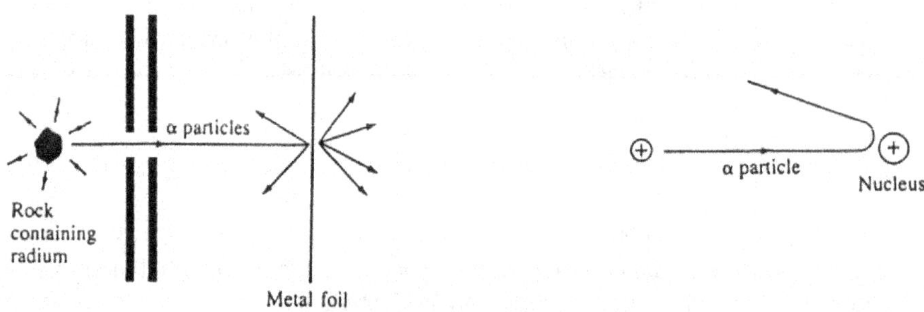

Based on observations, Rutherford concluded that most of the mass of the atom resided in a tiny region called the nucleus. The nucleus contains the positively charged protons, while the electrons travel in orbits around the nucleus.

Rutherford's model had two problems: (1) since protons repel one another, the positively charged nucleus should not exist, and (2) the orbiting electrons are undergoing centripetal acceleration and, according to electromagnetic theory, accelerated electric charges give rise to electromagnetic waves. Thus the electrons should give up their energy in the form of a continuous spectrum as they spiral into the nucleus.

Atomic Spectra

Emission spectra are produced by a high voltage placed across the electrodes of a tube containing a gas under low pressure. The light produced can be separated into its component colors by a diffraction grating. Such analysis reveals a spectrum of discrete lines and not a continuous spectrum.

In 1885, J. J. Balmer developed a mathematical equation that could be used to predict the wavelengths of the four visible lines in the hydrogen spectrum. Balmer's equation states

$$\frac{1}{\lambda} = R\left(\frac{1}{2^2} - \frac{1}{n^2}\right)$$

where λ is the wavelength of the spectral line in meters, and $R = 1.097 \times 10^7\ \text{m}^{-1}$ is known as the **Rydberg constant** and $n = 3, 4$, etc. For example, for the visible spectrum $n = 3$ (red light), $n = 4$ (blue light), $n = 5$ (violet light), and $n = 6$ (violet light).

Bohr's Model of the Hydrogen Atom

Based on the Rutherford model of the atom, line spectra, and the Balmer formula, Niels Bohr in 1913 proposed the following postulates for the hydrogen atom:

(1) The electron travels in circular orbits about the positively charged nucleus. However, only certain orbits are allowed. The electron does not radiate energy when it is in one of these orbits, thus violating classical theory.

(2) The allowed orbits have radii (r_n), where

$$r_n = n^2(0.53 \times 10^{-10}\text{ m}) = n^2(0.53\text{ nm}) \quad \text{and} \quad n = 1, 2, 3, \text{etc.}$$

If $n = 1$, then the electron is in its smallest orbit, which is known as its **ground state**.

(3) The orbits have angular momentum (L) given by

$L = mvr_n = \dfrac{nh}{2\pi}$, where $n = 1, 2, 3$, etc. The angular momentum has only discrete values; thus, it is **quantized**.

(4) If an electron falls from one orbit, also known as an **energy level**, to another, it loses energy in the form of a photon of light. The energy of the photon equals the difference between the energy of the orbits.

(5) The energy level of a particular orbit is given by $E = \dfrac{-13.6\text{ eV}}{n^2}$, where $n = 1, 2, 3,$ etc. If $n = 1$, then the electron is in its lowest energy level and it would take 13.6 eV to remove it from the atom (**ionization energy**).

(6) A hydrogen atom can absorb only those photons of light that will cause the electron to jump from a lower level to a higher level. Thus the energy of the photon must equal the difference in the energy between the two levels. Therefore, a continuous spectrum passing through a cool gas will exhibit dark absorption lines at the same wavelengths as the emission lines.

The Bohr model proved to be successful for the hydrogen atom and for one-electron ions; however, it did not work for multi-electron atoms.

De Broglie's Contribution to the Bohr Model

In 1923, de Broglie extended his idea of matter waves to the Bohr atom by stating that the electron wave must be a circular standing wave. The circumference $(2\pi r_n)$ of the standing wave contains a whole number of wavelengths $(n\lambda)$; thus, $2\pi r_n = n\lambda$, where $n = 1, 2, 3$, etc. So $\lambda = \dfrac{2\pi r_n}{n}$ but $\lambda = \dfrac{h}{mv}$. Then $\dfrac{2\pi r_n}{n} = \dfrac{h}{mv}$; rearranging gives

$$mvr_n = \frac{nh}{2\pi}$$

This formula is Bohr's postulate concerning the quantization of the electron's angular momentum. de Broglie provided an explanation for the key postulate in Bohr's model of the hydrogen atom; that is, the idea that the angular momentum (L) of the electron is quantized:

$L = mvr_n = \dfrac{nh}{2\pi}$ It is from this postulate that the equations for the discrete orbits and energy states are derived.

EXAMPLE PROBLEM 4. (Section 27-13) An electron and a proton are both traveling at 3.25×10^{6} m/s. Determine the de Broglie wavelength of each particle.

Part a. Step 1.

Use the de Broglie equation to determine the wavelength of each particle.

The mass of an electron is 9.1×10^{-31} kg, and that of a proton is 1.67×10^{-27} kg. Use the de Broglie equation to solve for the wavelength:

Electron: $\lambda = \dfrac{h}{mv}$

$$\lambda = \frac{6.63\times10^{-34}\ \text{J}\cdot\text{s}}{(9.1\times10^{-31}\ \text{kg})(3.25\times10^{6}\ \text{m/s})}$$

$\lambda = 2.24\times10^{-10}\ \text{m} = 0.224\ \text{nm}$

Proton: $\lambda = \dfrac{h}{mv}$

$$\lambda = \frac{6.63\times10^{-34}\ \text{J}\cdot\text{s}}{(1.67\times10^{-27}\ \text{kg})(3.25\times10^{6}\ \text{m/s})}$$

$\lambda = 1.22\times10^{-13}\ \text{m}$

The de Broglie wavelength of the electron described in this problem corresponds to the distance between the layers of atoms in a crystal.

Therefore, it would be possible to detect the wave nature of the electrons by using a diffraction experiment. The wavelength of the proton is such that the wave nature of these particles would be detected.

EXAMPLE PROBLEM 5. (Section 27-12) Determine the Bohr radius of the $n = 3$ orbit of the hydrogen atom. Determine the angular momentum of an electron in this orbit.

Part a. Step 1.

Determine the radius of the orbit.

The radius of orbit is given by

$r = n^{2}(0.53\times10^{-10}\ \text{m})$, where $n = 3$

$r = 3^{2}(0.53\times10^{-10}\ \text{m})$

$r = 4.77\times10^{-10}\ \text{m}$

Part a. Step 2.

Determine the angular momentum of the electron.

$$L = \frac{nh}{2\pi} = \frac{(3)(6.63 \times 10^{-34}\ \text{J}\cdot\text{s})}{2\pi}$$

$$L = 3.17 \times 10^{-34}\ \text{J}\cdot\text{s}$$

PROBLEM SOLVING SKILLS

For problems related to the photoelectric effect:

1. Construct a data table listing information both given and requested in the problem. For example, include the work function of the metal, threshold frequency and threshold wavelength for the metal, energy of the incident photons, wavelength of the incident photons, Planck's constant, and the speed of light.
2. Use the equation $E = \text{KE}_{\text{max}} + W_0$, where $W_0 = hf_0$ to solve the problem.

For problems related to the Compton effect:

1. Construct a data table listing information both given and requested in the problem. For example, include the wavelength of the incident and scattered photons, the angle between the incident and scattered photons, and the energy of the recoil electron.
2. Use the equation $\lambda' - \lambda = \frac{h}{m_e c}(1 - \cos\phi)$ to solve for the unknown quantity.
3. Use the law of conservation of energy to solve for the energy of the recoil electron.

For problems related to pair production:

1. List the mass of the particle and its antiparticle.
2. Determine the minimum energy required for pair production.
3. Use the law of conservation of energy to determine the kinetic energy of the particles after they are produced.

For problems related to the de Broglie wavelength of a particle:

1. List the mass and velocity of the particle.
2. Use $\lambda = \frac{h}{mv}$ to solve for the de Broglie wavelength.

For problems related to emission spectra:

1. List information both given and requested in the problem.
2. Solve for the wavelength of the light emitted when an electron transition occurs between two energy levels.
3. Use $f = \frac{c}{\lambda}$ to determine the frequency of the photon.
4. Use $E = hf$ to determine the energy of the photon.

SOLUTIONS TO SELECTED TEXTBOOK PROBLEMS

TEXTBOOK PROBLEM 5. (Section 27-2) Estimate the peak wavelength for radiation from (*a*) ice at 0°C, (*b*) a floodlamp at 3100 K, (*c*) helium at 4 K, assuming blackbody emission. In what region of the EM spectrum is each?

Part a. Step 1.

Use Wien's law to determine the peak wavelength for ice.

$$\lambda_p T = 2.90 \times 10^{-3}\,\text{m}\cdot\text{K} \quad 0°\text{C} = 273\ \text{K}$$

$$\lambda_p(273\ \text{K}) = 2.90 \times 10^{-3}\ \text{m}\cdot\text{K} = \frac{2.90 \times 10^{-3}\ \text{m}\cdot\text{K}}{273\ \text{K}}$$

$$\lambda_p = 1.06 \times 10^{-5}\,\text{m} = 10.6\ \mu\text{m} \quad \text{(infrared)}$$

Part b. Step 1.

Determine the peak wavelength for the floodlamp.

$$\lambda_p(3100\ \text{K}) = 2.90 \times 10^{-3}\ \text{m}\cdot\text{K}$$

$$\lambda_p = \frac{2.90 \times 10^{-3}\ \text{m}\cdot\text{K}}{3100\ \text{K}}$$

$$\lambda_p = 9.35 \times 10^{-7}\ \text{m} = 935\ \text{nm} \quad \text{(infrared)}$$

Part c. Step 1.

Determine the peak wavelength for helium.

$$\lambda_p(4\ \text{K}) = 2.90 \times 10^{-3}\ \text{m}\cdot\text{K}$$

$$\lambda_p = \frac{2.90 \times 10^{-3}\ \text{m}\cdot\text{K}}{4\ \text{K}}$$

$$\lambda_p = 7.3 \times 10^{-4}\ \text{m} = 7.3 \times 10^{5}\ \text{nm} = 0.73\ \text{mm} \quad \text{(microwave)}$$

Note: Visible light extends from 400 nm to 700 nm. Therefore, each wavelength is outside of the visible part of the EM spectrum.

TEXTBOOK PROBLEM 23. (Sections 27-3 and 27-4) Barium has a work function of 2.48 eV. What is the maximum kinetic energy of electrons if the metal is illuminated by UV light of wavelength 365 nm? What is their speed?

Part a. Step 1.

Determine the energy of the photons in eV.

$$E = hf = \frac{hc}{\lambda}$$

$$= \frac{(6.63 \times 10^{-34}\ \text{J}\cdot\text{s})(3.0 \times 10^{8}\ \text{m/s})}{(365\ \text{nm})\left(\frac{10^{-9}\ \text{m}}{1\ \text{nm}}\right)}$$

$$E = 5.45 \times 10^{-19}\ \text{J}$$

$$(5.45 \times 10^{-19}\ \text{J})\frac{1.0\ \text{eV}}{1.6 \times 10^{-19}\ \text{J}} = 3.41\ \text{eV}$$

Part a. Step 2.

Determine the maximum kinetic energy of the emitted electrons.

$$E = \text{KE} + W_0$$

$$3.41\ \text{eV} = \text{KE} + 2.48\ \text{eV}$$

$$\text{KE} = 0.926\ \text{eV}\,\frac{1.6 \times 10^{-19}\ \text{J}}{1\ \text{eV}} = 1.48 \times 10^{-19}\ \text{J}$$

Part a. Step 3.

Determine the maximum speed of the emitted electrons.

$$\text{KE} = \tfrac{1}{2}mv^2$$

$$1.48 \times 10^{-19}\ \text{J} = \tfrac{1}{2}(9.1 \times 10^{-31}\ \text{kg})v^2$$

$$v^2 = 3.26 \times 10^{11}\ \text{m}^2/\text{s}^2$$

$$v^2 = 5.7 \times 10^{5}\ \text{m/s}$$

TEXTBOOK PROBLEM 34. (Section 27-6) What is the longest wavelength photon that could produce a proton–antiproton pair? (Each has a mass of 1.67×10^{-27} kg.)

Part a. Step 1.

Use the law of conservation of energy to determine the minimum energy of the photon.

Using the law of conservation of energy, the minimum energy of the photon is equal to the sum of the rest energies of the two particles.

$$E = mc^2 = m_{\text{proton}}c^2 + m_{\text{antiproton}}c^2$$

$$E = (1.67 \times 10^{-27}\ \text{kg})(9.0 \times 10^{16}\ \text{m}^2/\text{s}^2) + (1.67 \times 10^{-27}\ \text{kg})(9.0 \times 10^{16}\ \text{m}^2/\text{s}^2)$$

$$E = 3.01 \times 10^{-10}\ \text{J} = 1.88 \times 10^{9}\ \text{eV}$$

Part a. Step 2.

Determine the wavelength of photon.

$$E = hf = \frac{hc}{\lambda}$$

$$\lambda = \frac{hc}{E} = \frac{(6.63 \times 10^{-34}\ \text{J}\cdot\text{s})(3.0 \times 10^{8}\ \text{m/s})}{3.01 \times 10^{-10}\ \text{J}}$$

$$\lambda = 6.61 \times 10^{-16}\ \text{m}$$

TEXTBOOK PROBLEM 37. (Section 27-6) A gamma-ray photon produces an electron and a positron, each with a kinetic energy of 285 keV. Determine the energy and wavelength of the photon.

Part a. Step 1.

Use the law of conservation of energy to determine the rest energy of the electron–positron pair.

Using the law of conservation of energy, the minimum energy required to produce the electron–positron pair is equal to the sum of the rest energies of the two particles.

$$E = m_{\text{electron}}c^2 + m_{\text{positron}}c^2$$

$$E = (9.1 \times 10^{-31}\ \text{kg})(9.0 \times 10^{16}\ \text{m}^2/\text{s}^2) + (9.1 \times 10^{-31}\ \text{kg})(9.0 \times 10^{16}\ \text{m}^2/\text{s}^2)$$

$$E = (1.64 \times 10^{13}\ \text{J})\left(\frac{1\ \text{MeV}}{1.6 \times 10^{-13}\ \text{J}}\right) = 1.02\ \text{MeV}$$

Part a. Step 2.

Determine the total energy of the photon.

$$\text{total energy} = \text{kinetic energy} + \text{rest energy}$$

$$\text{total energy} = 2(285\ \text{keV}) + 1.02\ \text{MeV}$$

$$= 570\ \text{keV}\left(\frac{1\ \text{MeV}}{1000\ \text{keV}}\right) + 1.02\ \text{MeV}$$

$$\text{total energy} = 1.59\ \text{MeV}\left(\frac{1.6 \times 10^{-13}\ \text{J}}{1\ \text{MeV}}\right) = 2.54 \times 10^{-13}\ \text{J}$$

Part a. Step 3.

Determine the wavelength of the gamma-ray photon.

$$E = hf = h\frac{c}{\lambda}$$

$$2.54 \times 10^{-13}\ \text{J} = \frac{(6.63 \times 10^{-34}\ \text{J}\cdot\text{s})(3.0 \times 10^{8}\ \text{m/s})}{\lambda}$$

$$\lambda = \frac{(6.63 \times 10^{-34}\ \text{J}\cdot\text{s})(3.0 \times 10^{8}\ \text{m/s})}{2.54 \times 10^{-13}\ \text{J}} = 7.83 \times 10^{-13}\ \text{m}$$

TEXTBOOK PROBLEM 42. (Section 27-8) An electron has a de Broglie wavelength $\lambda = 4.5 \times 10^{-10}$ m. (*a*) What is its momentum? (*b*) What is its speed? (*c*) What voltage was needed to accelerate it from rest to this speed?

Part a. Step 1.

Use Eq. 27-6 to determine the electron's momentum.

$$p = \frac{h}{\lambda} = \frac{6.63 \times 10^{-34}\ \text{J}\cdot\text{s}}{4.5 \times 10^{-10}\ \text{m}}$$

$$p = 1.47 \times 10^{-24}\ \text{kg}\cdot\text{m/s}$$

Part b. Step 1.

Use Eq. 27-8 to determine the electron's speed.

$$p = \frac{h}{\lambda} \quad \text{but} \quad p = mv$$

$$(9.1 \times 10^{-31}\ \text{kg})v = 1.47 \times 10^{-24}\ \text{kg}\cdot\text{m/s}$$

$$v = \frac{1.47 \times 10^{-24}\ \text{kg}\cdot\text{m/s}}{9.1 \times 10^{-31}\ \text{kg}}$$

$$v = 1.62 \times 10^{6}\ \text{m/s}$$

Part c. Step 1.

Determine the electron's kinetic energy in joules.

Note: The electron's speed is approximately $5.4 \times 10^{-3}c$. This speed is low, so the classical formula for kinetic energy can be used:

$$\text{KE} = \tfrac{1}{2}mv^2 = \tfrac{1}{2}(9.1 \times 10^{-31}\ \text{kg})(1.62 \times 10^{6}\ \text{m/s})^2$$

$$\text{KE} = 1.2 \times 10^{-18}\ \text{J}$$

Part c. Step 2.

Determine the potential difference to accelerate the electron to a speed of 1.62×10^6 m/s.

$W = qV$

$\Delta\text{KE} = qV$

$1.2 \times 10^{-18}\ \text{J} = (1.6 \times 10^{-19}\ \text{C})V$

$V = 7.5\ \text{J/C} = 7.5\ \text{V}$

TEXTBOOK PROBLEM 56. (Section 27-12) What wavelength photon would be required to ionize a hydrogen atom in the ground state and give the ejected electron a kinetic energy of 11.5 eV?

Part a. Step 1.

Determine the minimum energy required to ionize an electron in the ground state.

The energy of any particular level in the hydrogen atom is given by the formula

$$E = \frac{(-13.6\ \text{eV})(Z^2)}{n^2}$$

The difference in the energy between the ground state $(n = 1)$ and ionization $(n = \infty)$ for hydrogen $(Z = 1)$ can be found as follows:

$$\Delta E = E_2 - E_3 = \frac{(-13.6\ \text{eV})(1^2)}{1^2} - \frac{(-13.6\ \text{eV})(1^2)}{\infty^2}$$

$$= (-13.6\ \text{eV}) - (-0)$$

$$\Delta E = -13.6\ \text{eV}$$

The photon must provide 13.6 eV of energy to ionize the electron.

Part a. Step 2.

Determine the total energy of the photon needed to provide the electron with 11.5 eV of KE after ionization.

The photon must provide 13.6 eV to ionize the electron plus 11.5 eV to provide the electron with kinetic energy after ionization:

$$E_{\text{photon}} = 13.6\ \text{eV} + 11.5\ \text{eV} = 25.1\ \text{eV}$$

Part a. Step 3.

Determine the wavelength of the photon in meters and nanometers.

$$E_{\text{photon}} = hf = \frac{hc}{\lambda}$$

$$(25.1 \text{ eV})(1.6 \times 10^{-19} \text{ J/eV}) = \frac{(6.63 \times 10^{-34} \text{ J}\cdot\text{s})(3.0 \times 10^{8} \text{ m/s})}{\lambda}$$

$$4.02 \times 10^{-18} \text{ J} = \frac{(1.99 \times 10^{-25} \text{ J}\cdot\text{m})}{\lambda}$$

$$\lambda = \frac{(1.99 \times 10^{-25} \text{ J}\cdot\text{m})}{(4.02 \times 10^{-18} \text{ J})} = 4.96 \times 10^{-8} \text{ m}$$

TEXTBOOK PROBLEM 77. (Section 27-8) In some of Rutherford's experiments (Fig. 27-19) the α particles (mass $= 6.64 \times 10^{-27}$ kg) had a kinetic energy of 4.8 MeV. How close could they get to the surface of a gold nucleus (radius $\approx 7.0 \times 10^{-15}$ m, charge $= +79e$)? Ignore the recoil motion of the nucleus.

Part a. Step 1.

Complete a data table based on information both given and implied.

Note: $q_{\alpha} = +2e \qquad q_{\text{gold}} = +79e$

$e = 1.6 \times 10^{-19}$ C $\quad m_{\alpha} = 6.64 \times 10^{-27}$ kg

$$q_{\alpha}q_{\text{gold}} = (+2 \times 1.6 \times 10^{-19} \text{ C})(+79 \times 1.6 \times 10^{-19} \text{ C}) = 4.04 \times 10^{-36} \text{ (}$$

Part a. Step 2.

Determine the kinetic energy of the α particles in joules.

$$(4.8 \text{ MeV})\left(\frac{1.6 \times 10^{-13} \text{ J}}{1 \text{ MeV}}\right) = 7.7 \times 10^{-13} \text{ J}$$

Part a. Step 3.

Use the law of conservation of energy to determine the total distance r from the α particles to the gold nucleus.

The collision is head on; therefore, all of the kinetic energy of the α particles will be converted into the form of electrical potential energy at a total distance r:

$$\mathrm{KE} = \mathrm{PE}$$

$$7.7 \times 10^{-13}\ \mathrm{J} = \frac{k q_\alpha q_{\text{gold}}}{r}$$

$$7.7 \times 10^{-13}\ \mathrm{J} = \left(\frac{9 \times 10^9\ \mathrm{N \cdot m^2}}{\mathrm{C^2}}\right)\left(\frac{4.04 \times 10^{-36}\ \mathrm{C^2}}{r}\right)$$

$$r = 4.72 \times 10^{-14}\ \mathrm{m}$$

Part a. Step 4.

Subtract the radius of the gold nucleus to get the distance of closest approach to the surface.

Total distance r − radius of gold nucleus = distance to surface

$$(4.72 \times 10^{-14}\ \mathrm{m}) - (7.0 \times 10^{-15}\ \mathrm{m}) = 4.0 \times 10^{-14}\ \mathrm{m}$$

28

QUANTUM MECHANICS OF ATOMS

OBJECTIVES

After studying the material of this chapter, the student should be able to:

- distinguish between the Bohr model and the quantum-mechanical model of the atom.
- state two forms of the Heisenberg uncertainty principle and explain how the principle predicts an inherent unpredictability in nature. Use the uncertainty principle to compute the minimum uncertainty of a molecule's momentum or position.
- name the four quantum numbers required to describe the state of an electron in an atom. State the symbol used to represent each quantum number.
- given a value for the principle quantum number and list the range of values for the other quantum numbers.
- state the Pauli exclusion principle. Use this principle to determine the maximum number of electrons that fill the energy levels of atoms where $n = 1$ or $n = 2$.
- given the atomic number of a particular element, write the electronic configuration for the ground state of the atom.
- use the Periodic Table to identify an element whose outer electronic configuration is given.
- describe two ways X-ray photons can be produced.
- determine the cutoff frequency and wavelength of an X-ray photon produced by accelerating electrons through a known potential difference.
- determine the wavelength of a K_α X-ray of known energy.

KEY WORDS AND PHRASES

quantum mechanics or wave mechanics unified the wave–particle duality into a single consistent theory.

Heisenberg uncertainty principle is an important result of quantum mechanics. This principle results from the wave–particle duality and an intrinsic limit in our ability to make accurate measurements. One form of the uncertainty principle states that it is impossible to know simultaneously both the precise position and momentum of a particle.

quantum numbers The state of an electron in the hydrogen atom is governed by four quantum numbers. The quantum numbers are n, ℓ, m_ℓ, and m_s: n is the principle quantum number, where $n = 1, 2, 3, 4,$ etc.; ℓ is the orbital quantum number; m_ℓ is the magnetic quantum number; and m_s is the spin quantum number.

Pauli exclusion principle is used to explain the arrangement of electrons in multi-electron atoms. This principle states that no two electrons in an atom can occupy the same quantum state. Thus, each electron has a unique set of quantum numbers: n, ℓ, m_ℓ, and m_s.

electronic configuration of the elements listed in the **Periodic Table** of the elements can be specified using the n and ℓ quantum numbers. Electrons with the same value of ℓ are in the same subshell within the main shell designated by the letter n. The subshells are designated by the letters *s*, *p*, *d*, *f*, etc.

X-rays exhibit properties of electromagnetic waves of very short wavelength. X-rays can be produced in two ways. In one method, high-energy electrons knock an electron out of an inner energy level of certain atoms. When an electron drops from a higher level to a lower level, an X-ray photon is emitted. The second method is **bremsstrahlung**, or braking radiation. In this method, the electron is deflected as it passes near the nucleus of the atom.

SUMMARY OF MATHEMATICAL FORMULAS

Heisenberg uncertainty principle	$(\Delta x)(\Delta p) \geq \frac{h}{2\pi}$	It is impossible to know simultaneously both the precise position and momentum of a particle. The product of uncertainty of the position (Δx) and the uncertainty of the momentum (Δp) must be greater than or equal to Plank's constant divided by 2π.
	$(\Delta E)(\Delta t) \geq \frac{h}{2\pi}$	Another form of the uncertainty principle states that the product of the uncertainty of energy (ΔE) and the uncertainty in time (Δt) must be greater than or equal to Planck's constant divided by 2π.
principle quantum numbers	$n = 1, 2, 3,$ etc.	n is called the principle quantum number, where $n = 1, 2, 3, 4,$ etc.
	$\ell \leq n-1$	ℓ is the orbital quantum number. ℓ can take on integer values up to $n-1$. For example, if $n = 3$, then ℓ can have the following values: 0, 1, 2.
	$-\ell \leq m_\ell \leq +\ell$	m_ℓ is the magnetic quantum number. It is related to the direction of the electron's angular momentum. m_ℓ is an integer and can have values from $-\ell$ to $+\ell$.
	$m_s = +\frac{1}{2}$ or $-\frac{1}{2}$	m_s is the spin quantum number. It is related to the spin angular momentum

electronic configuration of the elements	The order of filling is as follows: $1s^2$, $2s^2$, $2p^6$, $3s^2$, $3p^6$, $4s^2$, $3d^{10}$, $4p^6$, $5s^2$, $4d^{10}$, $5p^6$, $6s^2$, $4f^{14}$, $5d^{10}$, etc.	The s orbital holds up to 2 electrons, the p orbital up to 6 electrons, the d orbital up to 10 electrons, and the f orbital up to 14 electrons.
cutoff wavelength of X-ray photons	$\lambda_0 = \frac{hc}{eV}$	The cutoff wavelength (λ_0) is the shortest wavelength X-ray produced, e is the charge on the electron, and V is the accelerating voltage.

CONCEPTS AND EQUATIONS

Quantum Mechanics

About 1925, Erwin Schrödinger and Werner Heisenberg produced a new theory, called wave mechanics or **quantum mechanics**, which unified the wave–particle duality into a single consistent theory.

Applied to the atom, quantum mechanics pictures the electron as spread out in space in the form of a cloud of negative charge. The shape and size of the electron cloud can be mathematically determined for a particular state of an atom. For the hydrogen atom, the shape of the electron cloud is spherically symmetric about the nucleus. The cloud model can be interpreted as an electron wave spread out in space or as a probability distribution for electrons as particles.

Quantum mechanics has been used to explain phenomena such as spectra emitted by complex atoms, the relative brightness of spectral lines, and even how atoms form molecules. Quantum mechanics reduces to classical physics in instances where classical physics applies. Newtonian mechanics is a special case of quantum mechanics. Even Bohr's postulate on the quantization of angular momentum of the electron in the hydrogen atom can be shown to be a special case of the more general quantum mechanics.

Heisenberg Uncertainty Principle

Newtonian mechanics implies that if an object's position and momentum are known at a particular moment in time, and if the forces that are or will be acting on the object are known, then its future position can be predicted. This idea is referred to as determinism.

However, an important result of quantum mechanics is the **uncertainty principle**. This principle results from the wave–particle duality and an intrinsic limit in our ability to make accurate measurements. One form of the uncertainty principle states that it is impossible to know simultaneously both the precise position and precise momentum of a particle. Expressed mathematically, the product of uncertainty of the position (Δx) and the uncertainty of the momentum (Δp) must be greater than or equal to Planck's constant divided by 2π:

$$(\Delta x)(\Delta p) \geq \frac{h}{2\pi}$$

Another form of the uncertainty principle states that the product of the uncertainty of energy (ΔE) and the uncertainty in time (Δt) must be greater than or equal to Plank's constant divided by 2π:

$$(\Delta E)(\Delta t) \geq \frac{h}{2\pi}$$

Thus, unlike Newtonian mechanics, quantum mechanics states that only approximate predictions are possible and that there is an inherent unpredictability in nature.

TEXTBOOK QUESTION 6. (Section 28-3) A cold thermometer is placed in a hot bowl of soup. Will the temperature reading of the thermometer be the same as the temperature of the hot soup before the measurement was made? Explain.

ANSWER: No, the temperature of the soup will decrease slightly when the thermometer is placed in the soup. As discussed in Section 14-4 of the textbook, the law of conservation of energy states that in a closed system the heat gained by one object (that is, the thermometer) equals the heat lost by the second object (that is, the soup). The heat absorbed by the thermometer will cause a slight decrease in the temperature of the hot soup. Therefore, an uncertainty in the true temperature of the soup is introduced when the thermometer is used.

EXAMPLE PROBLEM 1. (Section 28-3) An air molecule travels at 1350 miles per hour (603 m/s) at 0°C. Suppose the uncertainty in an experimental measurement of its speed is ±3.0%. Compute the minimum uncertainty in its position.

Part a. Step 1.

Determine the uncertainty in the molecule's speed in m/s.	The uncertainty in its speed is 3.0%; thus, $v = 3.0\% \times 603 \text{ m/s} = \pm 18.1 \text{ m/s}$.

Part a. Step 2.

Determine the mass of a nitrogen molecule.	Air is mainly nitrogen, and the molecular weight of diatomic nitrogen is 28 grams/mole: $m = \left(\frac{28\text{g}}{\text{mole}}\right)\left(\frac{1 \text{ mole}}{6.02 \times 10^{23} \text{molecules}}\right)$ $m = 4.7 \times 10^{-23} \text{ g} = 4.7 \times 10^{-26} \text{ kg}$

Part a. Step 3.

Determine the uncertainty in the molecule's momentum.	$\Delta p = m\,\Delta v = (4.7 \times 10^{-26} \text{ kg})(\pm 18.1 \text{ m/s})$ $\Delta p = \pm 8.5 \times 10^{-25} \text{ kg} \cdot \text{m/s}$

Part a. Step 4.

Determine the minimum uncertainty in the molecule's position.

Using the Heisenberg uncertainty principle:

$$(\Delta x)(\Delta p) \geq \frac{h}{2\pi}$$

$$(\Delta x)(\pm 8.5\times 10^{-25}\ \text{kg}\cdot\text{m/s}) \geq \frac{6.63\times 10^{-34}\ \text{J}\cdot\text{s}}{2\pi}$$

$$\Delta x \geq \pm 1.2\times 10^{-10}\ \text{m}$$

Quantum Mechanics of the Hydrogen Atom; Quantum Numbers

The state of an electron in the hydrogen atom is governed by four quantum numbers: $n,\ \ell,\ m_\ell,$ and m_s. The energy of a particular level in the hydrogen atom is related to n by the equation

$$E_n = \frac{-13.6\,\text{eV}}{n^2}$$

n is called the **principle quantum number** and $n = 1, 2, 3, 4$, etc.

Quantum number ℓ is the **orbital quantum number**. The angular momentum (L) is related to the orbital quantum number by the formula $L = \ell\sqrt{\ell+1}\left(\frac{h}{2\pi}\right)$. ℓ can take on integer values up to $n-1$. For example, if $n = 3$, then ℓ can have the following values: 0, 1, 2.

Quantum number m_ℓ is the **magnetic quantum number.** It is related to the direction of the electron's angular momentum. m_ℓ is an integer and can have values ranging from $-\ell$ to $+\ell$. The component of the angular momentum in an assigned direction, usually along the z axis, is given by $L_z = m_\ell\left(\frac{h}{2\pi}\right)$.

Quantum number m_s is the **spin quantum number**. This quantum number has values of only $+\frac{1}{2}$ or $-\frac{1}{2}$. The spin angular momentum in an assigned direction equals $m_s\left(\frac{h}{2\pi}\right)$. The word "spin" was originally given to this quantum number because it was thought to be associated with the electron spinning on its axis as it revolves around the nucleus. However, this is an oversimplification because the electron exhibits properties of waves as well as particles.

Pauli Exclusion Principle

In order to explain the arrangement of electrons in multi-electron atoms, the **Pauli exclusion principle** is used. This principle states that no two electrons in an atom can occupy the same quantum state. Thus, each electron has a unique set of quantum numbers: $n,\ \ell,\ m_\ell,$ and m_s. For example, helium has two electrons. Both electrons have $n = 1$, $\ell = 0$, and $m_\ell = 0$. However, one of the electrons has $m_s = +\frac{1}{2}$ and the other has $m_s = -\frac{1}{2}$. Thus each electron has a different set of quantum numbers.

Electronic Configuration of the Elements

The **electronic configuration** of the elements listed in the Periodic Table of the elements can be specified using the n and ℓ quantum numbers. Electrons with the same value of n are in the same shell. The shells are given letter symbols as shown in the following table. If $n = 1$, then the electrons are in the K shell.

Value of n	Symbol of shell	Value of ℓ	Symbol of subshell	Maximum number of electrons
1	K	0	s	2
2	L	1	p	6
3	M	2	d	10
4	N	3	f	14
.	.	4	g	18
.	.	5	h	32
.	.	.	.	.

As shown in the table, electrons with the same value of ℓ are in the same subshell within the main shell designated by the letter n. The subshells are designated by the letters s, p, d, f, etc.

The number of electrons in the subshell can be found by using the formula $2(2\ell+1)$. Thus, if $\ell = 3$, then $2[2(3)+1] = 14$, and the f subshell can hold 14 electrons.

The designation of an electron involves both n and ℓ plus a superscript that designates the number of electrons in the subshell. For example, if $n = 2$ and $\ell = 1$ and there are three electrons in the orbital, then the designation would be $2p^3$.

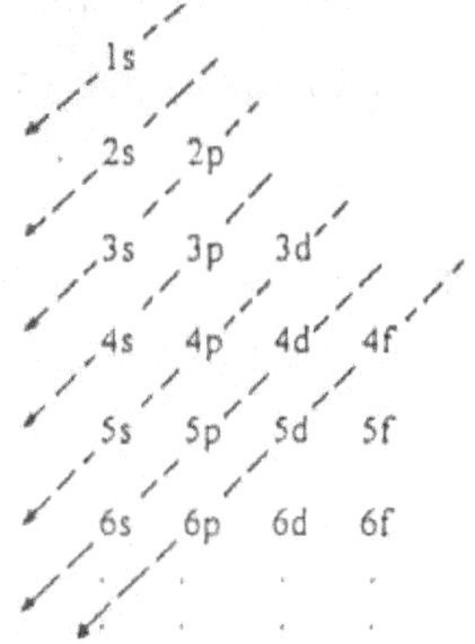

The order of filling is as follows: $1s^2$, $2s^2$, $2p^6$, $3s^2$, $3p^6$, $4s^2$, $3d^{10}$, $4p^6$, $5s^2$, $4d^{10}$, $5p^6$, $6s^2$, $4f^{14}$, $5d^{10}$, etc. A simplified way of determining the order of filling is shown in the diagram.

Write down the principle energy levels and their subshells and follow the diagonal lines. The diagonal lines follow the order of filling.

In filling the subshells, the lower energy subshells are filled first, as are the lower principle energy levels.

Each box in the Periodic Table contains the symbol of the element, its atomic number, its atomic mass, and the ground-state electronic configuration of the outermost electrons. For example, the symbol for calcium is Ca, 20 is the atomic number, 40.08 is the atomic mass, and $4s^2$ is the electronic configuration of the outermost electrons.

Ca	20
40.08	
$4s^2$	

TEXTBOOK QUESTION 15. (Section 28-8) Which of the following electron configurations are not allowed: (*a*) $1s^2 2s^2 2p^4 3s^2 4p^2$; (*b*) $1s^2 2s^2 2p^8 3s^1$; (*c*) $1s^2 2s^2 2p^6 3s^2 3p^5 4s^2 4d^5 4f^1$? If not allowed, explain why.

ANSWER: (*a*) This configuration could represent an excited state of a multi-electron atom. (*b*) The $2p$ orbital contains a maximum of six electrons. Therefore, this configuration is forbidden. (*c*) This configuration could represent an excited state of a multi-electron atom.

TEXTBOOK QUESTION 18. (Section 28-8) Explain why potassium and sodium exhibit similar properties.

ANSWER: From the Periodic Table, it can be seen that the outer electron configuration of both elements consists of an s^1 electron. Potassium is $4s^1$ while sodium is $3s^1$. The physical and chemical properties of atoms are primarily determined by the configuration of their outer electrons, also known as the valence electrons. Therefore, both elements exhibit similar properties and are said to belong to the same chemical family, in this case the alkali metal family.

TEXTBOOK QUESTION 20 (Section 28-8) The ionization energy for neon $(Z = 10)$ is 21.6 eV, and that for sodium $(Z = 11)$ is 5.1 eV. Explain the large difference.

ANSWER: Using the Periodic Table, it can be seen that neon is a noble gas (group VIII). The shells and subshells of elements in group VIII are completely filled and are not readily removed from the atom. As a result, they have high ionization energies.

The Periodic Table also shows that sodium is an alkali metal (group I). Sodium has a single s^1 electron in its outer shell. This electron is weakly held by the nucleus and is readily removed. As a result, sodium has a rather low ionization energy.

EXAMPLE PROBLEM 2. (Sections 28-6 to 28-8) List the possible quantum states for $_{10}\text{Ne}$.

Part a. Step 1.

Use the Pauli exclusion principle to list the possible quantum states.

The Pauli exclusion principle states that no two electrons in an atom can occupy the same quantum state. Each atom must have a unique set of quantum numbers n, ℓ, m_ℓ, and m_s.

Based on the exclusion principle, it is possible to construct a table of possible quantum states for $_{10}\text{Ne}$:

Sub-shell	n	ℓ	m_ℓ	m_s	Sub-shell	n	ℓ	m_ℓ	m_s
1s	1	0	0	$+\frac{1}{2}$	2p	2	1	–1	$-\frac{1}{2}$
1s	1	0	0	$-\frac{1}{2}$	2p	2	1	0	$+\frac{1}{2}$
2s	2	0	0	$+\frac{1}{2}$	2p	2	1	0	$-\frac{1}{2}$
2s	2	0	0	$-\frac{1}{2}$	2p	2	1	1	$+\frac{1}{2}$
2p	2	1	–1	$+\frac{1}{2}$	2p	2	1	1	$-\frac{1}{2}$

EXAMPLE PROBLEM 3. (Sections 28-6 to 28-8) Determine the values of m_ℓ that are allowed for the (*a*) 1*s* subshell and (*b*) 3*d* subshell.

Part a. Step 1.

Determine the values for the 1*s* subshell.	For the 1*s* subshell, $n = 1$, and since $\ell \le n-1$, $\ell = 0$. m_ℓ can have values from $-\ell$ to $+\ell$ but because $\ell = 0$, $m_\ell = 0$.

Part b. Step 1.

Determine the values for the 3*d* subshell.	For the 3*d* subshell, $n = 3$; therefore, $\ell = 1$ or 2, and since m_ℓ can have values from $-\ell$ to $+\ell$, m_ℓ can have the values $-2, -1, 0, +1, +2$.

EXAMPLE PROBLEM 4. (Section 28-6) What is the range of values of the angular momentum of an electron in the $n = 4$ state of the hydrogen atom?

Part a. Step 1.

Determine the possible values of the orbital quantum number. Then determine the range of values of the angular momentum.

$n = 4$ and ℓ can take on integer values from 0 to $n-1$; then ℓ can be 0, 1, 2, 3. The range of values for the angular momentum (L) can be determined by substituting the possible values of ℓ into the following equation:

$$L = \sqrt{\ell(\ell+1)}\frac{h}{2\pi}$$

$$\ell = 0: \quad L = \sqrt{0(0+1)}\frac{h}{2\pi} = 0$$

$$\ell = 1: \quad L = \sqrt{1(1+1)}\frac{6.63 \times 10^{-34}\ \text{J}\cdot\text{s}}{2\pi} = 1.49 \times 10^{-34}\ \text{J}\cdot\text{s}$$

$$\ell = 2: \quad L = \sqrt{2(2+1)}\frac{6.63 \times 10^{-34}\ \text{J}\cdot\text{s}}{2\pi} = 2.58 \times 10^{-34}\ \text{J}\cdot\text{s}$$

$$\ell = 3: \quad L = \sqrt{3(3+1)}\frac{6.63 \times 10^{-34}\ \text{J}\cdot\text{s}}{2\pi} = 3.66 \times 10^{-34}\ \text{J}\cdot\text{s}$$

EXAMPLE PROBLEM 5. (Section 28-8) Write the electronic configuration for each of the following elements: ${}_{8}O$, ${}_{18}Ar$, ${}_{32}Ge$, and ${}_{40}Zr$.

Part a. Step 1.

Write the electronic configuration for each element.	The order of filling of the subshells is $1s^2, 2s^2, 2p^6, 3s^2, 3p^6, 4s^2, 3d^{10}$, etc.: Oxygen (O) has 8 electrons, and its electronic configuration is $1s^2 2s^2 2p^4$ Argon (Ar) has 18 electrons, and its electronic configuration is $1s^2 2s^2 2p^6 3s^2 3p^6$ Germanium (Ge) has 32 electrons, and its electronic configuration is $1s^2 2s^2 2p^6 3s^2 3p^6 4s^2 3d^{10} 4p^2$ Zirconium (Zr) has 40 electrons, and its electronic configuration is $1s^2 2s^2 2p^6 3s^2 3p^6 4s^2 4p^6 3d^{10} 5s^2 4d^2$

EXAMPLE PROBLEM 6. (Section 28-8) Write the symbols for the elements whose outer electron configurations are as follows: $3d^{10}4s^1$ and $5p^2$.

Part a. Step 1.

Use the Periodic Table to determine the symbol for each element.	The Periodic Table of the elements specifies the configuration of the outermost electrons and any other unfilled subshells. Using the Periodic Table as a guide, it can be seen that copper (Cu) has an outer configuration of $3d^{10}4s^1$ and tin (Sn) has an outer configuration of $5p^2$.

X-Rays and X-Ray Production

X-rays were discovered in 1895 by Wilhelm Roentgen. The nature of X-rays was not determined until 1913, when it was shown that X-rays exhibit properties of electromagnetic waves of very short wavelength.

X-rays can be produced in two ways. In one method, high-energy electrons knock an electron out of an inner energy level of certain atoms. When an electron drops from a higher level to a lower level, an X-ray photon is emitted. The second method is a continuous spectrum called **bremsstrahlung**, or braking radiation. In this method, the electron is deflected as it passes near the nucleus of the atom. During the resulting deceleration, energy in the form of an X-ray is produced.

An X-ray tube produces a spectrum of wavelengths. The shortest wavelength X-ray is the result of the electron losing all of its kinetic energy during the collision. In this case,

$$\text{energy lost by electron} = \text{energy gained by X-ray photon, or } eV = hf_0$$

and because $f_0 = \frac{c}{\lambda_0}$,

$$\lambda_0 = \frac{hc}{eV}$$

Here f_0 is the cutoff frequency and is the highest frequency X-ray produced, λ_0 is the cutoff wavelength and is the shortest wavelength X-ray produced, e is the charge on the electron, and V is the accelerating voltage.

EXAMPLE PROBLEM 7. (Section 28-9) The wavelength of the strongest X-ray spectral line from an X-ray tube with a copper target is 0.200 nm. Determine the minimum accelerating voltage required to give an electron enough energy to produce this X-ray.

Part a. Step 1.

Determine the minimum accelerating voltage necessary for an electron to produce an X-ray photon of this wavelength.

Assume that all of the electron's energy is converted into the energy of the X-ray photon upon collision with the copper target:

kinetic energy of the electron = energy of the X-ray photon

$$qV = \frac{hc}{\lambda} = \frac{1240\,\text{eV}\cdot\text{nm}}{\lambda}$$

$$(1\text{ electron})V = \frac{1240\text{ eV}\cdot\text{nm}}{0.200\text{ nm}}$$

$$V = 6200\ V$$

X-rays and Atomic Number

One method of producing X-rays is for high-energy electrons to knock an electron out of an inner shell of certain atoms. If the missing electron was in the K shell $(n = 1)$ and is replaced by an electron that falls from the L shell $(n = 2)$, then the X-ray is referred to as an K_α X-ray.

In 1914, Henry Moseley determined that a graph of $\sqrt{\frac{1}{\lambda}}$ vs. Z produces a straight line, where λ is the wavelength of the K_α line. Because of this, the atomic number of a number of elements could be determined, and his research led to the arrangement of the Periodic Table on the basis of atomic number.

PROBLEM SOLVING SKILLS

For problems involving the uncertainty principle:

1. Determine the object's mass.
2. If necessary, determine the uncertainty in the object's speed.
3. Determine the uncertainty in the molecule's momentum.
4. Use the uncertainty principle to determine the uncertainty in the object's position.
5. If the uncertainty in position is given, then use the uncertainty principle to determine the uncertainty in the object's momentum and velocity.
6. If the problem involves uncertainty in energy, then note either the uncertainty in the energy or the time and solve the problem using the equation $(\Delta E)(\Delta t) \leq \frac{h}{2\pi}$.

For problems involving the possible quantum states of an atom:

1. Note the value of the principle quantum number (n) for the atom. For example, if an atom is in the third period of the Periodic Table, then the value of n is 3.
2. Determine the possible values of ℓ. Remember that ℓ can have positive values up to $n-1$.
3. Determine the possible values of m_ℓ. Remember that m_ℓ can have values of $-\ell$ to $+\ell$.
4. m_s can have values of $+\frac{1}{2}$ or $-\frac{1}{2}$.
5. Use the Pauli exclusion principle to construct a table for the possible quantum states.

For problems involving the range of values of the angular momentum of an electron:

1. Note the principle quantum number (n) of the particular orbital in which the electron is located.
2. Determine the range of values of the orbital quantum number (ℓ).
3. Use the equation $L = \sqrt{\ell(\ell+1)}\left(\frac{h}{2\pi}\right)$ to determine the possible values of the angular momentum (L).

For problems involving the electronic configuration of an element:

1. Write down the order of filling of the *s*, *p*, *d*, and *f* subshells and the maximum number of electrons in each subshell.
2. Take note of the atomic number of the element.
3. Write down the electronic configuration until the number of electrons in the atom is reached.
4. If the outer electron configuration is given, then the element can be identified by using the Periodic Table.

For problems involving X-ray spectra and atomic number:

1. If the accelerating voltage is given, then determine the electron's kinetic energy in electron volts ($W = \Delta\text{KE} = q\,\Delta V$).
2. The highest energy and shortest wavelength X-ray photon is produced if all of the electron's kinetic energy is used in producing the photon. Use $q\,\Delta V = \frac{hc}{\lambda}$ to determine the wavelength of the X-ray.

3. If the wavelength of the shortest X-rays is given, then the equation $q\,\Delta V = \frac{hc}{\lambda}$ can be used to determine the voltage needed to give an electron the energy to produce the X-ray.

SOLUTIONS TO SELECTED TEXTBOOK PROBLEMS

TEXTBOOK PROBLEM 5. (Section 28-3) An electron remains in an excited state of an atom for typically 10^{-8} s. What is the minimum uncertainty in the energy of the state (in eV)?

Part a. Step 1.

Use Eq. 28-2 to determine the minimum uncertainty in the energy.

$$(\Delta E)(\Delta t) \geq \frac{h}{2\pi}$$

$$(\Delta E)(10^{-8}\text{ s}) \geq \frac{6.626 \times 10^{-34}\text{ J}\cdot\text{s}}{2\pi}$$

$$\Delta E \geq \frac{1.055 \times 10^{-34}\text{ J}\cdot\text{s}}{10^{-8}\text{ s}}$$

$$\Delta E = (1.1 \times 10^{-26}\text{ J})\left(\frac{1\text{ eV}}{1.6 \times 10^{-19}\text{ J}}\right)$$

$$\Delta E = 6.6 \times 10^{-8}\text{ eV}$$

TEXTBOOK PROBLEM 16. (Section 28-6) For $n = 6,\ \ell = 3,$ what are the possible values of m_ℓ and m_s?

Part a. Step 1.

Determine the possible values of m_ℓ.

The values of m_ℓ range from $-\ell$ to $+\ell$. Therefore, for $\ell = 3$, $m_\ell = -3, -2, -1, 0, +1, +2, +3$.

Part a. Step 2.

Determine the possible values of m_s.

The values of m_s are $-\frac{1}{2}$ and $+\frac{1}{2}$.

TEXTBOOK PROBLEM 26. (Section 28-8) What is the full electronic configuration for (*a*) silver (Ag), (*b*) gold (Au), (*c*) uranium (U)? [*Hint*: See the Periodic Table inside the back cover.]

Part a. Step 1.

Write the electronic configuration of Ag.	The atomic number (Z) of silver = 47. Therefore, the electronic configuration of silver is as follows: $1s^2 2s^2 2p^6 3s^2 3p^6 3d^{10} 4s^2 4p^6 4d^{10} 5s^1$

Part b. Step 1.

Write the electronic configuration of Au.	The atomic number (Z) of gold = 79. Therefore, the electronic configuration of gold is as follows: $1s^2 2s^2 2p^6 3s^2 3p^6 3d^{10} 4s^2 4p^6 4d^{10} 4f^{14} 5s^2 5p^6 6s^1 5d^{10}$

Part c. Step 1.

Write the electronic configuration of U.	The atomic number (Z) of uranium = 92. Therefore, the electronic configuration of uranium is as follows:

$$1s^2 2s^2 2p^6 3s^2 3p^6 4s^2 3d^{10} 4p^6 5s^2 4d^{10} 5p^6 6s^2 5d^{10} 6p^6 4f^{14} 7s^2 5f^3 6d^1$$

TEXTBOOK PROBLEM 34. (Section 28-9) If the shortest-wavelength bremsstrahlung X-rays emitted from an X-ray tube have a $\lambda = 0.035$ nm, what is the voltage across the tube?

Part a. Step 1.

Use Eq. 27-4 determine the energy of the electrons in eV.	The energy of the shortest wavelength X-rays equals the maximum kinetic energy of the electrons striking the X-ray tube. Therefore, $E = hf = \frac{hc}{\lambda}, \quad \text{where} \quad hc = 1240 \text{ eV}$ $E = \frac{1240 \text{ eV} \cdot \text{nm}}{0.035 \text{ nm}}$ $E = 3.54 \times 10^4 \text{ eV}$

Part a. Step 2.

Determine the potential difference across the X-ray tube.

$$3.54 \times 10^4 \text{ eV} = (1 \text{ electron})V$$

$$V = 3.54 \times 10^4 \ V \approx 35 \text{ kV}$$

TEXTBOOK PROBLEM 38. (Section 28-9) Estimate the wavelength for an $n = 3$ to $n = 2$ transition in iron $(Z = 26)$.

Part a. Step 1.

Use the Bohr formula Eq. 27-16 of the textbook), with Z replaced by $Z - 1$, to determine the wavelength of the X-ray.

Using the Bohr formula, we take n to be 3 and n' to be 2. Then

$$\frac{1}{\lambda} = \left(\frac{2\pi^2 e^4 mk^2}{h^3 c}\right)(Z-1)^2\left(\frac{1}{n'^2} - \frac{1}{n^2}\right)$$

$$\frac{1}{\lambda} = (1.097 \times 10^7 \text{ m}^{-1})(26-1)^2\left(\frac{1}{2^2} - \frac{1}{3^2}\right)$$

$$\frac{1}{\lambda} = 9.523 \times 10^9 \text{ m}^{-1}$$

$$\lambda = \frac{1}{9.523 \times 10^9 \text{m}^{-1}}$$

$$\lambda = 1.050 \times 10^{-9} \text{ m}$$

TEXTBOOK PROBLEM 41. (Section 28-11) A laser used to weld detached retinas puts out 25-ms-long pulses of 640-nm light which average 0.68-W output during a pulse. How much energy can be deposited per pulse and how many photons does each pulse contain? [*Hint*: See Example 27-4.]

Part a. Step 1.

Determine the energy deposited during a pulse.
Note: 25 ms = 0.025 s.

$$W = \Delta E = Pt$$

$$\Delta E = (0.68 \text{ J/s})(0.025 \text{ s})$$

$$\Delta E = 0.0170 \text{ J}$$

Part a. Step 2.

Determine the number of photons in a single pulse.

$$\Delta E = nhf = nf\frac{c}{\lambda}, \quad \text{where } n = \text{the number of photons}$$

$$0.0170\text{ J} = n(6.63\times10^{-34}\text{J s})\frac{3.0\times10^{8}\text{ m/s}}{(640\text{ nm})(1\text{m}/10^{9}\text{ nm})}$$

$$0.0170\text{ J} = n\frac{(1.99\times10^{-25}\text{J}\cdot\text{m})}{(6.40\times10^{-7}\text{ m})}$$

$$0.0170\text{ J} = n(3.11\times10^{-19}\text{ J})$$

$$n = 5.5\times10^{16}\text{ photons}$$

MOLECULES AND SOLIDS

OBJECTIVES

After studying the material of this chapter, the student should be able to:

- distinguish between a covalent bond, polar covalent bond, and an ionic bond.
- distinguish between a strong bond and a weak bond.
- explain what is meant by bond length and binding energy.
- calculate the potential energy between ions separated by a given distance and determine the magnitude of the electrostatic force acting between them.
- draw a diagram showing potential energy versus separation distance for point charges when the charges are either of the same sign or the opposite sign.
- explain what is meant by activation energy.
- draw a diagram showing potential energy versus separation distance for two atoms that come together to form a covalent bond or an ionic bond. The diagram should include a section to represent the activation energy.
- identify the point on the graph of potential energy versus separation distance where the potential energy is a minimum, and explain what is meant by bond length and binding energy.
- explain why molecular spectra appear in the form of band spectra rather than line spectra.
- determine the difference in energy between rotational and/or vibrational energy states of a diatomic molecule. Also, determine the moment of inertia and the bond length between the atoms of the molecule.
- use the band theory of solids to explain the classification of solids into conductors, insulators, and semiconductors.
- calculate the energy gap between the valence band and the conduction band for a semiconductor or insulator if the wavelength of the longest wavelength photon that causes a transition is given.
- explain why semiconductors become better electrical conductors as the temperature increases.
- use the idea of electronic configuration to explain how a doped semiconductor can become highly conducting.
- explain the principle of the junction diode and how the junction diode can be used as a half-wave rectifier.

KEY TERMS AND PHRASES

chemical **bonds** refer to the forces that hold the atoms of a molecule together.

covalent bond is the type of chemical bond in which the electrons are shared equally.

partial ionic character describes a covalent bond in which the electrons that form the bond are not shared equally. One end of the molecule is charged positively, while the other end is charged negatively; the molecule is called a **polar** molecule.

ionic bond is formed when one or more electrons are transferred from one atom to another. The bond formed is based on the electrostatic attraction of the negatively charged ion for the positively charged ion.

bond energy, or **binding energy**, refers to the energy required to break a chemical bond that holds the atoms of a molecule together. The bonds that hold atoms of a molecule together are called **strong** bonds.

weak bond, or **van der Waals bond**, usually refers to electrostatic attraction between molecules. An example of a weak bond is that between two dipoles; such a bond is often called a **dipole–dipole bond.**

activation energy refers to the energy that must be added in order to force atoms together to form a molecule.

band spectra, or molecular spectra, are exhibited by molecules. This is because molecules have additional energy levels due to the vibration of the atoms of the molecule with respect to each other and the rotational energy of the molecule.

***n*-type semiconductors** are semiconductors where electrons (negative charge) carry the current.

***p*-type** semiconductors are semiconductors where positive **holes** appear to carry the current.

***pn* junction diode** is produced when a *p*-type and an *n*-type semiconductor are joined.

forward biased refers to a diode that is connected to a battery from which the voltage is large enough to overcome the internal potential difference. The result is a current flow through the diode.

reversed biased refers to a diode that is connected to a battery from which the voltage causes the holes and electrons to be separated, which means that the negative charge tends to be separated from the positive charge. The result is a diode that is essentially nonconducting.

half-wave rectifier allows current from an ac source to flow only in one direction. A *pn* junction diode acts as a half-wave rectifier. It can be used to change an ac current to a dc current.

SUMMARY OF MATHEMATICAL FORMULAS

potential energy between two point charges	$PE = \left(\frac{1}{4\pi\varepsilon_0}\right)\frac{q_1q_2}{r}$ $\frac{1}{4\pi\varepsilon_0} = k,$ where $k = 9\times10^9\ \text{N}\cdot\text{m}^2/\text{C}^2$	The potential energy between two point charges is related to the product of the charges (q_1q_2) and is inversely proportional to the distance r between their centers. If both charges are positive or both are negative, then the PE is positive and decreases with increasing r. If the charges are of opposite sign, then PE is negative and increases with increasing r.

molecular rotational energy	$E_{rot} = \ell(\ell+1)\hbar^2/2I$ where $\ell = 0, 1, 2,$ etc.	Molecular rotational energy (E_{rot}) depends on the quantum number (ℓ), the moment of inertia (I), and $\hbar$, where $\hbar = h/2\pi$.
molecular vibrational energy	$E_{vib} = (v+\frac{1}{2})hf,$ where $v = 0, 1, 2,$ etc.	Energy levels for vibrational motion depend on the frequency of vibration (f) of the molecule and the vibrational quantum number (v).

CONCEPT SUMMARY

Chemical Bonds

The force that holds the atoms of a molecule together is referred to as a **bond**. The following is a summary of the two main types of chemical bonds: covalent and ionic.

Covalent Bonds

In the case of diatomic molecules, such as H_2, O_2, and N_2, the outermost electrons are shared equally by both atoms. The type of bond in which the electrons are shared is called a **covalent bond**.

The cloud model from quantum mechanics is useful in attempting to explain chemical bonding. A simple molecule to consider is hydrogen. When two hydrogen atoms are at a distance, the electron clouds repel, and the positively charged nuclei repel. There is no unbalanced force between the atoms.

As the atoms approach, the positively charged nucleus of one atom attracts the electron cloud of the other atom, and the shapes of the electron clouds become distorted. The nuclear charge is concentrated, and as a result the attraction of one nucleus for the electron cloud of the other atom is greater than the repulsion between the clouds.

As the electron clouds overlap, the overlapping regions cause the repulsion between the clouds to be further reduced. However, as the atoms come closer, the repulsion between the positively charged nuclei increases until the forces balance. The distance between the nuclei when the balance point is reached is called the bond length.

Covalent Bonds with Partial Ionic Character

The chemical bond formed by molecules such as H_2, O_2, and N_2 is a covalent bond. This is because the electrons that form the chemical bond are shared equally by the atoms that form the molecule. When the atoms involved are from different elements, the electrons that form the bond are not shared equally.

The water molecule contains two atoms of hydrogen and one atom of oxygen. Hydrogen has one proton in its nucleus, while oxygen has eight. The result is that oxygen's large nuclear charge tends to pull the electron cloud of the hydrogen toward it so that the region near the oxygen atom is negatively charged, while the region near each hydrogen is positively charged. Because one end of the molecule is charged positively while the other end is charged negatively, the molecule is called a **polar** molecule. This type of covalent bond is said to have a **partial ionic character**.

Ionic Bonds

An **ionic bond** is formed when one or more electrons are transferred from one atom to another. The bond formed is based on the electrostatic attraction of the negatively charged ion for the positively charged ion.

An example of a compound exhibiting ionic bonding is sodium chloride, NaCl. The nucleus of the sodium atom contains 11 protons, while that of chlorine contains 17 protons. The nuclear charge of the chlorine exerts a greater force on the outer electron of the sodium ion than does the sodium nucleus. The result is the transfer of the outer electron of the sodium ion to the chlorine atom. The sodium ion (Na^+) exerts a force of electrostatic attraction on the chlorine (Cl^-). Because the force is between two ions, the bond is called an ionic bond.

Strong Versus Weak Bonds

Energy is required in order to break a chemical bond that holds the atoms of a molecule together. The energy required to break a bond is called the **bond energy**, or **binding energy**. The binding energy for covalent and ionic bonds is usually in the range of 2 eV to 5 eV. The bonds that hold atoms of a molecule together are called **strong bonds**.

The term **weak bond**, or **van der Waals bond**, usually refers to electrostatic attraction between molecules. An example of a weak bond is between two dipoles; such a bond is often called a **dipole–dipole bond**. When one of the atoms in a dipole–dipole bond is hydrogen, the bond is usually referred to as a **hydrogen bond**. Another type of weak bond is a **dipole-induced dipole bond**. This type of bond results when a polar molecule with a permanent dipole moment induces a dipole moment in an electrically balanced, nonpolar molecule.

The strength of a weak bond is in the range of 0.04 to 0.3 eV. In a biological cell the average kinetic energy of molecules is in the same range. A weak bond can be broken during molecular collisions and therefore weak bonds are not permanent. Strong bonds are almost never broken by molecular collisions and are therefore relatively permanent. They can be broken by chemical action in a biological cell with the aid of an enzyme.

TEXTBOOK QUESTION 1. (Sections 29-1 and 29-5) What type of bond would you expect for (*a*) the N_2 molecule, (*b*) the HCl molecule, (*c*) Fe atoms in a solid?

ANSWER: (*a*) The electrons that form the N_2 bond are equally shared by the N atoms. The molecule does not have a permanent dipole moment, and the bond is a covalent bond.

(*b*) HCl consists of one hydrogen atom and one chlorine atom. As can be seen in the Periodic Table, the electronic configuration of chlorine is $1s^2 2s^2 2p^6 3s^2 3p^5$, while the electronic configuration of hydrogen is $1s^1$. The chlorine atom pulls the $1s^1$ of hydrogen toward it, with the result that the region near the chlorine atom is negatively charged while the region near each hydrogen atom is positively charged. Overall, the molecule is electrically neutral. However, one end of the molecule is charged positively while the other end is charged negatively, and the bond is an ionic bond.

(*c*) Fe is a metal, and the nuclei have a weak attraction for the outer electron. As a result, the outer electrons of the Fe atoms tend to move freely through the solid and are shared by all of the atoms in the metal. This type of bond is referred to as a metallic bond.

Potential Energy Diagrams for Molecules

As discussed in Chapter 17, the potential energy between two point charges separated by a distance *r* is given by

$$\text{PE} = \left(\frac{1}{4\pi\varepsilon_0}\right)\frac{q_1 q_2}{r}, \quad \text{where } \frac{1}{4\pi\varepsilon_0} = k \text{ and } k = 9\times 10^9 \text{ N}\cdot\text{m}^2/\text{C}^2$$

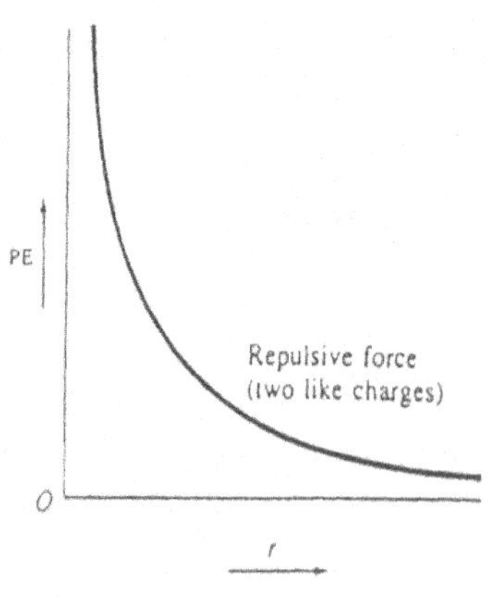

Figure A

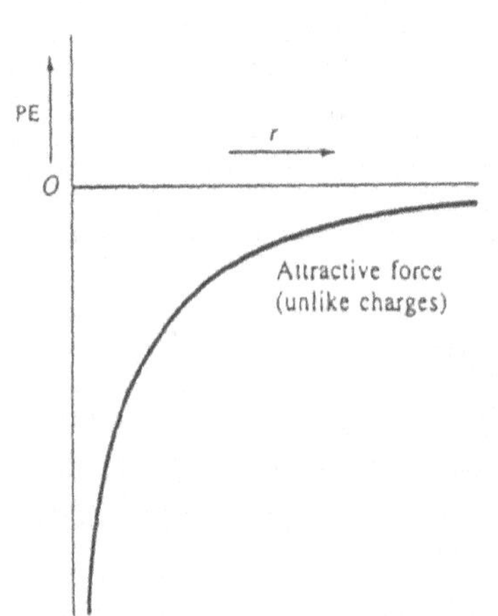

Figure B

If the charges are both positive or both negative, then the PE is positive and decreases with increasing r, as shown in Figure A. If the charges are of opposite sign, then PE is negative and increases with increasing r, as shown in Figure B.

EXAMPLE PROBLEM 1. (Section 29-2) (*a*) Determine the electrostatic potential energy of the Rb^+ and Cl^- ions in an RbCl molecule that are at a stable separation of 0.329 nm. Assume each ion carries a charge of +1.0*e*. Express your answer in both joules and electron volts. (*b*) What is the magnitude of the electrostatic attractive force acting between the ions?

Part a. Step 1.

Calculate the magnitude of the electrostatic potential energy in joules.

$$\text{PE} = \left(\frac{1}{4\pi\varepsilon_0}\right)\frac{q_1 q_2}{r}$$

but

$$q_1 q_2 = (1.6\times10^{-19}\text{ C})(-1.6\times10^{-19}\text{ C}) = -2.6\times10^{-38}\text{ C}^2$$

$$\text{PE} = (9\times10^{9}\text{ N}\cdot\text{m}^2/\text{kg}^2)\left(\frac{2.6\times10^{-38}\text{ C}^2}{0.329\times10^{-9}\text{ m}}\right)$$

$$\text{PE} = -7.0\times10^{-19}\text{ J}$$

Part a. Step 2.

Determine the energy in eV.

$$\text{PE} = (-7.0\times10^{-19}\text{ J})\left(\frac{1.0\text{ eV}}{1.6\times10^{-19}\text{J}}\right)$$

$$\text{PE} = -4.37\text{ eV}$$

Part b. Step 1.

Determine the magnitude of the electrostatic force attraction between the two ions.

$$F = \left(\frac{1}{4\pi\varepsilon_0}\right)\frac{q_1 q_2}{r^2}$$

$$= (9\times10^9\ \text{N}\cdot\text{m}^2/\text{kg}^2)\frac{(-2.6\times10^{-38}\ \text{C}^2)}{(0.329\times10^{-9}\ \text{m})^2}$$

$$F = -2.12\times10^{-9}\ \text{N}$$

Binding Energy

The potential energy function of a covalent bond, such as H_2 or NaCl, is shown in the accompanying Figure C. As discussed previously, as the two atoms approach, they tend to attract and share their valence electrons. The value of the PE decreases to a minimum value at a certain optimum distance between their nuclei (r_0). This distance is known as the bond length. However, if the distance between the nuclei becomes less than r_0, then the nuclei repel and the PE increases. In Figure C, r_0 is the approximate point of greatest stability for the molecule and the approximate point of lowest energy. The energy at this point is called the **binding energy**. The binding energy is the amount of energy required to separate the two atoms to infinity; at infinity, PE = 0.

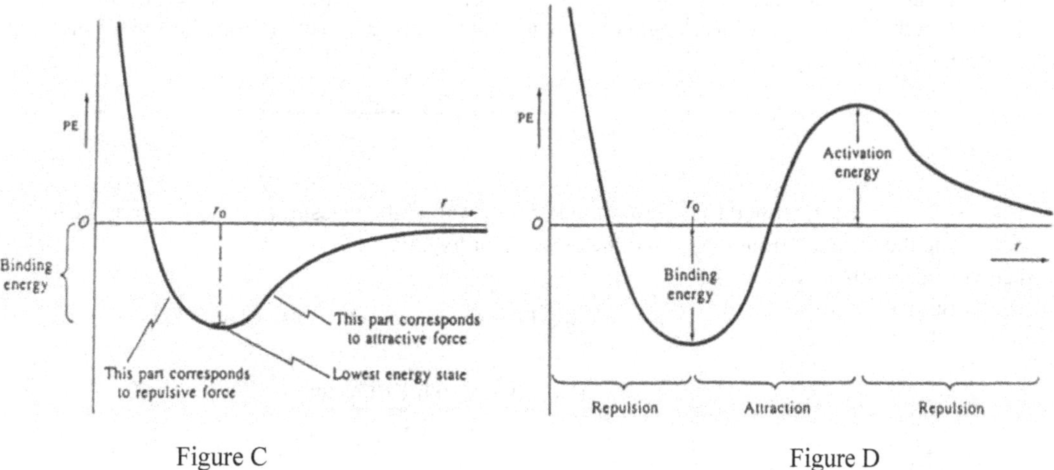

Figure C

Figure D

Activation Energy

For many molecules, the force between the atoms as they approach is repulsive. In order for the atoms to form a molecule, additional energy called the **activation energy** must be added to force them together. Figure D is a graph of the PE curve for a situation where the activation energy must be considered.

Molecular Spectra

As stated in Section 29-4 of the textbook, "When atoms combine to form molecules, the energy levels of the outer electrons are altered because they now interact with each other. Additional energy levels also become possible because the atoms can vibrate with respect to each other, and the atom as a whole can rotate. The energy levels for both vibrational and rotational motion are quantized, and are very close together (typically 10^{-1} to 10^{-3} eV apart). Each atomic energy level thus becomes a set of closely spaced levels corresponding to the vibrational and rotational motions. Transitions from one level to another appear as many very closely spaced lines." The simple line spectra associated with the Bohr model of the hydrogen atom are not observed in molecules. Instead, molecules exhibit **band spectra**.

The quantized rotational energy levels are given by

$$E_{\text{rot}} = \ell(\ell+1)\hbar^2/2I$$

where $\ell = 0, 1, 2$, etc., I is the moment of inertia, and $\hbar = h/2\pi$. Transitions from one rotational energy level to another are subject to the selection rule $\Delta\ell = \pm 1$.

The energy levels for vibrational motion are given by

$$E_{\text{vib}} = (v + \tfrac{1}{2})hf$$

where v is the vibrational quantum number, $v = 0, 1, 2$, etc., and f is the classical frequency of vibration of the molecule. Transitions from one vibrational energy level to another are subject to the selection rule $\Delta v = \pm 1$.

EXAMPLE PROBLEM 2. (Section 29-4) The bond length between the carbon and oxygen atoms in a carbon monoxide molecule is 0.113 nm. The mass of the carbon atom is 2.00×10^{-26} kg and the mass of the oxygen atom is 2.67×10^{-26} kg. Determine the (*a*) location of the center of mass as measured from the carbon atom, (*b*) moment of inertia of the molecule about its center of mass, and (*c*) wavelength of the longest wavelength photon that could cause a transition from the $\ell = 0$ rotational energy state to the $\ell = 1$ state.

Part a. Step 1.

Use Eq. 7-9a of the textbook to determine the distance from the center of the carbon atom to the center of mass of the molecule.

From the Periodic Table, the mass of a carbon atom is 12 u and the mass of an oxygen atom is 16 u.

Note: 1 u = 1 atomic mass unit $= 1.66 \times 10^{-27}$ kg.

$$x_{\text{CM}} = \frac{(12\text{ u})(0) + (16\text{ u})(0.113\text{ nm})}{12\text{ u} + 16\text{ u}}$$

$$x_{\text{CM}} = 0.0646\text{ nm}$$

Part b. Step 1.

Use Eq. 8-13 of the textbook to determine the moment of inertia of the CO molecule.

As determined in Part a, the center of mass of the molecule is located 0.0646 nm from the carbon atom. The distance from the center of mass to the oxygen atom =
0.113 nm − 0.0646 nm = 0.0484 nm.

Note: $1\text{ nm} = 10^{-9}$ m.

The moment of inertia equals the sum of the masses of each atom multiplied by the square of the distance of each atom as measured from the center of mass of the molecule:

$$I = m_1 r_1^2 + m_2 r_2^2$$

$$I = (12)(1.66 \times 10^{-27}\ \text{kg})(0.0646 \times 10^{-9}\ \text{m})^2$$

$$+ (16)(1.66 \times 10^{-27}\ \text{kg})(0.0484 \times 10^{-9}\ \text{m})^2$$

$$I = 8.32 \times 10^{-47}\ \text{kg}\cdot\text{m}^2 + 6.22 \times 10^{-47}\ \text{kg}\cdot\text{m}^2$$

$$I = 1.46 \times 10^{-46}\ \text{kg}\cdot\text{m}^2$$

Part c. Step 1.

Use Eq. 29-2 of the textbook to determine the rotational energy difference between $\ell = 1$ and $\ell = 0$.

$$\Delta E_{\text{rot}} = \hbar^2 \frac{\ell}{I} = \frac{h^2 \ell}{4\pi^2 I} = \frac{(6.63 \times 10^{-34}\ \text{J}\cdot\text{s})^2(1)}{4\pi(1.46 \times 10^{-46}\ \text{kg}\cdot\text{m}^2)}$$

$$\Delta E_{\text{rot}} = 7.63 \times 10^{-23}\ \text{J}$$

Part c. Step 2.

Determine the wavelength of the photon required for the transition from the $\ell = 0$ to $\ell = 1$ state.

$$\Delta E_{\text{rot}} = E_{\text{photon}} = hf = h\frac{c}{\lambda}$$

$$7.63 \times 10^{-23}\ \text{J} = (6.63 \times 10^{-34}\ \text{J}\cdot\text{s})\left(\frac{3.0 \times 10^8\ \text{m/s}}{\lambda}\right)$$

$$\lambda = (6.63 \times 10^{-34}\ \text{J}\cdot\text{s})\left(\frac{3.0 \times 10^8\ \text{m/s}}{7.63 \times 10^{-23}\ \text{J}}\right)$$

$$\lambda = 2.60 \times 10^{-3}\ \text{m} = 2.60\ \text{mm}$$

EXAMPLE PROBLEM 3. (Section 29-4) It is determined that a photon of wavelength 0.00698 m causes the O_2 molecule to make a transition from the lowest rotational energy state ($\ell = 0$) to the first excited state ($\ell = 1$). Determine the (*a*) energy of the photon, (*b*) difference in energy between the rotational energy states, (*c*) moment of inertia of the O_2 molecule, and (*d*) bond length of the molecule. Note: The mass of an oxygen atom is 2.67×10^{-26} kg.

Part a. Step 1.

Determine the energy of the photon.

$E = hf$ but $f = \frac{c}{\lambda}$

Therefore,

$$E = h\frac{c}{\lambda}$$

$$E = (6.63 \times 10^{-34}\ \text{J} \cdot \text{s})\left(\frac{3 \times 10^{8}\ \text{m/s}}{0.00698\ \text{m}}\right) = 2.85 \times 10^{-23}\ \text{J}$$

Part b. Step 1.

Determine the difference in energy between the states where $\ell = 0$ and $\ell = 1$.

The photon caused the transition from $\ell = 0$ to $\ell = 1$; therefore, the energy difference between the two states equals the energy of the photon:

$$\Delta E = E_1 - E_0 = 2.85 \times 10^{-23}\ \text{J}$$

Part c. Step 1.

Determine the moment of inertia of the O_2 molecule.

The formula for the quantized rotational energy level is

$$E_{\text{rot}} = \ell(\ell + 1)\frac{\hbar^2}{2I},$$

where $\hbar^2 = \left(\frac{6.63 \times 10^{-34}\ \text{J} \cdot \text{s}}{2\pi}\right)^2 = 1.11 \times 10^{-68}\ \text{J}^2 \cdot \text{s}^2$

If $\ell = 0$, then $E_0 = 0(0+1)\frac{\hbar^2}{2I} = 0.$

If $\ell = 1$, then $E_1 = 1(1+1)\frac{\hbar^2}{2I} = 2\frac{\hbar^2}{2I} = \frac{\hbar^2}{I}.$

$$\Delta E = E_1 - E_0 = \frac{\hbar^2}{I} - 0 = \frac{\hbar^2}{I}$$

$2.85 \times 10^{-23}\ \text{J} = \frac{\hbar^2}{I}$ and

$$I = \frac{1.11 \times 10^{-68}\ \text{J}^2 \cdot \text{s}^2}{2.85 \times 10^{-23}\ \text{J}}$$

$$I = 3.91 \times 10^{-46}\ \text{J} \cdot \text{s}^2 = 3.91 \times 10^{-46}\ \text{kg} \cdot \text{m}^2$$

Part d. Step 1.

Determine the bond length of the molecule.

The oxygen atoms have the same mass, and each atom can be considered to be a point mass. The center of mass of the molecule lies midway between the two atoms. The moment of inertia about the center of mass is given by

$$I = m_1 r_1^2 + m_2 r_2^2$$

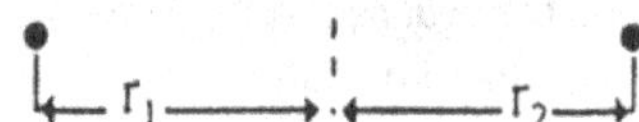

But $m_1 = m_2 = m$ and $r_1 = r_2 = r$. Therefore,

$$I = 2mr^2$$

$$3.91 \times 10^{-46}\ \text{kg} \cdot \text{m}^2 = 2(2.67 \times 10^{-26}\ \text{kg})r^2$$

$$r^2 = 7.3 \times 10^{-21}\ \text{m}^2$$

$$r = 8.6 \times 10^{-11}\ \text{m}$$

The bond length is the distance between the oxygen atoms.

$$\text{Bond length} = 2r = 1.71 \times 10^{-10}\ \text{m}$$

Bonding in Solids

In Chapter 16, the ability of a solid to conduct an electrical current resulted in it being classified as a conductor, semiconductor, or insulator. This classification can now be discussed in terms of what is referred to as the band theory of solids.

As stated in the text, "If a large number of atoms come together to form a solid, then each of the original atomic levels becomes a **band**." As shown in the accompanying diagrams, "The energy levels are so close together in each band that they seem essentially continuous."

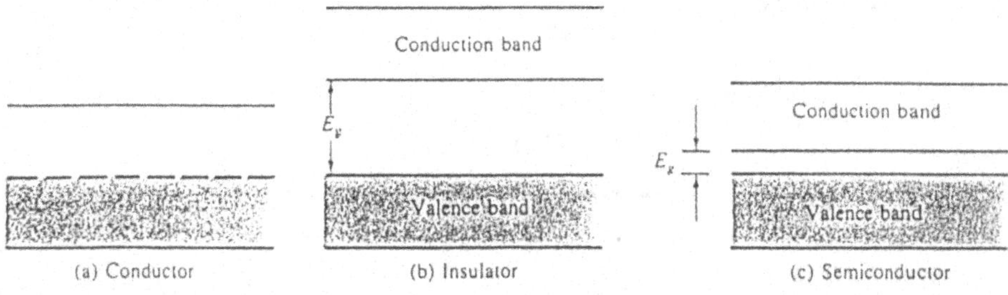

As described in the textbook, in a good conductor, such as a metal, the highest energy band (valence band) containing electrons is only partially filled. As a result, many electrons are relatively free to move throughout the volume of the metal and the metal can carry an electric current.

In a good **insulator**, the highest band (**valence band**) is completely filled with electrons. The next highest energy band, called the **conduction band**, is separated from the valence band by a large **energy gap** (E_g) of 5 to 10 eV. At room temperature, molecular kinetic energy available due to collisions is only about 0.04 eV, so almost no electrons can jump from the valence to the conduction band. When a potential difference is applied across the material, there are no available states accessible to the electrons, and no current flows. Hence the material is a good insulator.

The bands for a pure **semiconductor**, such as silicon or germanium, are like those for an insulator, except that the unfilled conduction band is separated from the filled valence band by a much smaller energy gap (E_g), typically on the order of 1 eV. At room temperature, there will be a few electrons that can acquire enough energy to reach the conduction band and so a very small current can flow when a voltage is applied. At higher temperatures, more electrons will have enough energy to jump the gap. This effect can often more than offset the effects of more frequent collisions due to increased disorder at higher temperature, so that the resistivity of semiconductors can decrease with temperature.

EXAMPLE PROBLEM 4. (Section 29-7) The longest wavelength radiation absorbed by a certain semiconductor is 2000 nm. Determine the energy gap (E_g) for this semiconductor in joules and in electron volts.

Part a. Step 1.

Determine the energy gap in joules.

E_g is the minimum energy required to cause an electron to jump from the valence band to the conduction band. The energy of the wavelength radiation must equal E_g:

$$E_g = E_{\text{photon}} = h\frac{c}{\lambda}$$

$$= (6.63 \times 10^{-34}\ \text{J}\cdot\text{s})\left(\frac{3 \times 10^8\ \text{m/s}}{(2000\ \text{nm})(1.0 \times 10^{-9}\ \text{m/1 nm})}\right)$$

$$E_g = 9.95 \times 10^{-20}\ \text{J}$$

Part a. Step 2.

Determine the energy in eV.

$$E_g = (9.95 \times 10^{-20}\ \text{J})\left(\frac{1.0\ \text{eV}}{1.6 \times 10^{-19}\ \text{J}}\right)$$

$$E_g = 0.621\ \text{eV}$$

EXAMPLE PROBLEM 5. (Section 29-7) The energy gap for silicon is 1.14 eV at room temperature. Determine the (*a*) lowest frequency photon that will cause an electron to jump from the valence band to the conduction band, and (*b*) wavelength of the photon found in part (*a*).

Part a. Step 1.

Express the energy in joules.

$$E_g = (1.14\text{ eV})\left(\frac{1.6\times 10^{-19}\text{ J}}{1.0\text{ eV}}\right)$$

$$E_g = 1.82\times 10^{-19}\text{ J}$$

Part a. Step 2.

Determine the frequency of the photon.

$$E_g = E_{\text{photon}} = hf$$

$$1.82\times 10^{-19}\text{ J} = (6.63\times 10^{-34}\text{ J}\cdot\text{s})f$$

$$f = 2.75\times 10^{14}\text{ Hz}$$

Part b. Step 1.

Determine the wavelength of the photon.

$$c = f\lambda$$

$$3.0\times 10^{8}\text{ m/s} = (2.75\times 10^{14}\text{ Hz})\lambda$$

$$\lambda = 1.09\times 10^{-6}\text{ m}$$

Semiconductors and Doping

The electronic configuration of the valence electrons of silicon is $3s^2 3p^2$ and that for germanium is $4s^2 4p^2$, which means that each element has four outer electrons and is a relatively poor conductor of electricity. Silicon and germanium are examples of semiconductors.

However, if a small amount of an impurity such as arsenic $(4s^2 4p^3)$ is introduced into the crystal structure of germanium, then arsenic's fifth electron is not bound and is free to move about. As a result, the **doped** semiconductor becomes highly conducting. An arsenic-doped germanium crystal is called an ***n*-type semiconductor** because electrons (negative charge) carry the current.

If a small amount of gallium $(4s^2 4p^1)$ is added to the germanium crystal, then an empty place or **hole** is introduced because gallium has only three outer electrons. An electron from the germanium atom can jump into this hole, but as a result the hole moves to a new location. Because most of the atoms of the crystal are germanium, this new location is invariably next to a germanium atom that is now positively charged because it has lost an electron. An electron can jump from another germanium atom to fill the previous hole, and thus the hole can move through the crystal. The flow of electricity in this instance is called a hole current. A germanium crystal doped with gallium is called a ***p*-type semiconductor** since it is the positive holes that appear to carry the current.

TEXTBOOK QUESTION 8. (Section 29-5) If conduction electrons are free to roam about in a metal, why don't they leave the metal entirely?

ANSWER: The nuclei of metals have a weak attraction for the outer electrons. As a result, the outer electrons of the metal atoms tend to move freely through the solid and are shared by all of the atoms in the metal. As a result, the metal is electrically neutral. If electrons left the metal, then the metal would have an excess positive charge. The resulting electrostatic attraction between the positively charged metal and the escaping electrons would tend to cause the electrons to be attracted back to the metal.

TEXTBOOK QUESTION 9. (Section 29-5) Explain why the resistivity of metals increases with temperature whereas the resistivity of semiconductors may decrease with increasing temperature.

ANSWER: The nuclei of metals have a weak attraction for the outer electrons. As a result, at room temperature the outer electrons of the metal atoms tend to move freely through the solid. As the temperature increases, the atoms that make up the metal are vibrating more rapidly. The electric field of the individual atoms then has a greater probability of interfering with the electrons as they move through the metal. As a result, the resistivity of the metal increases.

As the temperature of a semiconductor increases, more electrons obtain enough energy to overcome the small energy gap between the valence band and the conduction band. As a result, the resistivity of a semiconductor may decrease with increasing temperature.

Semiconductor Diodes

A semiconductor **diode** is produced when a *p*-type and an *n*-type semiconductor are joined. This combination is called a ***pn* junction diode.** At the junction, a few electrons from the *n*-type diffuse into holes in the *p*-type, and an internal difference in potential develops between the sections.

If a battery is connected as shown in Figure G, then the diode is said to be **forward biased**. If the battery voltage is large enough to overcome the internal potential difference, then a current will flow through the diode.

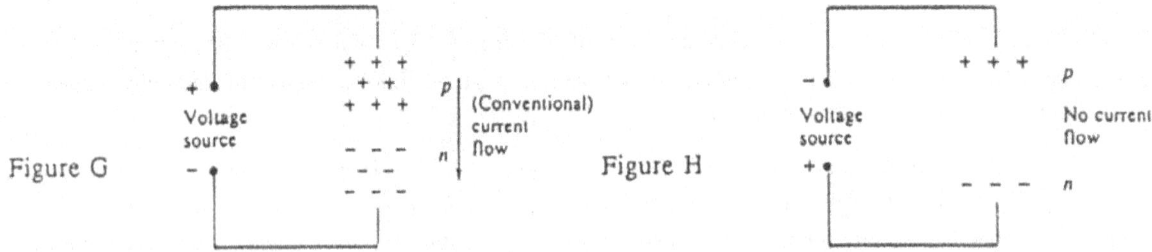

If the battery is connected as shown in Figure H, then the diode is **reverse biased**. This causes the holes and electrons to be separated, which means that the negative charge tends to be separated from the positive charge. The result is a diode that is essentially nonconducting. A graph of current versus voltage for a typical diode is shown next.

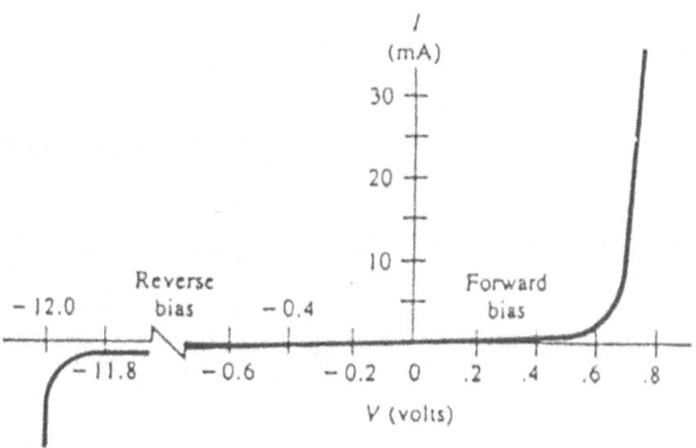

Since a *pn* junction diode allows current to flow only in one direction, it will allow current from an ac source to flow only in one direction through the circuit. Therefore, it can be used to change an ac current to a dc current. This is called **rectification**, and in the simple circuit shown below, the diode is acting as a **half-wave rectifier**.

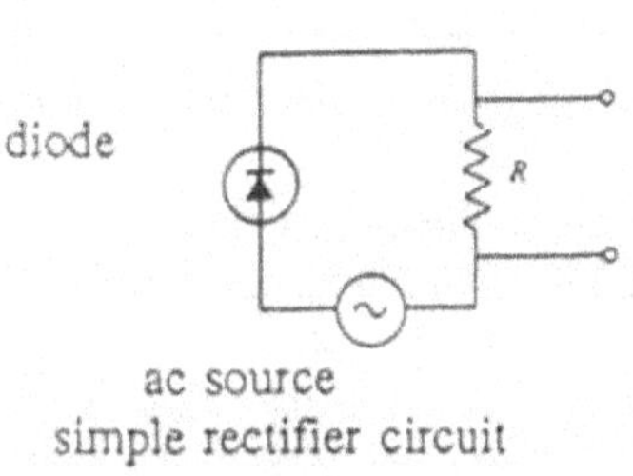

simple rectifier circuit

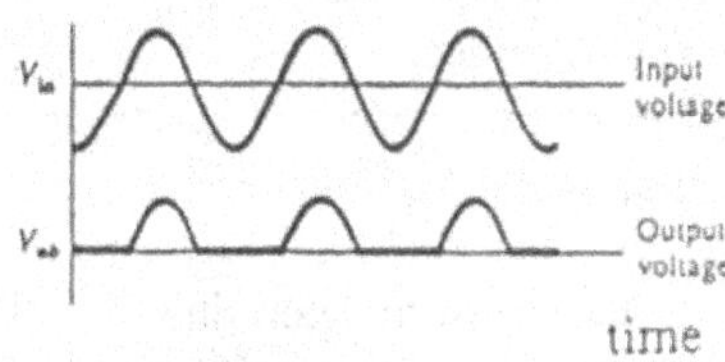

output voltage across R as a function of time

Note that the symbol for a diode includes an arrowhead. The diode allows current to flow in the direction of the arrowhead but not in the reverse direction.

TEXTBOOK QUESTION 10. (Section 28-9) Compare the resistance of a *pn* junction diode connected in forward bias to its resistance when connected in reverse bias.

ANSWER: A diode is forward biased when connected to a battery as shown in Figure G above. A battery voltage large enough to overcome the internal potential difference can cause a current to flow through the diode, and the resistance to current flow is small.

A diode is reverse biased when connected to a battery as shown in Figure H. This causes the holes and electrons to be separated. The result is that the diode is essentially nonconducting, and the electrical resistance to current flow is very large.

EXAMPLE PROBLEM 6. (Section 29-9) A diode whose current voltage characteristics are in the accompanying figure is connected in series with a voltage source and a 200-Ω resistor. What voltage is needed to cause a 10-mA current to flow through the circuit?

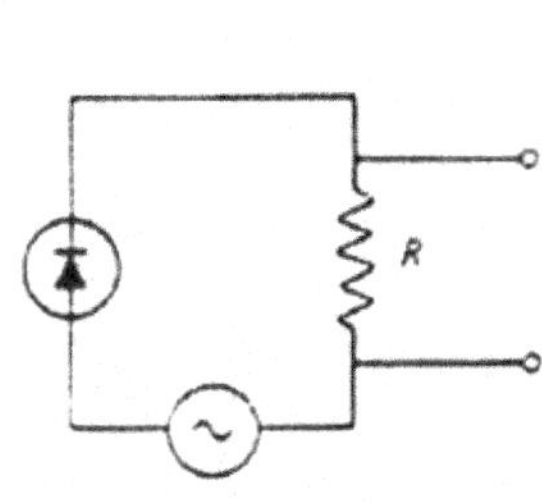

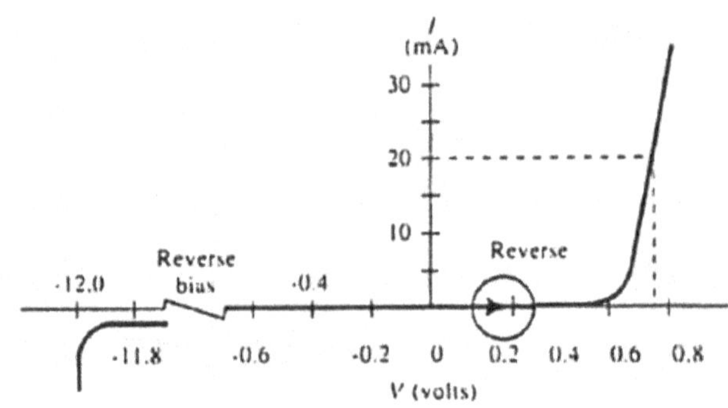

Part a. Step 1.

Use the diagram to determine the approximate voltage drop across the diode.	Using the figure, it can be determined that a voltage of approximately 0.68 V is needed to produce a 10-mA current.

Part a. Step 2.

Use Ohm's law to determine the voltage drop across the resistor.	The circuit is a series circuit; therefore, the diode current equals the current flowing through the 200-Ω resistor: $V_R = IR = (10\text{ mA})(200\ \Omega)$ $V_R = (10\times10^{-3}\text{ A})(200\ \Omega) = 2.0\text{ V}$

Part a. Step 3.

Use Kirchhoff's voltage rule to determine the battery voltage.	The resistor is in series with the diode and the voltage source, $V_{\text{battery}} = V_{\text{diode}} + V_R = 0.68\text{ V} + 2.0\text{ V}$ $V_{\text{battery}} = 2.68\text{ V}$

PROBLEM SOLVING SKILLS

For problems related to electrostatic potential energy:

1. Determine the charge on each ion and the distance between the ions.
2. Use $\text{PE} = \left(\dfrac{1}{4\pi\varepsilon_0}\right)\dfrac{q_1q_2}{r}$ to solve for the potential energy.

For problems related to rotational energy states:

1. Determine the difference in energy between the rotational energy states.
2. Use the energy difference to determine the moment of inertia of the molecule.
3. Use the moment of inertia to determine the bond length.

For problems related to the energy gap between the valence band and conduction band for a semiconductor or insulator:

1. If the energy for the electron to travel from the valence band to the conduction band is provided by a photon, determine the energy of the photon. The energy of a photon can be determined by using $E = hf$ or $E = h\frac{c}{\lambda}$.
2. The energy difference (energy gap) between the valence band and conduction band is equal to the lowest frequency (or longest wavelength) photon capable of causing the electron's transition.

For problems related to a series circuit containing a voltage source, a diode, and an additional resistor:

1. Use a figure giving the current–voltage characteristics of the diode to determine the approximate voltage drop across the diode.
2. Use Ohm's law to determine the voltage drop across the series resistor.
3. Use Kirchhoff's voltage rule to determine the voltage of the source.

SOLUTIONS TO SELECTED TEXTBOOK PROBLEMS

TEXTBOOK PROBLEM 9. (Section 29-4) The "characteristic rotational energy," $\frac{\hbar^2}{2I}$, for N_2 is 2.48×10^{-4} eV. Calculate the N_2 bond length.

Part a. Step 1.

Determine the moment of inertia of the N_2 molecule.

$$\Delta E = \frac{\hbar^2}{2I},$$

$$\text{where } \hbar = \frac{h}{2\pi} = \frac{6.63 \times 10^{-34}\ \text{J}\cdot\text{s}}{2\pi} = 1.05 \times 10^{-34}\ \text{J}\cdot\text{s}$$

$$I = \frac{\hbar^2}{2(\Delta E)} = \frac{(1.05 \times 10^{-34}\ \text{J}\cdot\text{s})^2}{2(2.48 \times 10^{-4}\ \text{eV})(1.6 \times 10^{-19}\ \text{J})/(1\ \text{eV})}$$

$$I = \frac{1.1 \times 10^{-68}\ \text{J}^2\cdot\text{s}^2}{7.94 \times 10^{-23}\ \text{J}}$$

$$I = 1.39 \times 10^{-46}\ \text{kg}\cdot\text{m}^2$$

Part a. Step 2.

Determine the bond length of the molecule.

The nitrogen atoms have the same mass, and each atom can be considered to be a point mass. The center of mass of the molecule lies midway between the two atoms. The moment of inertia about the center of mass is given by

$I = m_1 r_1^2 + m_2 r_2^2$ $\quad |\leftarrow r_1 \rightarrow . \leftarrow r_2 \rightarrow|$

But $m_1 = m_2 = m$ and $r_1 = r_2 = r$. Therefore,

$I = 2mr^2$ Note: The atomic mass of a nitrogen atom = 14 amu.

$$1.39 \times 10^{-46}\ \text{kg} \cdot \text{m}^2 = 2(14\ \text{amu})\left(\frac{1.66 \times 10^{-27}\ \text{kg}}{1\ \text{amu}}\right) r^2$$

$$r^2 = 2.99 \times 10^{-21}\ \text{m}^2$$

$$r = 5.47 \times 10^{-11}\ \text{m}$$

The bond length is the distance between the nitrogen atoms.

$$\text{Bond length} = 2r = 2(5.47 \times 10^{-11}\ \text{m}) = 1.09 \times 10^{-10}\ \text{m}$$

TEXTBOOK PROBLEM 19. (Section 29-7) The energy gap between valence and conduction bands in germanium is 0.72 eV. What range of wavelengths can a photon have to excite the electron from the top of the valence band into the conduction band?

Part a. Step 1.

Express the energy in joules.

$$E_g = (0.72\ \text{eV})\left(\frac{1.6 \times 10^{-19}\ \text{J}}{1.0\ \text{eV}}\right)$$

$$E_g = 1.15 \times 10^{-19}\ \text{J}$$

Part a. Step 2.

Determine the frequency of the photon.

$$E_g = E_{\text{photon}} = hf$$

$$1.15 \times 10^{-19}\ \text{J} = (6.63 \times 10^{-34}\ \text{J} \cdot \text{s})f$$

$$f = 1.74 \times 10^{14}\ \text{Hz}$$

Part a. Step 3.

Determine the longest wavelength the photon.

$c = f\lambda$

$3.0 \times 10^8 \text{ m/s} = (1.74 \times 10^{14} \text{ Hz})\lambda$

$\lambda = 1.72 \times 10^{-6} \text{ m}$

Part a. Step 4.

Determine the range of wavelengths to bridge the energy gap.

The range of wavelengths is $\lambda \le 1.72 \times 10^{-6}$ m.

TEXTBOOK PROBLEM 21. (Section 29-7) The energy gap E_g in germanium is 0.72 eV. When used as a photon detector, roughly how many electrons can be made to jump from the valence to the conduction band by the passage of an 830-keV photon that loses all of its energy in this fashion?

Part a. Step 1.

Express the photon energy in eV.

$E_{\text{photon}} = (830 \text{ keV})\left(\dfrac{1000 \text{ eV}}{1 \text{ keV}}\right) = 8.30 \times 10^5 \text{ eV}$

Part a. Step 2.

Determine the maximum number of electrons that can be made to jump to the conduction band.

E_g = energy required for one electron to jump to the conduction band

n = maximum number of electrons that can make the jump

$E_{\text{photon}} = nE_g$

$8.30 \times 10^5 \text{ eV} = n(0.72 \text{ eV})$

$n \approx 1.15 \times 10^6$ electrons

TEXTBOOK PROBLEM 24. (Section 29-9) At what wavelength will an LED radiate if made from a material with an energy gap $E_g = 1.3$ eV?

Part a. Step 1.

Express the energy in joules.

$$E_g = (1.3\text{ eV})\left(\frac{1.6\times 10\text{ J}}{1.0\text{ eV}}\right)$$

$$E_g = 2.08\times 10^{-19}\text{ J}$$

Part a. Step 2.

Determine the frequency of the photon.

The photon's energy is equal to the energy gap:

$$E_{\text{photon}} = E_g \text{ and } E_{\text{photon}} = hf$$

$$2.08\times 10^{-19}\text{ J} = (6.63\times 10^{-34}\text{ J}\cdot\text{s})f$$

$$f = 3.14\times 10^{14}\text{ Hz}$$

Part a. Step 3.

Determine the wavelength of the photon.

$$c = f\lambda$$

$$3.0\times 10^{8}\text{ m/s} = (3.14\times 10^{14}\text{ Hz})\lambda$$

$$\lambda = 9.56\times 10^{-7}\text{ m} \approx 0.96\ \mu\text{m}$$

TEXTBOOK PROBLEM 27. (Section 29-9) A silicon diode, whose current–voltage characteristics are given in Fig. 29-30, is connected in series with a battery and a $960\text{-}\Omega$ resistor. What battery voltage is needed to produce a 14-mA current?

Part a. Step 1.

Use Fig. 29-30 to determine the voltage necessary to produce a 14-mA current.

Based on Fig. 29-30, a current of 14 mA is produced by an applied voltage of approximately 0.70 V.

Part a. Step 2.

Determine the approximate battery voltage.

$$V_{\text{battery}} \approx V_{\text{diode}} + V_{\text{R}}$$

$$\approx 0.70\text{ V} + (0.014\text{ A})(960\ \Omega)$$

$$\approx 0.70\text{ V} + 13.4\text{ V}$$

$$V_{\text{battery}} \approx 14.1\text{ V}$$

TEXTBOOK PROBLEM 28. (Section 29-9) An ac voltage of 120-V rms is to be rectified. Estimate very roughly the average current in the output resistor $R(=31\ \text{k}\Omega)$ for (*a*) a half-wave rectifier (Fig. 29-31), and (*b*) a full-wave rectifier (Fig. 29-32) without capacitor.

Part a. Step 1.

Determine the average current for a half-wave rectifier without a capacitor.

For a half-wave rectifier without a capacitor, the current is zero half of the time:

$$I_{\text{average}} = \frac{1}{2}\frac{V_{\text{rms}}}{R} = \frac{1}{2}\left(\frac{120\ \text{V}}{31{,}000\ \Omega}\right)$$

$$I_{\text{average}} = 1.94 \times 10^{-3}\ \text{A} = 1.94\ \text{mA}$$

Part b. Step 1.

Determine the average for current for a full-wave rectifier without a capacitor.

For a full-wave rectifier without a capacitor, the current is positive all of the time.

$$I_{\text{average}} = \frac{V_{\text{rms}}}{R} = \left(\frac{120\ \text{V}}{31{,}000\ \Omega}\right)$$

$$I_{\text{average}} = 3.87 \times 10^{-3}\ \text{A} = 3.87\ \text{mA}$$

TEXTBOOK PROBLEM 46. (Section 29-6) When EM radiation is incident on diamond, it is found that light with wavelengths shorter than 226 nm will cause the diamond to conduct. What is the energy gap between the valence band and the conduction band for diamond?

Part a. Step 1.

Use Eq. 27-4 to determine the energy of the 226-nm photon.

$$E_{\text{photon}} = hf = h\frac{c}{\lambda}$$

$$= \frac{(6.63 \times 10^{-34}\ \text{J}\cdot\text{s})(3.0\times 10^{8}\ \text{m/s})}{(226\ \text{nm})(1.0 \times 10^{-9}\ \text{m/1 nm})}$$

$$E_{\text{photon}} = 8.80 \times 10^{-19}\ \text{J}$$

Part a. Step 2.

Determine the photon's energy in eV.

$$(8.80 \times 10^{-19}\ \text{J})\left(\frac{1\ \text{eV}}{1.6 \times 10^{-19}\ \text{J}}\right) = 5.50\ \text{eV}$$

Part a. Step 3.

Determine the energy gap between the valence band and the conduction band.	The wavelength of the photon for conduction to occur must be less than 226 nm. Therefore, the energy gap between the valence band and the conduction band must be slightly greater than 5.50 eV.

30

NUCLEAR PHYSICS AND RADIOACTIVITY

OBJECTIVES

After studying the material of this chapter, the student should be able to:

- determine the number of neutrons in a nuclide of known atomic number and mass number.
- explain what is meant by an isotope of an element. State how isotopes of an element differ and state the properties they have in common.
- explain what is meant by the unified atomic mass unit. Calculate the energy equivalent in MeV of an atomic mass unit.
- given a table of nuclear masses, calculate the binding energy of a nucleus and the binding energy per nucleon.
- identify the three kinds of radiation emitted by radioactive substances. State which radiations are deflected by electric and magnetic fields.
- give the symbol used to represent each of the following: alpha particle, beta^- particle, beta^+ particle, gamma ray.
- write a general equation to represent each of the following possible radiation decays: alpha decay, beta^+ decay, beta^- decay, gamma decay.
- distinguish between the parent nucleus and daughter nucleus in a nuclear transmutation.
- calculate the disintegration energy (Q) for a given alpha decay.
- list the four conservation laws that apply to radioactive decays.
- write the equation that relates the half-life of a substance to its decay constant.
- write the equation for the law of radioactive decay. Explain the meaning of each symbol in the equation. Solve problems related to the law of radioactive decay.

KEY TERMS AND PHRASES

nucleus of an atom contains **protons** and **neutrons**. These particles are called **nucleons**.

1 **unified atomic mass unit** $(\text{u}) = 1.660 \times 10^{-27}$ kg.

atomic number is the number of protons contained in the nucleus.

atomic mass number is the total number of protons and neutrons in the nucleus.

isotopes are atoms that have the same number of protons but different number of neutrons in the nucleus.

total binding energy is the energy required to break apart a nucleus into its constituent protons and neutrons.

strong nuclear force refers to the attractive force that holds the nucleus together. The strong nuclear force acts between all nucleons, protons, and neutrons alike. This force is much greater than the force of electrostatic repulsion that exists between the protons. The strong nuclear force is a **short-range** force. It acts between nucleons if they are less than 10^{-15} m apart but is essentially zero if the separation distance is greater than 10^{-15} m.

weak nuclear force is much weaker than the strong nuclear force and appears in a type of radioactive decay called **beta decay**.

radioactive decay results from the instability of certain nuclei. There are three different radiations produced by radioactive decay: alpha, beta, and gamma.

parent nucleus refers to the nucleus before radioactive decay.

daughter nucleus refers to the nucleus after radioactive decay.

alpha decay occurs when a helium nucleus is spontaneously emitted from the nucleus. An alpha particle (α) consists of two protons and two neutrons but no electrons.

disintegration energy (Q) is the total energy released during alpha decay.

beta decay occurs when an electron (e^-) and an antineutrino ($\overline{\nu}$) are spontaneously emitted from the nucleus. Beta (β^-) decay is observed in nuclei that have a high ratio of neutrons to protons.

electron capture occurs when the nucleus absorbs one of the inner orbital electrons in the atom to form a neutron.

transmutation is the changing of one element into a new element. Transmutation is the result of alpha or beta decay.

gamma decay occurs when a nucleus in an excited state drops to a lower energy state. In the process a photon called a gamma (γ) ray is emitted.

law of conservation of nucleon number states that the total number of nucleons before decay equals the total number of nucleons after decay.

half-life of an isotope is the time required for half of the radioactive nuclei present in the sample to decay.

radioactive decay series is a series of successive decays that starts with one parent isotope and proceeds through a number of daughter isotopes. The series ends when a stable, nonradioactive isotope is produced.

radioactive dating refers to a method of estimating the age of an object based on the object's half-life and the amount of the isotope present in the sample being analyzed.

SUMMARY OF MATHEMATICAL FORMULAS

atomic mass number	${}^{A}_{Z}\mathrm{X}$ $A = Z + N$	The nucleus of a chemical element is designated by ${}^{A}_{Z}\mathrm{X}$. X is the chemical symbol of the element, Z is the atomic number, which is the number of protons contained in the nucleus, A is the atomic mass number, and N is the number of neutrons in the nucleus.

atomic radius	$r \approx (1.2\times10^{-15}\text{ m})(A^{\frac{1}{3}})$	The approximate atomic radius (r) of the nucleus increases with the mass number (A).
binding energy	$E=(\Delta m)c^2$ $E=(\text{u})(931.5\text{ MeV/u})$	The energy equivalent of the mass difference (Δm) is the binding energy (E) of the nucleus. The binding energy is determined by multiplying the mass difference, expressed in u, by the conversion factor 931.5 MeV/u.
alpha decay	${}_{Z}^{A}\text{N} \rightarrow {}_{Z-2}^{A-4}\text{N}' + {}_{2}^{4}\text{He}$	An alpha decay (α) particle consists of a helium nucleus. ${}_{Z}^{A}\text{N}$ is the parent nucleus (N) with atomic number (Z) and mass number (A). ${}_{Z-2}^{A-4}\text{N}'$ is the daughter nucleus, and ${}_{2}^{4}\text{He}$ is the alpha particle.
disintegration energy	$Q=(M_\text{P}-M_\text{D}-m_\alpha)c^2$	The total energy released during alpha decay is called the disintegration energy (Q). M_P is the mass of the parent nucleus, M_D is the mass of the daughter nucleus, and m_α is the mass of the alpha particle.
beta decay	${}_{Z}^{A}\text{N} \rightarrow {}_{Z+1}^{A}\text{N}' + {}_{-1}^{0}\text{e} + \overline{v}$	Beta decay occurs when an electron (e^-) and an antineutrino ($\overline{v}$) are spontaneously emitted from the nucleus. The electron carries one negative charge, and as a result the atomic number of the daughter nucleus is one greater than the atomic number of the parent nucleus. The antineutrino carries no electric charge and has zero mass.
electron capture	${}_{Z}^{A}\text{N} + {}_{-1}^{0}\text{e} \rightarrow {}_{Z-1}^{A}\text{N}' + v$	Electron capture occurs when the nucleus absorbs one of the inner orbital electrons in the atom to form a neutron. When electron capture occurs, a neutrino (v) is spontaneously emitted from the nucleus.

gamma decay	${}^{A}_{Z}\text{N}^{*} \rightarrow {}^{A}_{Z}\text{N} + \gamma$	Gamma decay occurs when a nucleus in an excited state drops to a lower energy state. In the process a photon called a gamma (γ) ray is emitted. Since the gamma ray is a photon, it carries no electric charge and has no mass. No change in nucleon number or atomic number occurs due to a gamma decay. ${}^{A}_{Z}\text{N}^{*}$ is the nucleus in an excited state.
radioactive decay rate	$\Delta N = -\lambda N \Delta t$ $R = \dfrac{\Delta N}{\Delta t} = \lambda N$	The number of radioactive decays (ΔN) that occur in a short time interval (Δt) is proportional to the length of the time interval and the total number (N) of radioactive nuclei present. λ is the decay constant, which is different for different isotopes. $R = \dfrac{\Delta N}{\Delta t}$ is the rate of decay or the activity of the isotope.
law of radioactive decay	$N = N_0 e^{-\lambda t}$	Based on the law of radioactive decay, the number of radioactive nuclei present (N) after time (t) depends on the number of radioactive nuclei present in the original sample (N_0) and the decay constant (λ) for the particular isotope.
half-life	$T_{\frac{1}{2}} = \dfrac{\ln 2}{\lambda} = \dfrac{0.693}{\lambda}$	The half-life $(T_{\frac{1}{2}})$ of an isotope is the time required for half of the radioactive nuclei present in the sample to decay. The half-life is related to the decay constant (λ).

CONCEPT SUMMARY

Structure of the Nucleus

The nucleus of an atom contains **protons** and **neutrons**, which are referred to as **nucleons**. The mass of a proton (m_p) is 1.673×10^{-27} kg, or 1.0073 u. The proton carries a single positive charge of magnitude 1.60×10^{-19} C. The mass of the electrically neutral neutron (m_n) is 1.675×10^{-27} kg or 1.0087 u. Note: $1\text{ u} = 1.660 \times 10^{-27}$ kg = 1 **unified atomic mass unit**.

The nucleus of a chemical element is designated by ${}^{A}_{Z}\text{X}$. X is the chemical symbol of the element, for example, H for hydrogen and Ca for calcium. Z is the **atomic number**, which is the number of protons contained in the nucleus. A is the **atomic mass number**. The atomic mass number is the total number of protons and neutrons in the nucleus.

Experiments indicate that the nuclei have a roughly spherical shape. The radius (r) of the nucleus increases with the mass number (A), and the approximate radius is given by

$$r \approx (1.2 \times 10^{-15}\text{ m})(A^{\frac{1}{3}})$$

EXAMPLE PROBLEM 1. (Section 30-1) Determine the approximate density of the nucleus of an ${}^{27}_{13}\text{Al}$ atom in kg/m^3.

Part a. Step 1.

Determine the radius of the nucleus.

The atomic mass number (A) of an aluminum atom is 27.

The approximate radius of the nucleus may be determined as follows:

$$r \approx (1.2 \times 10^{-15}\text{ m})(A^{\frac{1}{3}})$$

$$r \approx (1.2 \times 10^{-15}\text{ m})(27^{\frac{1}{3}}) \approx 3.6 \times 10^{-15}\text{ m}$$

Part a. Step 2.

Determine the approximate volume of the nucleus.

$$V = \frac{4}{3}\pi r^3$$

$$\approx \frac{4}{3}(3.14)(3.6 \times 10^{-15}\text{ m})^3$$

$$V \approx 2.0 \times 10^{-43}\text{ m}^3$$

Part a. Step 3.

Determine the mass of the nucleus in kg.

$$m = (27\text{ u})(1.66 \times 10^{-27}\text{ kg/u})$$

$$m = 4.48 \times 10^{-26}\text{ kg}$$

Part a. Step 4.

Determine the approximate density of the nucleus.

$$\rho = \frac{m}{V} \approx \frac{4.48 \times 10^{-26}\text{ kg}}{2.0 \times 10^{-43}\text{ m}^3}$$

$$\rho \approx 2.2 \times 10^{17}\text{ kg/m}^3$$

Isotopes

Atoms that have the same atomic number but different mass numbers are called **isotopes**. Isotopes of the same element have the same (1) atomic number, (2) number of electrons, (3) electronic configuration, and (4) chemical properties. Isotopes differ in the number of neutrons in the nucleus. For example, the two isotopes of lithium are ${}^{6}_{3}\text{Li}$ and ${}^{7}_{3}\text{Li}$. Each isotope contains three protons; however, one contains three neutrons while the other contains four neutrons.

TEXTBOOK QUESTION 1. (Section 30-1) What do different isotopes of a given element have in common? How are they different?

ANSWER: Isotopes are atoms that have the same number of protons but different number of neutrons in the nucleus. Isotopes of the same element have the same chemical properties and the neutral atoms of the same element contain the same number of electrons. For example, ${}^{63}_{29}\text{Cu}$ and ${}^{65}_{29}\text{Cu}$ are isotopes. Both have 29 protons in the nucleus, but ${}^{63}_{29}\text{Cu}$ has 34 neutrons while ${}^{65}_{29}\text{Cu}$ has 36 neutrons. As shown in Appendix B, the atomic mass number as well as atomic mass of isotopes of the same element differ.

TEXTBOOK QUESTION 2. (Section 30-1) What are the elements represented by X in the following: (*a*) ${}^{232}_{92}\text{X}$; (*b*) ${}^{18}_{7}\text{X}$; (*c*) ${}^{1}_{1}\text{X}$; (*d*) ${}^{86}_{38}\text{X}$; (*e*) ${}^{252}_{100}\text{X}$?

ANSWER: The identity of the element is determined by its atomic number. Using the Periodic Table, (*a*) the atomic number of uranium is 92; (*b*) the atomic number of nitrogen is 7; (*c*) the atomic number of hydrogen is 1; (*d*) the atomic number of strontium is 38; (*e*) the atomic number of fermium is 100.

TEXTBOOK QUESTION 5. (Section 30-1) Why are the atomic masses of many elements (see the Periodic Table) not close to whole numbers?

ANSWER: Isotopes of the same element contain the same number of protons but different number of neutrons. The weighted average of the percent abundance of the isotopes is used to determine the atomic mass of the element. For most elements this weighted average is not close to a whole number. For example, in Appendix B chlorine is listed as having 17 protons, but one isotope ($^{35}_{17}\text{Cl}$) has 18 neutrons and $^{37}_{17}\text{Cl}$ has 20 neutrons. The percent abundance of $^{35}_{17}\text{Cl}$ is 75.76% and $^{37}_{17}\text{Cl}$ is 24.24%. The atomic mass is determined as follows:

$$(0.7576)(35)+(0.2424)37=35.48$$

Binding Energy and Nuclear Forces

The total mass of the nucleus is always less than the sum of the masses of the protons and neutrons of which it is composed. The energy equivalent of the mass difference (Δm) is the **total binding energy** of the nucleus and can be determined by multiplying the mass difference, expressed in u, by the conversion factor 931.5 MeV/u.

The average binding energy per nucleon is defined as the total binding energy divided by the mass number (A). Figure 30-1 in the text shows that the average binding energy per nucleon rises sharply and reaches a plateau at about 8.7 MeV per nucleon above $A \approx 40$, levels off at until $A \approx 80$, and then slowly decreases.

The nucleus is held together by an attractive force that acts between all nucleons, protons and neutrons alike. This force is called the **strong nuclear force** and is much greater than the force of electrostatic repulsion that exists between the protons. The strong nuclear force is a **short-range** force. It acts between nucleons if they are less than $10^{-15}\,\text{m}$ apart but is essentially zero if the separation distance is greater than $10^{-15}\,\text{m}$.

A second type of nuclear force is the **weak nuclear force**. This force is much weaker than the strong nuclear force and appears in a type of radioactive decay called **beta decay**.

EXAMPLE PROBLEM 2. (Section 30-2) The exact atomic mass of a $^{65}_{29}\text{Cu}$ atom is 64.92779 u. Determine the total mass of its constituent particles and calculate the binding energy of the atom. The mass of an electron, proton, and neutron is 0.00055 u, 1.00728 u, and 1.00867 u, respectively.

Part a. Step 1.

Determine the total mass of the constituent particles.	The atomic number is 29; therefore, the Cu atom has 29 electrons and 29 protons. The mass number is 65; therefore, Cu has $65-29=36$ neutrons:
	Electrons $29\times 0.00055\text{ u} = 0.01595\text{ u}$
	Protons $29\times 1.00728\text{ u} = 29.21112\text{ u}$
	Neutrons $36\times 1.00867\text{ u} = 36.31212\text{ u}$
	Total mass of nucleons 65.53919 u

Part a. Step 2.

Determine the difference in mass between $^{65}_{29}\text{Cu}$ and its constituent particles.	Mass of nucleons	65.53919 u
	Mass of Cu	64.92779 u
	Mass difference	0.61140 u

Part a. Step 3.

Determine the binding energy.

$$E = (\Delta m)c^2$$

$$E = (0.61140\ \text{u})(931.5\ \text{MeV/u}) = 569.5\ \text{MeV}$$

EXAMPLE PROBLEM 3. (Section 30-2) (*a*) Use Fig. 30-1 from the textbook to determine the binding energy per nucleon for $^{120}_{50}\text{Sn}$. (*b*) Use this value to determine the total binding energy for the $^{120}_{50}\text{Sn}$ nucleus.

Part a. Step 1.

Determine the binding energy per nucleon.

It is not possible to determine an exact value from Fig. 30-1. An approximate value for the binding energy per nucleon for $^{120}_{50}\text{Sn}$ is 8.5 MeV per nucleon.

Part b. Step 1.

Determine the total binding energy.

$$\left(\frac{8.5\ \text{MeV}}{\text{nucleon}}\right)(120\ \text{nucleons}) \approx 1020\ \text{MeV}$$

Radioactive Decay Mode

Certain nuclei are unstable and undergo **radioactive decay**. There are three different radiations produced by radioactive decay: alpha, beta, and gamma.

Alpha Decay

An **alpha decay** particle consists of a helium nucleus. Thus, an alpha (α) particle consists of two protons and two neutrons but no electrons. An alpha decay can be represented by the following equation:

$$^{A}_{Z}\text{N} \rightarrow {}^{A-4}_{Z-2}\text{N}' + {}^{4}_{2}\text{He}$$

$^{A}_{Z}\text{N}$ is the **parent** nucleus (**N**) with atomic number (Z) and mass number (A). $^{A-4}_{Z-2}\text{N}'$ is the **daughter** nucleus, and $^{4}_{2}\text{He}$ is the alpha particle.

When alpha decay occurs, the atomic number of the element decreases by two while the mass number decreases by four. Alpha decay occurs in large nuclei in which the strong nuclear force is not strong enough to hold the nucleus together. It occurs when the mass of the parent nucleus is greater than the mass of the daughter plus the mass of the alpha particle; that is, $M_P > M_D + \alpha$. The mass difference appears in the form of kinetic energy of the daughter nucleus and the alpha particle (but mainly in the alpha particle).

The total energy released during alpha decay is called the disintegration energy Q:

$$Q = (M_P - M_D - m_\alpha)c^2$$

EXAMPLE PROBLEM 4. (Section 30-4) (*a*) Determine the disintegration energy Q when the following reaction occurs:

$${}^{226}_{88}\text{Ra} \to {}^{222}_{86}\text{Rn} + {}^{4}_{2}\text{He}.$$

(*b*) Determine the kinetic energy of each of the products.

Part a. Step 1.	Before		After	
Determine the mass before and after the decay.	${}^{226}_{88}\text{Ra}$	226.0254 u	${}^{222}_{86}\text{Rn}$	222.0176 u
			${}^{4}_{2}\text{He}$	4.0026 u
			Total mass of products	226.0202 u

Part a. Step 2.

Determine the mass difference	Mass difference = 226.0254 u − 226.0202 u
	Mass difference = 0.0052 u

Part a. Step 3.

Determine the disintegration energy Q.	$Q = (0.0052 \text{ u})(931.5 \text{ MeV/u}) = 4.84 \text{ MeV}$

Part b. Step 1.

Use the law of conservation of momentum to express the final velocity of the alpha particle in terms of the final velocity of the radon nucleus.	In any reaction, both energy and momentum must be conserved. Assuming that the radium nucleus was initially at rest, then the initial momentum was zero and the total momentum after the decay must also be zero. This means that the alpha particle's momentum must be equal to but opposite that of the radon nucleus.

Note: Assume that the initial momentum of the system was zero.

Initial momentum = Final momentum

$$0 = m_{\alpha}v_{\alpha} + m_{\text{Rn}}v_{\text{Rn}}$$

$$m_{\alpha}v_{\alpha} = -m_{\text{Rn}}v_{\text{Rn}} \quad \text{and} \quad v_{\alpha} = -\frac{m_{\text{Rn}}}{m_{\alpha}}v_{\text{Rn}}$$

$$v_{\alpha} = -\left(\frac{222.0176\text{ u}}{4.0026\text{ u}}\right)v_{\text{Rn}}$$

$$v_{\alpha} = -55.47\,v_{\text{Rn}}$$

Part b. Step 2.

Determine the ratio of the KE of the radon nucleus to the KE of the alpha particle.

$$\frac{\text{KE}_{\text{Rn}}}{\text{KE}_{\alpha}} = \frac{\frac{1}{2}m_{\text{Rn}}v_{\text{Rn}}^2}{\frac{1}{2}m_{\alpha}v_{\alpha}^2} \quad \text{but} \quad v_{\alpha} = -55.47\,v_{\text{Rn}}$$

$$\frac{\text{KE}_{\text{Rn}}}{\text{KE}_{\alpha}} = \frac{\frac{1}{2}(222.0176\text{ u})v_{\text{Rn}}^2}{\frac{1}{2}(4.002\text{ u})(-55.47\,v_{\text{Rn}})^2}$$

Upon simplifying,

$$\frac{\text{KE}_{\text{Rn}}}{\text{KE}_{\alpha}} = 0.01803$$

Part b. Step 3.

Determine the kinetic energy of the alpha particle and the daughter nucleus.

The total kinetic energy of the daughter nucleus and the alpha particle is 4.84 MeV.

$$\text{KE}_{\alpha} + \text{KE}_{\text{Rn}} = 4.84\text{ MeV} \quad \text{but} \quad \text{KE}_{\text{Rn}} = 0.01803\,\text{KE}_{\alpha}$$

Substituting gives

$$\text{KE}_{\alpha} + 0.01803\text{KE}_{\alpha} = 4.84\text{ MeV}$$

$$1.01803\,\text{KE}_{\alpha} = 4.84\text{ MeV}$$

$$\text{KE}_{\alpha} = 4.75\text{ MeV}$$

$$\text{KE}_{\text{Rn}} = 4.84\text{ MeV} - 4.75\text{ MeV} = 0.086\text{ MeV}$$

TEXTBOOK QUESTION 17. (Section 30-4) Can hydrogen or deuterium emit an α particle? Explain.

ANSWER: An alpha (α) particle contains four nucleons, two protons plus two neutrons. Hydrogen has one nucleon (one proton), while deuterium has two nucleons (one proton and one neutron). Neither hydrogen nor deuterium has the necessary number of nucleons to emit an α particle.

TEXTBOOK QUESTION 21. (Section 30-4) A proton strikes a ${}^{6}_{3}\text{Li}$ nucleus. As a result, an α particle and another particle are released. What is the other particle?

ANSWER: The reaction can be written as ${}^{6}_{3}\text{Li} + {}^{1}_{1}\text{H} \rightarrow {}^{A}_{Z}\text{X} + {}^{4}_{2}\text{He}$.

The sum of the atomic numbers of the reactant particles must equal the sum of the atomic numbers of the product particles. The atomic number (Z) of the unknown particle can be found as follows: $3+1 = Z+2$. Therefore, the atomic number of the unknown particle is 2.
The sum of the mass numbers of the reactant particles must equal the sum of the mass numbers of the product particles. The mass number (A) of the unknown particle can be found as follows: $6+1 = A+4$. Therefore, the mass number (A) of the unknown particle $= 3$.
The unknown particle must be ${}^{3}_{2}\text{He}$.

Beta Decay

Beta decay occurs when an electron (e^-) and an antineutrino ($\overline{v}$) are spontaneously emitted from the nucleus. Beta (β^-) decay is observed in nuclei that have a high ratio of neutrons to protons. The electron is not a nucleon; therefore, there is no change in the mass number of the daughter nucleus. However, the electron carries one negative charge, and as a result the atomic number of the daughter nucleus is one greater than the atomic number of the parent nucleus. Beta decay is represented by the following equation:

$${}^{A}_{Z}\text{N} \rightarrow {}_{Z+1}^{\;\;A}\text{N}' + {}^{\;\;0}_{-1}\text{e} + \overline{v}$$

The **antineutrino** ($\overline{v}$) carries no electric charge and has zero mass.

Certain unstable isotopes have a low neutron to proton ratio. In such a situation, a **positron** and a **neutrino** (v) may be emitted from the nucleus. The charge on a positron (${}^{\;\;0}_{+1}\text{e}$) is opposite that of the electron, but the mass of both particles is the same. The positron is represented as e^+ or β^+ and is the antiparticle to the electron. As in the case of the antineutrino, the neutrino carries no electric charge and has zero mass. The equation for positron decay is as follows:

$${}^{A}_{Z}\text{N} \rightarrow {}_{Z-1}^{\;\;A}\text{N}' + {}^{\;\;0}_{+1}\text{e} + v$$

Electron Capture

Electron capture occurs when the nucleus absorbs one of the inner orbital electrons in the atom to form a neutron. When electron capture occurs, a neutrino is spontaneously emitted from the nucleus. Electron capture may occur if the neutron–proton ratio is low. Electron capture is represented by the following equation:

$$^{A}_{Z}\mathrm{N} + {}^{0}_{-1}\mathrm{e} \rightarrow {}_{Z-1}^{A}\mathrm{N}' + \nu$$

Transmutation

A new element is formed when alpha or beta decay occurs. The changing of one element (parent nucleus) into a new element (daughter nucleus) is called **transmutation**.

Gamma Decay

Gamma decay occurs when a nucleus in an excited state drops to a lower energy state. In the process a photon called a gamma (γ) ray is emitted. The process is analogous to photon emission when an orbital electron drops from a higher energy level to a lower energy level.

The nucleus may be in the excited state as a result of either a violent collision with another particle or a previous radioactive decay.

Since the gamma ray is a photon, it carries no electric charge and has no mass. Therefore, no change in nucleon number or atomic number occurs due to a gamma decay. A gamma decay can be represented as follows:

$$^{A}_{Z}\mathrm{N}^* \rightarrow {}^{A}_{Z}\mathrm{N} + \gamma$$

EXAMPLE PROBLEM 5. (Sections 30-4, 30-5, and 30-7) Cite the type of reaction and then complete the following:

(*a*) $^{239}_{92}\mathrm{U} \rightarrow \mathrm{X} + {}^{0}_{-1}\mathrm{e} + \bar{\nu}_{\mathrm{e}}$

(*b*) $^{105}_{48}\mathrm{Cd} + {}^{0}_{-1}\mathrm{e} \rightarrow \mathrm{X} + \nu_{\mathrm{e}}$

(*c*) $^{226}_{88}\mathrm{Ra} \rightarrow \mathrm{X} + {}^{4}_{2}\mathrm{He}$

(*d*) $^{15}_{8}\mathrm{O} \rightarrow \mathrm{X} + {}^{0}_{+1}\mathrm{e} + \nu_{\mathrm{e}}$

Part a. Step 1.

Cite the type of reaction, then use the law of conservation of charge and nucleon number to solve for X.

This reaction is a beta decay. The product nucleus must have an atomic number of $92-(-1)=93$ in order to satisfy the law of conservation of charge. It must have a mass number of 239 in order to conserve nucleon number. Using the Periodic Table, it can be determined that X is neptunium; thus

X is $^{239}_{93}\mathrm{Np}$

Part b. Step 1.	
Cite the type of reaction and repeat Part a.	This is an example of electron capture. The atomic number of the product nucleus is $48+(-1)=47$. The mass number is 105. Using the Periodic Table, the unknown nucleus is silver; thus X is $^{105}_{47}\text{Ag}$
Part c. Step 1.	
Cite the type of reaction and repeat Part a.	This is an example of alpha decay. The atomic number of the product nucleus is $88-2=86$. The mass number is $226-4=222$. The unknown nucleus is radon; thus, X is $^{222}_{86}\text{Ra}$
Part d. Step 1.	
Cite the type of reaction and repeat Part a.	This is an example of positron emission. The atomic number of the product nucleus is $8-(+1)=7$. The mass number is $15-0=15$. The element is nitrogen; thus, X is $^{15}_{7}\text{N}$

Conservation Laws

In addition to conservation of energy, linear momentum, angular momentum, and electric charge, a radioactive decay also obeys the **law of conservation of nucleon number**. This law states that the total number of nucleons before decay equals the total number of nucleons after decay.

Half-Life and Rate of Decay

The number of radioactive decays (ΔN) that occur in a short time interval (Δt) is proportional to the length of the time interval and the total number (N) of radioactive nuclei present:

$$\Delta N = -\lambda N\,\Delta t \quad \text{and} \quad R = \frac{\Delta N}{\Delta T} = \lambda N$$

Here λ is the decay constant, which is different for different isotopes. The negative sign in the equation indicates that the number of radioactive nuclei present is decreasing. $R = \frac{\Delta N}{\Delta T}$ is the rate of decay or the **activity** of the isotope.

Based on the law of radioactive decay, the number of radioactive nuclei present is given by the equation

$$N = N_0 e^{-\lambda t}$$

Here N_0 is the number of radioactive nuclei present in the original sample—that is, the number at $t = 0$. N is the number of radioactive nuclei present at time t and $e = 2.718$.

The rate of decay of any isotope is usually given by its **half-life**. The half-life $(T_{\frac{1}{2}})$ of an isotope is the time required for half of the radioactive nuclei present in the sample to decay. The relationship between the half-life and the decay constant is given by

$$T_{\frac{1}{2}} = \frac{0.693}{\lambda}$$

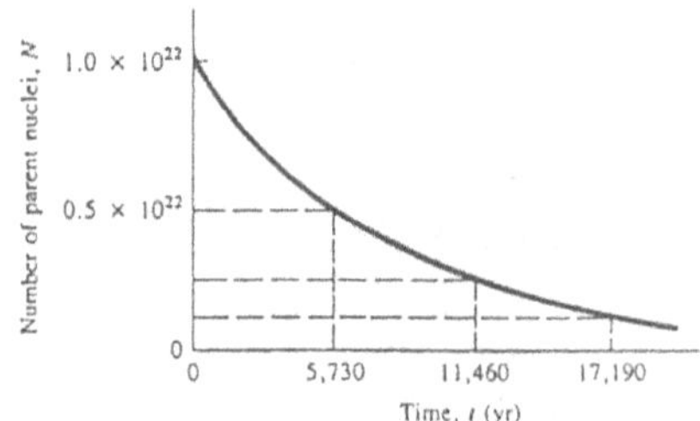

The graph shown here indicates the number of undecayed nuclei (parent nuclei) present as a function of time, where the time is expressed in terms of half-lives. This type of curve is known as an exponential decay curve.
Note: In this diagram, the time is given in terms of the half-life of an isotope of carbon, carbon-14.

EXAMPLE PROBLEM 6. (Section 30-8) A 2.00-microgram sample of pure $^{201}_{81}\text{Tl}$ is to be used in a laboratory experiment. The half-life of the thallium isotope is 73 hours. Determine the (*a*) decay constant of the isotope, and (*b*) number of original nuclei remaining after 292 hours. Note: This radioactive isotope of thallium is commonly used in thallium cardiac stress tests.

Part a. Step 1.

Determine the half-life in seconds.

$$T_{\frac{1}{2}} = (73\text{ h})\left(\frac{3600\text{ s}}{1\text{ h}}\right) = 2.63 \times 10^5\text{ s}$$

Part a. Step 2.

Determine the decay constant λ for this element.

$$\lambda = \frac{0.693}{T_{\frac{1}{2}}} = \frac{0.693}{2.63 \times 10^5\text{ s}}$$

$$\lambda = 2.64 \times 10^{-6}\text{ s}$$

Part b. Step 1.

Determine the number of half-lives that have passed in 292 h.

$$(292\text{ h})\left(\frac{1\text{ half-life}}{73\text{ h}}\right) = 4\text{ half-lives}$$

Part b. Step 2.

Determine the number of moles of the original sample that is present after 4 half-lives.

The number of moles of the original present after 4 half-lives should be $\left(\frac{1}{2}\right)^4$ mole of the original. Thus,

$$\left(\frac{1}{2}\right)^4 (2.00 \times 10^{-6}\text{ g})\left(\frac{1\text{ mole}}{201\text{ g}}\right) = 6.22 \times 10^{-10}\text{ moles}$$

Part b. Step 3.	
Determine the number of nuclei of ${}^{201}_{81}\text{Tl}$ present after 292 h.	$N = (6.22 \times 10^{-10} \text{ moles})(6.02 \times 10^{23}/\text{mole})$ $N = 3.74 \times 10^{14}$ nuclei

ALTERNATE SOLUTION: Use the law of radioactive decay to solve Part b.

Part b. Step 1.	
Determine the number of moles of ${}^{201}_{81}\text{Tl}$ in the original sample.	$(2.0 \times 10^{-6} \text{ g})\left(\dfrac{1 \text{ mole}}{201 \text{ g}}\right) = 9.95 \times 10^{-9}$ moles
Part b. Step 2.	
Determine the number of nuclei present in the original sample.	$N_0 = (9.95 \times 10^{-9} \text{ moles})(6.02 \times 10^{23}/\text{mole})$ $N_0 = 5.99 \times 10^{15}$ nuclei
Part b. Step 3.	
Use the law of radioactive decay to solve for the number of ${}^{201}_{81}\text{Tl}$ nuclei remaining after 292 h.	$N = N_0 e^{-\lambda t}$, where $\lambda = 2.63 \times 10^{-6}$ s, $t = 292 \text{ h} = 1.05 \times 10^{6}$ s, and $\lambda t = (2.64 \times 10^{-6} \text{ s})(1.05 \times 10^{6} \text{ s}) = 2.77$ $N = (5.99 \times 10^{15} \text{ nuclei})e^{-2.77} = (5.99 \times 10^{15} \text{ nuclei})(0.0627)$ $N = 3.74 \times 10^{14}$ nuclei

Decay Series

A radioactive decay often results in a daughter nucleus that is also radioactive. A **radioactive decay series** is a series of successive decays that starts with one parent isotope and proceeds through a number of daughter isotopes. The series ends when a stable, nonradioactive isotope is produced.

Radioactive Dating

If the half-life of a radioactive isotope is known, then an estimate of the age of an object can often be made. For example, the ratio of ${}^{14}_{6}\text{C}$ to ${}^{12}_{6}\text{C}$ in a living object is relatively constant. However, a living object stops absorbing ${}^{14}_{6}\text{C}$ when it dies. Therefore, by knowing the half-life of ${}^{14}_{6}\text{C}$ (5700 years) and the object's ${}^{14}_{6}\text{C}$ to ${}^{12}_{6}\text{C}$ ratio, an estimate of the object's age can be made.

If a rock contains uranium, geologists are able to estimate the age of the rock by determining the amount of ${}^{238}_{92}\text{U}$ in the rock relative to the amount of daughter nuclei that are present.

TEXTBOOK QUESTION 22. (Section 30-11) Can $^{14}_{6}C$ dating be used to measure the age of stone walls and tablets of ancient civilizations?

ANSWER: As stated in the textbook, plants absorb carbon from the air in the form of CO_2. While most of the carbon in the carbon dioxide molecules contain $^{12}_{6}C$, a small fraction contains radioactive $^{14}_{6}C$. Animals eat plants and absorb carbon from the plants they have eaten. Once the animal or plant dies, it stops absorbing $^{14}_{6}C$. The half-life of $^{14}_{6}C$ is about 5730 years, so it is possible to use carbon dating to measure the age of dead plants or animals.

Stone walls and tablets are not made from substances that absorb carbon. Therefore, $^{14}_{6}C$ dating is not useful in determining the age of stone walls or tablets.

PROBLEM SOLVING SKILLS

For problems involving nuclear density:

1. Use the equation $r \approx (1.2 \times 10^{-15}\ \text{m})(A^{\frac{1}{3}})$ to determine the approximate radius of the nucleus.
2. Use the equation $V = \frac{4}{3}\pi r^3$ to determine the approximate volume of the nucleus.
3. Convert the mass of the nucleus from unified atomic mass units to kg. Note: The mass of the nucleus expressed in atomic mass units can be determined by noting the mass number.
4. Use the equation $\rho = \frac{m}{V}$ to determine the approximate density.

For problems related to the binding energy of the nucleus:

1. Use a table of nuclear masses to determine the total mass of the constituent particles that make up the nucleus. Note also the exact atomic mass of the nucleus.
2. Determine the difference between the mass of the nucleus and the mass of the constituent particles.
3. Use the equation $\Delta E = (\Delta m)c^2$ to determine the binding energy.

For questions related to a particular type of nuclear decay when the parent nucleus is known:

1. Take note whether the decay is alpha, beta, or gamma decay.
2. Write down the general equation for the particular type of decay.
3. Use the conservation laws to determine the atomic number and mass number of the daughter nucleus.
4. Use the Periodic Table to identify the daughter element.

For problems related to the calculation of the disintegration energy of an alpha decay:

1. Use a table of nuclear masses to determine the mass of the parent nucleus, daughter nucleus, and alpha particle in atomic mass units.

2. Determine the difference between the mass of the parent nucleus and the mass of the daughter nucleus plus alpha particle. Express this difference in atomic mass units.
3. Determine the disintegration energy by multiplying the mass difference expressed in atomic mass units by 931.5 MeV/u.
4. If the problem asks for the kinetic energy of the daughter nucleus and alpha particle:
   a. Use the law of conservation of momentum to determine the ratio of the velocity of the alpha particle to that of the daughter nucleus.
   b. Determine the ratio of the kinetic energy of the alpha particle to that of the daughter nucleus.
   c. Knowing the total kinetic energy of the daughter nucleus and alpha particle, and the ratio of the two kinetic energies, algebraically solve for the kinetic energy of each particle.

For problems related to radioactive decay:

1. Determine the number of radioactive nuclei in the sample.
2. Use $\Delta N = -\lambda N\,\Delta t$ to determine the activity of the sample.
3. If the problem involves carbon dating, then determine the ratio of decays per second in the artifact to the number of decays per second in the sample.
4. Knowing the half-life of ${}^{14}_{6}\text{C}$, determine the age of the artifact.

For problems related to half-life, decay constant, and the law of radioactive decay.

1. If either the half-life or decay constant is given, use $T = \dfrac{0.693}{\lambda}$ to determine either the half-life or decay constant.
2. Use the law of radioactive decay $N = N_0 e^{-\lambda t}$ to determine the number of nuclei remaining after a certain time has passed.

SOLUTIONS TO SELECTED TEXTBOOK PROBLEMS

TEXTBOOK PROBLEM 2. (Section 30-1) What is the approximate radius of an α particle (${}^{4}_{2}\text{He}$)?

Part a. Step 1.

Use Eq. 30-1 to determine the radius of the nucleus.

The mass number (A) of the alpha particle is 4. The approximate radius of the nucleus may be determined as follows:

$$r \approx (1.2 \times 10^{-15}\,\text{m})(A^{\frac{1}{3}})$$

$$r \approx (1.2 \times 10^{-15}\,\text{m})(4^{\frac{1}{3}})$$

$$r \approx 1.9 \times 10^{-15}\,\text{m} \approx 1.9\,\text{fm}$$

TEXTBOOK PROBLEM 18. (Section 30-2) Calculate the total binding energy, and the binding energy per nucleon, for (*a*) ${}^{7}_{3}\text{Li}$, (*b*) ${}^{195}_{78}\text{Pt}$. Use Appendix B.

Part a. Step 1.

Use Appendix B to determine the masses of a neutron, an ${}^{1}\text{H}$ atom, and a ${}^{7}_{3}\text{Li}$ nucleus.

From Appendix B, $m_{\text{neutron}} = 1.008665\ \text{u}$, $m_{1\text{H}} = 1.007825\ \text{u}$, and $m({}^{7}_{3}\text{Li}) = 7.016003\,\text{u}$.

Part a. Step 2.

Determine the mass difference (Δm) between the nucleons and the ${}^{7}_{3}\text{Li}$ nucleus.

${}^{7}_{3}\text{Li}$ contains 4 neutrons and 3 protons:

$$\Delta m = 4(1.008665\ \text{u}) + 3(1.007825\ \text{u}) - 7.016003\ \text{u}$$

$$\Delta m = 4.03466\ \text{u} + 3.023475\ \text{u} - 7.016003\ \text{u}$$

$$\Delta m = 0.042132\ \text{u}$$

Part a. Step 3.

Determine the total binding energy for ${}^{7}_{3}\text{Li}$.

Note: The binding energy per nucleon $= 931.5\ \text{MeV}$, and $1\ \text{u} = 931.5\ \text{MeV}/c^2$:

$$E = (\Delta m)c^2$$

$$= [(0.042132\ \text{u})(931.5\ \text{MeV})/c^2/\text{u})]c^2$$

$$E = 39.2\ \text{MeV}$$

Part a. Step 4.

Determine the binding energy per nucleon.

$$\text{Binding energy per nucleon} = \frac{\text{total binding energy}}{\text{number of nucleons}}$$

$$= \frac{39.2\ \text{MeV}}{7}$$

$$\text{Binding energy per nucleon} = 5.61\ \text{MeV}$$

Part b. Step 1.

Use Appendix B to determine the mass of a neutron, an ${}^{1}\text{H}$ atom, and a ${}^{195}_{78}\text{Pt}$ atom.

From Appendix B, $m_{\text{neutron}} = 1.008665\ \text{u}$, $m_{1\text{H}} = 1.007825\ \text{u}$, and $m({}^{195}_{78}\text{Pt}) = 194.964792\ \text{u}$.

Part b. Step 2.

Determine the mass difference (Δm) between the nucleons and the $^{195}_{78}\text{Pt}$ atom.	$^{195}_{78}\text{Pt}$ contains 117 neutrons and 78 protons: $\Delta m = (117)(1.008665\text{ u}) + (78)(1.007825\text{ u}) - 194.964792\text{ u}$ $= 118.01381\text{ u} + 78.61035\text{ u} - 194.964774\text{ u}$ $\Delta m = 1.65937\text{ u}$

Part b. Step 3.

Determine the total binding energy.	Note: The binding energy per nucleon $= 931.5$ MeV: $E = (\Delta m)c^2 = [(1.65937\text{ u})(931.5\text{ MeV}/c^2/\text{u})]c^2$ $E = 1545.70\text{ MeV}$

Part a. Step 4.

Determine the binding energy per nucleon.	$\text{Binding energy per nucleon} = \dfrac{\text{total binding energy}}{\text{number of nucleons}}$ $= \dfrac{1545.70\text{ MeV}}{195}$ Binding energy per nucleon = 7.93 MeV

TEXTBOOK PROBLEM 26. (Section 30-3 to 30-7) Give the result of a calculation that shows whether or not the following decays are possible:

(*a*) $^{233}_{92}\text{U} \rightarrow {}^{232}_{92}\text{U} + \text{n};$

(*b*) $^{14}_{7}\text{N} \rightarrow {}^{13}_{7}\text{N} + \text{n};$

(*c*) $^{40}_{19}\text{K} \rightarrow {}^{39}_{19}\text{K} + \text{n}.$

Part a. Step 1.

Use Appendix B to determine the masses of the neutron and the $^{233}_{92}\text{U}$ and $^{232}_{92}\text{U}$ nuclei.	From Appendix B, $m_{\text{neutron}} = 1.008665\text{ u}$, $^{233}_{92}\text{U} = 233.039636\text{ u}$, and $^{232}_{92}\text{U} = 232.037156\text{ u}$.

Part a. Step 2.

Determine the mass difference (Δm) between the ${}^{233}_{92}\text{U}$ nuclei and ${}^{232}_{92}\text{U}$ nuclei plus neutron.	$\Delta m = 233.039636\text{ u} - 232.037156\text{ u} - 1.008665\text{ u}$ $\Delta m = -0.006185\text{ u}$ In order for the decay to occur, Δm must be positive. Since Δm is negative, the decay is not possible.

Part b. Step 1.

Use Appendix B to determine the masses of the neutron and the ${}^{14}_{7}\text{N}$ and ${}^{13}_{7}\text{N}$ nuclei.	From Appendix B, $m_{\text{neutron}} = 1.008665\text{ u}$, ${}^{14}_{7}\text{N} = 14.003074\text{ u}$, and ${}^{13}_{7}\text{N} = 13.005739\text{ u}$.

Part b. Step 2.

Determine the mass difference (Δm) between the ${}^{14}_{7}\text{N}$ nuclei and ${}^{13}_{7}\text{N}$ nuclei plus neutron.	$\Delta m = 14.003074\text{ u} - 13.005739\text{ u} - 1.008665\text{ u}$ $\Delta m = -0.011330\text{ u}$ In order for the decay to occur, Δm must be positive. Since Δm is negative, the decay is not possible.

Part c. Step 1.

Use Appendix B to determine the masses of the neutron and the ${}^{40}_{19}\text{K}$ and ${}^{39}_{19}\text{K}$ nuclei.	From Appendix B, $m_{\text{neutron}} = 1.008665\text{ u}$, ${}^{40}_{19}\text{K} = 39.963998\text{ u}$, and ${}^{39}_{19}\text{K} = 38.963706\text{ u}$.

Part c. Step 2.

Determine the mass difference (Δm) between the ${}^{40}_{19}\text{K}$ and the ${}^{39}_{19}\text{K}$ nuclei plus neutron.	$\Delta m = 39.963998\text{ u} - 38.963706\text{ u} - 1.008665\text{ u}$ $\Delta m = -0.008373\text{ u}$ In order for the decay to occur, Δm must be positive. Since Δm is negative, the decay is not possible.

TEXTBOOK PROBLEM 28. (Sections 30-3 to 30-7) A $^{238}_{92}\text{U}$ nucleus emits an α particle with kinetic energy = 4.20 MeV. (*a*) What is the daughter nucleus, and (*b*) what is the approximate atomic mass (in u) of the daughter atom? Ignore recoil of the daughter nucleus.

Part a. Step 1.

Determine the atomic number and mass number of the unknown nucleus.	An α particle has an atomic number of 2 and a mass number of 4. The $^{238}_{92}\text{U}$ nucleus has an atomic number of 92 and a mass number of 238. The atomic number of the unknown element $(\text{X}) = 92 - 2 = 90.$ The mass number of the unknown element $(\text{X}) = 238 - 4 = 234.$

Part a. Step 2.

Use the Periodic Table to identify the unknown nucleus.	The unknown element is $^{234}_{90}\text{Th},$ or thorium 234.

Part b. Step 1.

Use Appendix B to determine the masses of $^{238}_{92}\text{U}$ and the α particle.	$m(^{238}_{92}\text{U}) = 238.050788\ \text{u}$ $m_\alpha = 4.002603\ \text{u}$

Part b. Step 2.

Determine the mass difference between $^{238}_{92}\text{U},\ \alpha,$ and $^{234}_{90}\text{Th}.$

$$\text{KE} = (\Delta m)c^2$$

$$4.20\ \text{MeV} = (\Delta m)c^2$$

$$\Delta m = (4.20\ \text{MeV})/c^2 = (4.20\ \text{MeV})\left(\frac{1.6\times10^{-13}\ \text{J/MeV}}{9\times10^{16}\ \text{m}^2/\text{s}^2}\right)$$

$$\Delta m = (7.47\times10^{-30}\ \text{kg})\left(\frac{1\ \text{u}}{1.66\times10^{-27}\ \text{kg}}\right) = 0.00450\ \text{u}$$

Part b. Step 3.

Determine the mass of the daughter atom, $^{234}_{90}\text{Th}.$

$$\Delta m = m(^{238}_{92}\text{U}) - (^{234}_{90}\text{Th}) - m_\alpha$$

$$0.00450\ \text{u} = 238.050788\ \text{u} - m(^{234}_{90}\text{Th}) - 4.002603\ \text{u}$$

$$0.00450\ \text{u} = 234.048185\ \text{u} - m(^{234}_{90}\text{Th})$$

$$m(^{234}_{90}\text{Th}) = 234.043685\ \text{u}$$

TEXTBOOK PROBLEM 39. (Sections 30-8 to 30-11) A radioactive material produces 1120 decays per minute at one time, and 3.6 h later produces 140 decays per minute. What is its half-life?

Part a. Step 1.

Determine the decay constant.

$$\frac{\Delta N}{\Delta t} = \left(\frac{\Delta N}{\Delta t}\right)_0 e^{-\lambda t}$$

$$140 \text{ decays/min} = (1120 \text{ decays/min})\, e^{-\lambda(3.6\text{ h})}$$

$$0.125 = e^{-\lambda(3.6\text{ h})}$$

$$\ln(0.125) = \ln e^{-\lambda(3.6\text{ h})}$$

$$-2.08 = -\lambda(3.6\text{ h}) \ln e \quad \text{but} \quad \ln e = 1$$

$$\lambda = 0.578/\text{h}$$

Part a. Step 2.

Determine the half-life.

$$T_{\frac{1}{2}} = (0.693)/\lambda = (0.693)/(0.578/\text{h})$$

$$T_{\frac{1}{2}} = 1.20 \text{ hours}$$

TEXTBOOK PROBLEM 42. (Section 30-8) What fraction of a radioactive sample is left after exactly 5 half-lives?

Part a. Step 1.

Complete a data table listing the information provided.

Let N_0 = original number of nuclei:

N = number of nuclei remaining after the number of half-lives

n = number of half-lives that have passed

Part a. Step 2.

Determine the percent remaining after $n = 5$.

$$\frac{N}{N_0} = (\tfrac{1}{2})^n$$

$$\frac{N}{N_0} = (\tfrac{1}{2})^5 = 0.0313$$

Only 3.13% of the original nuclei remain after 5 half-lives.

TEXTBOOK PROBLEM 43. (Sections 30-8 to 30-11) The iodine isotope ${}^{131}_{53}\text{I}$ is used in hospitals for diagnosis of thyroid function. If 782 μg are ingested by a patient, determine the activity (*a*) immediately, (*b*) 1.5 h later when the thyroid is being tested, and (*c*) 3.0 months later. Use Appendix B.

Part a. Step 1.

Use Appendix B to determine the half-life of ${}^{131}_{53}\text{I}$.

From Appendix B, $T_{\frac{1}{2}} = 8.0252$ days. Converting the half-life to seconds gives

$$(8.0252\text{ days})\left(\frac{24\text{ hours}}{1\text{ day}}\right)\left(\frac{3600\text{ s}}{1\text{ hour}}\right) = 6.93 \times 10^{5}\text{ s}$$

Part a. Step 2.

Determine the decay constant.

$$\lambda = (0.693)/T_{\frac{1}{2}}$$

$$\lambda = \frac{0.693}{6.93 \times 10^{5}\text{ s}} = 1.00 \times 10^{-6}\text{ s}$$

Part a. Step 3.

Determine the number of nuclei in the initial sample. Note: $782\ \mu\text{g} = 7.82 \times 10^{-4}$ g.

1 mole of ${}^{131}_{53}\text{I} = 130.906$ g

$$N_0 = (7.82 \times 10^{-4}\text{ g})\left(\frac{1\text{ mole}}{130.906\text{ g}}\right)\left(\frac{6.02 \times 10^{23}\text{ atoms}}{1\text{ mole}}\right)$$

$$N_0 = 3.60 \times 10^{18}\text{ nuclei}$$

Part a. Step 4.

Determine the number of decays at $t = 0$.

$$\frac{\Delta N}{\Delta t} = -\lambda N_0 e^{-\lambda t}$$

$$= -(1.00 \times 10^{-6}/\text{s})(3.60 \times 10^{18}\text{ nuclei})e^{-(1.00 \times 10^{-6}/\text{s})(0)}$$

$$\frac{\Delta N}{\Delta t} = -3.60 \times 10^{12}\text{ decays/s}$$

Part b. Step 1.

Determine λt at $t = 1.5$ hour.

1.5 h = 5400 s

$$\lambda t = (1.00 \times 10^{-6}/\text{s})(5400\text{ s}) = 0.00540$$

Part b. Step 2.

Determine the number of decays per second at 1.5 h.

$$\frac{\Delta N}{\Delta t} = \left(\frac{\Delta N}{\Delta t}\right)_0 e^{-\lambda t}$$

$$= (3.60 \times 10^{12} \text{ decays/s})\, e^{-0.00540}$$

$$\frac{\Delta N}{\Delta t} = 3.60 \times 10^{12} \text{ decays/s}$$

Part c. Step 1.

Determine λt at 3.0 months.

$$t \approx (3 \text{ months})\left(\frac{1 \text{ year}}{12 \text{ months}}\right)\left(\frac{3.156 \times 10^7 \text{ s}}{1 \text{ year}}\right)$$

$$t \approx 7.89 \times 10^6 \text{ s}$$

$$\lambda t \approx (1.00 \times 10^{-6} \text{ s})(7.89 \times 10^6 \text{ s}) \approx 7.89$$

Part c. Step 2.

Determine the approximate number of decays after 3.0 months.

$$\frac{\Delta N}{\Delta t} = \left(\frac{\Delta N}{\Delta t}\right)_0 e^{-\lambda t}$$

$$\approx (3.60 \times 10^{12} \text{ decays/s}) e^{-7.89}$$

$$\frac{\Delta N}{\Delta t} \approx (3.60 \times 10^{12} \text{ decays/s})(3.74 \times 10^{-4})$$

$$\frac{\Delta N}{\Delta t} \approx 1.35 \times 10^9 \text{ decays/s}$$

TEXTBOOK PROBLEM 55. (Section 30-11) An ancient wooden club is found that contains 73 g of carbon and has an activity of 7.0 decays per second. Determine its age assuming that in living trees the ratio of $^{14}\text{C}/^{12}\text{C}$ atoms is about 1.3×10^{-12}.

Part a. Step 1.

Determine the approximate number of ^{12}C nuclei in the sample.

The percentage of ^{14}C is extremely small compared with the ^{12}C; therefore,

$$N(^{12}\text{C}) \approx (73 \text{ g C})\left(\frac{1 \text{ mol } ^{12}\text{C}}{12.0 \text{ g } ^{12}\text{C}}\right)\left(\frac{6.02 \times 10^{23} \text{ atoms}}{1 \text{ mole}}\right)$$

$$N(^{12}\text{C}) \approx 3.66 \times 10^{24} \text{ nuclei}$$

Part a. Step 2.

Determine the number of ^{14}C atoms in 73 g of carbon when the club was first made.

$$\frac{N(^{14}\text{C})}{N(^{12}\text{C})} \approx 1.3 \times 10^{-12}$$

$$\frac{N(^{14}\text{C})}{3.66 \times 10^{24}\ \text{nuclei}} \approx 1.3 \times 10^{-12}$$

$$N(^{14}\text{C}) \approx 4.76 \times 10^{12}\ \text{nuclei}$$

Part a. Step 3.

Determine the decay constant for carbon-14.

$$\lambda = \frac{0.693}{T_{\frac{1}{2}}} \quad \text{but the half-life of} \quad {}^{14}_{6}\text{C} = 5700\ \text{years:}$$

$$\lambda = \frac{0.693}{5700\ \text{years}}$$

$$\lambda = 1.21 \times 10^{-4}\ \text{decays/year} = 3.83 \times 10^{-12}\ \text{decays/s}$$

Part a. Step 4.

Determine the approximate age of the club.

$$\lambda N = \lambda N_0 e^{-\lambda t}, \quad \text{where} \quad N_0 = 4.76 \times 10^{12}\ \text{nuclei}$$

$$7.0\ \text{decays/s} = (3.83 \times 10^{-12}\ \text{decays/s})(4.76 \times 10^{12}\ \text{nuclei})\, e^{-\lambda t}$$

$$7.0\ \text{decays/s} = (18.2\ \text{decays/s})\, e^{-\lambda t}$$

$$e^{-\lambda t} = 0.385$$

$$\ln e^{-\lambda t} = \ln(0.385) \quad \text{but} \quad \ln e^{-\lambda t} = -\lambda t \ln e = -\lambda t$$

$$-\lambda t = -0.955$$

$$t = \frac{-0.955}{-3.83 \times 10^{-12}\ \text{decays/s}}$$

$$t \approx (2.49 \times 10^{11}\ \text{s}) \approx 7900\ \text{years}$$

31

NUCLEAR ENERGY; EFFECTS AND USES OF RADIATION

OBJECTIVES

After studying the material of this chapter, the student should be able to:

- explain what is meant by a nuclear reaction.
- write the general equation for a nuclear reaction in both the long form and the short form. Explain what each symbol in the equation represents.
- given a problem involving a nuclear reaction that is written in the short form, determine the missing particle or nucleus.
- calculate the reaction energy (Q-value) for a nuclear reaction and state whether the reaction is exothermic or endothermic.
- distinguish between nuclear fission and fusion and give an example of each process.
- explain what is meant by a self-sustaining chain reaction and how this reaction is kept under control in a nuclear reactor.
- explain what is meant by dosimetry.
- define the curie, roentgen, rad, gray, and sievert. Explain what is meant by the term rem.
- given the level of activity of a radioactive sample in curies, calculate the number of decays per second and the yearly dosage in rem absorbed by a person in contact with the sample. Determine whether the dosage exceeds the recommended standard.
- describe three devices that can be used to detect the presence and level of activity of radiation.

KEY TERMS AND PHRASES

nuclear reaction occurs when two nuclei collide and two (or more) nuclei are produced. In the process, the nucleus of one element is changed into a different element; therefore, a **transmutation** of elements has occurred.

threshold energy is the minimum energy for a reaction to occur.

nuclear fission is the process in which a nucleus is split into approximately equal parts along with the release of neutrons and a large amount of energy.

nuclear chain reaction occurs when the neutrons released in a typical fission cause other nuclei to undergo fission.

critical mass is the minimum amount of fissionable material required for a self-sustaining chain reaction to occur.

nuclear reactor is a device in which the chain reaction is controlled and the energy is released gradually.

nuclear fusion is the process by which small nuclei combine to form heavier nuclei.

thermonuclear device, or hydrogen bomb, is a device in which uncontrolled fusion reactions release enormous amount of energy in a few moments of time.

dosimetry refers to measuring the quantity, or **dose**, of radiation that passes through a material.

1 **rad** is the amount of radiation that deposits 10^{-2} J of energy per kg of absorbing material. The SI unit for absorbed dose is the **gray** (Gy), where $1\ \text{Gy} = 1\ \text{J/kg} = 100\ \text{rad}$.

rem (rad equivalent man) is a measure of the biological damage caused by radiation.

relative biological effectiveness (RBE) of a certain type of radiation is defined as the number of rad of X-ray or γ radiation that produces the same biological damage as 1 rad of the given radiation.

SUMMARY OF MATHEMATICAL FORMULAS

nuclear reaction	$\text{a} + \text{X} \rightarrow \text{Y} + \text{b}$	A nuclear reaction occurs when two nuclei collide and two (or more) nuclei are produced. In the process, the nucleus of one element is changed into a different element; therefore, a transmutation of elements has occurred. Here a is the bombarding particle, sometimes called the bullet; X is the nucleus hit by the bullet, usually referred to as the target nucleus; b is a small particle, or possibly a photon, and is called the product particle; and Y is the recoil nucleus.
reaction energy or Q-value	$Q = (M_\text{a} + M_\text{X} - M_\text{b} - M_\text{Y})c^2$ $Q = (\Delta m)c^2$ or $Q = \text{KE}_\text{b} + \text{KE}_\text{Y} - \text{KE}_\text{a} - \text{KE}_\text{X}$	The reaction energy (or Q-value) is related to the mass defect (Δm) by Einstein's equation. Here is a projectile particle or small nucleus that strikes nucleus X, producing nucleus Y and particle b. Q can also be written in terms of the change in kinetic energy.
nuclear fission	A typical fission reaction is ${}^{1}_{0}\text{n} + {}^{235}_{92}\text{U} \rightarrow$ ${}^{90}_{38}\text{Sr} + {}^{136}_{54}\text{Xe} + 10\,{}^{1}_{0}\text{n}$	Nuclear fission is the process in which a nucleus is split into approximately equal parts along with the release of neutrons and a large amount of energy. Fission occurs only in the nuclei of certain elements. These elements have a large nucleon number, such as ${}^{235}_{92}\text{U}$. The nucleus may undergo fission if struck by a slow-moving neutron.

dosimetry: curie	$1\ \text{Ci} = 3.70 \times 10^{10}$ disintegrations per second $1\ \mu\text{Ci} = 10^{-6}\ \text{Ci}$	Dosimetry refers to measuring the quantity, or dose, of radiation that passes through a material. The activity of a radioactive isotope is measured in terms of curies (Ci) or microcuries (μCi).
dosimetry: roentgen	One roentgen (1 R) is the amount of radiation that will produce 1.6×10^{12} ion pairs/g of dry air at STP. Today the roentgen is defined as that amount of X-ray or γ radiation that deposits 0.878×10^{-2} J of energy per kg of air.	The roentgen (R) is a unit of dosage that was previously defined in terms of the amount of ionized air produced by the radiation.
dosimetry: rad radiation equivalent man (rem) and relative biological effectiveness (RBE)	1 rad is the amount of radiation that deposits 10^{-2} J of energy per kg of absorbing material. Effective dose (in rem) = dose (in rad) × RBE effective dose (in sievert) = dose (Gy) × RBE	The roentgen has been replaced by the rad. The rem (rad equivalent man) is the unit that refers to the biological damage caused by radiation. RBE is the relative biological effectiveness of a certain type of radiation. It is defined as the number of rad of X-ray or γ radiation that produces the same biological damage as 1 rad of the given radiation. The RBE of X-rays and γ rays is 1.0; β particles, 1.0; α particles, 10–20; slow neutrons, 3–5; and fast neutrons and protons, 10.

CONCEPT SUMMARY

Nuclear Reactions and Transmutation of Elements

A **nuclear reaction** occurs when two nuclei collide and two (or more) nuclei are produced. In the process, the nucleus of one element is changed into a different element; therefore, a **transmutation** of elements has occurred.

A nuclear reaction can be written either as $a + X \rightarrow Y + b$ or as $X\ (a, b)\ Y$

Here a is the bombarding particle, sometimes called the bullet. X is the nucleus hit by the bullet, usually referred to as the target nucleus. b is a small particle, or possibly a photon, and is called the product particle. Y is the recoil nucleus.

For example, in the reaction ${}^{1}_{1}\text{H} + {}^{23}_{11}\text{Na} \rightarrow {}^{20}_{10}\text{Ne} + {}^{4}_{2}\text{He}$, a is ${}^{1}_{1}\text{H}$, X is ${}^{23}_{11}\text{Na}$, Y is ${}^{20}_{10}\text{Ne}$, and b is ${}^{4}_{2}\text{He}$.

The proton ${}^{1}_{1}\text{H}$ may be written as p; the alpha particle ${}^{4}_{2}\text{He}$ may be written as α.

TEXTBOOK QUESTION 1b. (Section 31-1) Fill in the missing particles or nuclei:

$\text{n} + {}^{137}_{56}\text{Ba} \rightarrow {}^{137}_{55}\text{Cs} + ?$ [n stands for a neutron, ${}^{1}_{0}\text{n}$]

ANSWER: The equation of the reaction is $\text{n} + {}^{137}_{56}\text{Ba} \rightarrow {}^{137}_{55}\text{Cs} + {}^{A}_{Z}\text{X}$

Conservation of nucleon number	Conservation of charge
$1 + 137 = 137 + A$	$0 + 56 = 55 + Z$
$138 = 137 + A$	$Z = +1$
$A = 1$	

Therefore, $A = 1$ and $Z = 1$. Using the Periodic Table, it is found that the atomic number of hydrogen is 1. Therefore, the resulting nuclide is a proton, ${}^{1}_{1}\text{H}$.

TEXTBOOK QUESTION 1d. (Section 31-1) Fill in the missing particles or nuclei: $\alpha + {}^{197}_{79}\text{Au} \rightarrow ? + \text{d}$. [d stands for deuterium]

The equation of the reaction is ${}^{4}_{2}\text{He} + {}^{197}_{79}\text{Au} \rightarrow {}^{A}_{Z}\text{X} + {}^{2}_{1}\text{H}$

Conservation of nucleon number	Conservation of charge
$197 + 4 = A + 2$	$79 + 2 = Z + 1$
$201 = A + 2$	$81 = Z + 1$
$A = 199$	$Z = 80$

Therefore, $A = 199$ and $Z = 80$. Using the Periodic Table, it is found that the atomic number of mercury is 80. Therefore, the resulting nuclide is ${}^{199}_{80}\text{Hg}$.

TEXTBOOK QUESTION 2. (Section 31-1) When ${}^{22}_{11}\text{Na}$ is bombarded by deuterons (${}^{2}_{1}\text{H}$), an α particle is emitted. What is the resulting nuclide? Write down the reaction equation.

ANSWER: The reaction is written as follows: ${}^{22}_{11}\text{Na} + {}^{2}_{1}\text{H} \rightarrow {}^{A}_{Z}\text{X} + {}^{4}_{2}\text{He}$.

Conservation of nucleon number	Conservation of charge
$22 + 2 = A + 4$	$11 + 1 = Z + 2$
$24 = A + 4$	$12 = Z + 2$
$A = 20$	$Z = 10$

Therefore, $A = 20$ and $Z = 10$. Using the Periodic Table, we find that the atomic number of neon is 10. Therefore, the resulting nuclide is ${}^{20}_{10}\text{Ne}$.

EXAMPLE PROBLEM 1. (Section 31-1) Fill in the missing particle or nucleus. (*a*) ${}^{35}_{17}\text{Cl}$ (?, α) ${}^{32}_{16}\text{S}$ and (*b*) ${}^{200}_{80}\text{Hg}$ (p, ?) ${}^{197}_{79}\text{Au}$

Part a. Step 1.	
Write the reaction in the long form.	${}^{35}_{17}\text{Cl} + ? \to {}^{32}_{16}\text{S} + {}^{4}_{2}\text{He}$
Part a. Step 2.	
Apply conservation of charge and determine the atomic number of the particle.	$17 + ? = 16 + 2$; therefore, the atomic number of the bombarding particle is 1, and the element is hydrogen (H).
Part a. Step 3.	
Apply conservation of nucleon number and identify the nuclide.	$35 + ? = 32 + 4$; therefore, the mass number is 1, and the bombarding particle is a proton, ${}^{1}_{1}\text{H}$.
Part b. Step 1.	
Write the reaction in the long form.	${}^{200}_{80}\text{Hg} + {}^{1}_{1}\text{H} \to {}^{197}_{79}\text{Au} + ?$
Part b. Step 2.	
Apply conservation of charge and determine the atomic number of the particle.	$80 + 1 = 79 + ?$; therefore, the atomic number of the product particle is 2, and the element is helium (He).
Part b. Step 3.	
Apply conservation of nucleon number and identify the nuclide.	$200 + 1 = 197 + ?$; therefore, the mass number of the product particle is 4. The product particle is an alpha particle, ${}^{4}_{2}\text{He}$.

EXAMPLE PROBLEM 2. (Section 31-1) Fill in the missing particle or nucleus. (*a*) ${}^{24}_{12}\text{Mg}+\text{d}\rightarrow ?+\alpha$, (*b*) ${}^{12}_{6}\text{C}+\text{d}\rightarrow \text{n}+?$

Part a. Step 1.	
Rewrite the reaction.	Here d represents a deuteron, ${}^{2}_{1}\text{H}$, while α represents an alpha particle, ${}^{4}_{2}\text{He}$: ${}^{24}_{12}\text{Mg}+{}^{2}_{1}\text{H}\rightarrow ?+{}^{4}_{2}\text{He}$
Part a. Step 2.	
Apply the law of conservation of charge to determine the atomic number.	$12+1=?+2$; thus the atomic number of the unknown nucleus is 11, and the element is sodium (Na).
Part a. Step 3.	
Apply the law of conservation of nucleon number and use the Periodic Table to identify the unknown particle.	$24+2=?+4$; thus the mass number is 22. The unknown nucleus is ${}^{22}_{11}\text{Na}$.
Part b. Step 1.	
Rewrite the reaction.	${}^{12}_{6}\text{C}+{}^{2}_{1}\text{H}\rightarrow ?+{}^{1}_{0}\text{n}$
Part b. Step 2.	
Use conservation of charge.	$6+1=?+0$; therefore, the atomic number of the nucleus is 7, and the element is nitrogen (N).
Part b. Step 3.	
Use conservation of nucleon number to identify the unknown nuclide.	$12+2=?+1$, and the mass number of the nucleus is 13. The unknown nucleus is ${}^{13}_{7}\text{N}$.

Reaction Energy, or Q-Value

In any nuclear reaction, all the conservation laws hold; therefore, it is possible to determine whether a particular reaction can occur. If, in a reaction, the mass of the products is less than the mass of the reactants, then it is possible for the reaction to occur and energy to be released to account for the missing mass. However, in some reactions, the mass of the products may be greater than that of the reactants. In such a case, the reaction will not occur unless the bombarding particle has sufficient kinetic energy. The **reaction energy** (or ***Q*-value**) can be determined as follows:

$$Q = (M_a + M_X - M_b - M_Y)c^2$$

where a is a projectile particle or small nucleus that strikes nucleus X, producing nucleus Y and particle b. Q can also be written in terms of the change in kinetic energy as follows:

$$Q = KE_b + KE_Y - KE_a - KE_X$$

If $Q > 0$, then the reaction is exothermic or exoergic, and energy is released in the reaction. If $Q < 0$, then the reaction is endothermic or endoergic, and the total kinetic energy of the projectile particle and the target particle must be greater than or equal to a minimum amount for the reaction to occur. This minimum energy is called the **threshold energy**. The threshold energy, when added to Q, results in an energy great enough to allow the final products to have velocities that obey the law of conservation of momentum and the law of conservation of energy.

EXAMPLE PROBLEM 3. (Section 31-1) Determine the Q-value of the following reaction. State whether the reaction is exothermic or endothermic. ${}^3_2\text{He} + {}^1_0\text{n} \rightarrow {}^2_1\text{H} + {}^2_1\text{H}$

Part a. Step 1.

Determine the total mass of the reactants and the total mass of the products.

Reactants		Products	
${}^3_2\text{He}$	3.01603 u	${}^2_1\text{H}$	2.01410 u
${}^1_0\text{n}$	1.00867 u	${}^2_1\text{H}$	2.01410 u
	4.02470 u		4.02820 u

Part a. Step 2.

Determine the mass difference between the products and the reactants.

Mass difference = 4.02470 u − 4.02820 u

Mass difference = −0.00350 u

The products have greater mass than the reactants; therefore, the reaction is endothermic. Energy must be added for the reaction to happen.

Part a. Step 3.

Determine the value of Q in joules and MeV.

$$Q = (M_a + M_X - M_b - M_Y)c^2$$

$$Q = (\Delta m)c^2$$

$$\Delta m = (-0.00350 \text{ u})(1.66 \times 10^{-27} \text{ kg/u}) = -5.81 \times 10^{-30} \text{ kg}$$

$$Q = (-5.81 \times 10^{-30} \text{ kg})(3 \times 10^8 \text{ m/s})^2 = -5.229 \times 10^{-13} \text{ J}$$

$$Q = (-5.229 \times 10^{-13} \text{ J})(1 \text{ MeV}/1.6 \times 10^{-13} \text{ J}) = -3.26 \text{ MeV}$$

Alternate solution:

$Q = (-0.00350\text{ u})(931.5\text{ MeV/u}) = -3.26\text{ MeV}$

Since $Q < 0,$ the reaction is endothermic or endoergic, and the total kinetic energy of the projectile particle and the target particle must be greater than or equal to a minimum amount for the reaction to occur.

Nuclear Fission

Nuclear fission is the process in which a nucleus is split into approximately equal parts along with the release of neutrons and a large amount of energy. Fission occurs only in the nuclei of certain elements with a large nucleon number, for example, ${}^{235}_{92}\text{U}$ and ${}^{239}_{94}\text{Pu}.$ Both of these nuclei may undergo fission if struck by a slow-moving neutron. A typical fission reaction is

$${}^{1}_{0}\text{n} + {}^{235}_{92}\text{U} \rightarrow {}^{90}_{38}\text{Sr} + {}^{136}_{54}\text{Xe} + 10\,{}^{1}_{0}\text{n}$$

It can be shown that in this reaction, the sum of the masses of the reactants is greater than the sum of the masses of the products. The missing mass is converted to energy, $Q > 0,$ and the amount of energy released can be found by using $E = (\Delta m)c^2.$

EXAMPLE PROBLEM 4. (Section 31-2) Determine the energy released by the following fission reaction: ${}^{235}_{92}\text{U} + {}^{1}_{0}\text{n} \rightarrow {}^{138}_{56}\text{Ba} + {}^{93}_{41}\text{Nb} + 5\,{}^{1}_{0}\text{n} + 5\,{}^{0}_{-1}\text{e}.$

Part a. Step 1.

Determine the total mass of the reactants and the total mass of the products.	Reactants		Products	
	${}^{235}_{92}\text{U}$	235.0439 u	${}^{138}_{56}\text{Ba}$	137.9052 u
	${}^{1}_{0}\text{n}$	1.0087 u	${}^{93}_{41}\text{Nb}$	92.9064 u
		236.0526 u	$5\,{}^{1}_{0}\text{n}$	5.0435 u
			$5\,{}^{0}_{-1}\text{e}$	0.00274 u
				235.8578 u

Part a. Step 2.

Determine the mass difference.	Mass difference = 236.0526 u − 235.8578 u
	Mass difference = 0.1948 u

Part a. Step 3.

Determine the energy released in the reaction.	The reactants have a greater mass than the products; therefore, the reaction can occur spontaneously.
	$(0.1948\text{ u})(931.5\text{ MeV/u}) = 181\text{ MeV}$

Nuclear Chain Reaction

The neutrons released in a typical fission can be used to cause other nuclei to undergo fission. If enough ${}^{235}_{92}\text{U}$ or ${}^{239}_{94}\text{Pu}$ is available, it is possible to create a **self-sustaining chain reaction**. The minimum amount of fissionable material required for a self-sustaining chain reaction is known as the **critical mass**. The critical mass depends on a number of factors, including the moderator used to slow the neutrons since the probability of causing ${}^{235}_{92}\text{U}$ or ${}^{239}_{94}\text{Pu}$ to fission increases with slow-moving neutrons; the type of fuel used, such as ${}^{235}_{92}\text{U}$ or ${}^{239}_{94}\text{Pu}$, and whether or not the fuel is enriched. Only 0.7% of naturally occurring uranium is the fissionable ${}^{235}_{92}\text{U}$, and in order to have a self-sustaining chain reaction, the percentage of ${}^{235}_{92}\text{U}$ must be increased.

In a **nuclear reactor**, the chain reaction is controlled and the energy is released gradually. In an atomic bomb, the chain reaction is uncontrolled and the energy release occurs in a few moments of time.

TEXTBOOK QUESTION 11. (Section 31-2) Why must the fission process release neutrons if it is to be useful?

ANSWER: In order for a nucleus such as ${}^{235}_{92}\text{U}$ to undergo fission, it must absorb a neutron. In order for a reaction to be self-sustaining, the ${}^{235}_{92}\text{U}$ nucleus that fissions must produce neutrons that can then be absorbed by other ${}^{235}_{92}\text{U}$ nuclei. Without this release of at least one or more neutrons produced per fission, the fission would not be self-sustaining and would not be useful.

TEXTBOOK QUESTION 3. (Section 31-1) Why are neutrons such good projectiles for producing nuclear reactions?

ANSWER: Neutrons, which are electrically neutral, are not repelled by the positive charges located in the nucleus of the atom. As a result they can approach the nucleus to the point where the strong nuclear force plays a dominant role.

Nuclear Fusion

Nuclear fusion is the process by which small nuclei combine to form heavier nuclei. In the reaction, large amounts of energy are released. The force of electrostatic repulsion keeps the nuclei from combining; therefore, the fusion of light nuclei does not occur spontaneously. In order for fusion to happen, it is necessary for the nuclei to collide while traveling at high velocity. Such velocities do occur if the nuclei are part of a hot gas in which the temperature approximates 100 million kelvins. Temperatures of this magnitude are found in the interior of stars, and thus fusion accounts for the enormous energy released by the Sun and other stellar objects.

Uncontrolled fusion reactions have been achieved in the form of the thermonuclear, or hydrogen, bomb. This is because the temperature required for the fusion is achieved by detonating a fission or atomic bomb. The fission bomb creates the necessary temperatures and pressures for the fusion process to occur. However, the difficulty of achieving and sustaining the very high temperatures needed for fusion has thus far prevented the construction of a controlled fusion reactor.

TEXTBOOK QUESTION 17. (Section 31-3) Light energy emitted by the Sun and stars comes from the fusion process. What conditions in the interior of stars make this possible?

ANSWER: The Sun has a very large mass, and because of the large gravitational force exerted by this mass, it has a very high density. In addition, the temperature in the Sun's interior is very high. As a result of these conditions, the strong nuclear force of attraction is able to overcome Coulomb repulsion between adjacent nuclei and nuclear fusion can occur.

TEXTBOOK QUESTION 16. (Section 31-4) Why do gamma particles penetrate matter more easily than beta particles do?

ANSWER: Gamma particles are electrically neutral and have no mass; therefore, they do not interact with matter as readily as electrically charged particles such alpha particles and beta particles. As a result, gamma particles pass through matter more readily.

EXAMPLE PROBLEM 5. (Section 31-3) (*a*) Determine the energy released when the following fusion reaction occurs: ${}_{1}^{3}\text{H}+{}_{1}^{3}\text{H}\rightarrow{}_{2}^{4}\text{He}+2\,{}_{0}^{1}\text{n}$. (*b*) Determine the amount of energy in joules produced per u of tritium consumed in the reaction. (*c*) Determine the amount of energy in joules produced per kg of tritium consumed in the reaction.

Part a. Step 1.

	Reactants		Products	
Determine the total mass of the reactants and the total mass of the products.	${}_{1}^{3}\text{H}$	3.016049 u	${}_{2}^{4}\text{He}$	4.002603 u
	${}_{1}^{3}\text{H}$	3.016049 u	$2\,{}_{0}^{1}\text{n}$	2.017330 u
		6.032098 u		6.019933 u

Part a. Step 2.	
Determine the mass difference and the energy released per reaction.	Mass difference $= 6.03098\ \text{u} - 6.019933\ \text{u}$ Mass difference $= 0.012165\ \text{u}$ $(0.012165\ \text{u})(931.5\ \text{MeV/u}) = 11.3\ \text{MeV}$
Part b. Step 1.	
Express the energy released per reaction in terms of joules.	$(11.3\ \text{MeV})(1.6 \times 10^{-13}\ \text{J/MeV}) = 1.81 \times 10^{-12}\ \text{J}$
Part b. Step 2.	
Determine the energy released per atomic mass unit of tritium consumed in the reaction.	$\dfrac{1.81 \times 10^{-12}\ \text{J}}{6.032098\ \text{u}} = 3.00 \times 10^{-13}\ \text{J/u}$
Part c. Step 1.	
Determine the energy released per kg of tritium consumed in the reaction.	$1\ \text{u} = 1.66 \times 10^{-27}\ \text{kg}$ $(3.00 \times 10^{-13}\ \text{J/u})\left(\dfrac{1\ \text{u}}{1.66 \times 10^{-27}\ \text{kg}}\right) = 1.81 \times 10^{14}\ \text{J/kg}$

Dosimetry

Dosimetry refers to measuring the quantity, or **dose**, of radiation that passes through a material. There are several units used for this purpose. The activity of a radioactive isotope is measured in terms of **curies** (Ci) or microcuries (μCi), where $1\ \text{Ci} = 3.70 \times 10^{10}$ disintegrations per second, and $1\ \mu\text{Ci} = 10^{-6}\ \text{Ci}$.

The **roentgen** (R) is a unit of dosage that was previously defined in terms of the amount of ionized air produced by the radiation. One roentgen (1 R) is the amount of radiation that will produce 1.6×10^{12} ion pairs/g of dry air at STP. Today the roentgen is defined as that amount of X-ray or γ radiation that deposits $0.878 \times 10^{-2}\ \text{J}$ of energy per kg of air. The roentgen has been replaced by the **rad,** where 1 rad is the amount of radiation that deposits $10^{-2}\ \text{J}$ of energy per kg of absorbing material. The SI unit for absorbed dose is the **gray** (Gy), where $1\,\text{Gy} = 1\ \text{J/kg} = 100\ \text{rad}$.

Units have been developed that refer to the biological damage caused by radiation. The **rem** (rad equivalent man) is such a unit and can be used for all types of radiation.

$$\text{rem} = \text{rad} \times \text{RBE}$$

Relative biological effectiveness (RBE) is defined as the number of rad of X-ray or γ radiation that produces the same biological damage as 1 rad of the given radiation. The RBE of X-rays and γ rays is 1.0; β particles, 1.0;

α particles, 10–20; slow neutrons, 3–5; and fast neutrons and protons, 10. While we are all subject to about 0.30 rem/year of **natural background radiation** from our environment, large doses of radiation can cause **radiation sickness** leading to death. However, the length of time over which the body receives the dose as well as the size of the dose are important.

The effective dose (in rem) = dose (in rad) × RBE. The SI unit for effective dose is measured in **sieverts**. The effect dose (in Sv) = dose (Gy) × RBE.

TEXTBOOK QUESTION 21. (Section 31-5) Radiation is sometimes used to sterilize medical supplies and even food. Explain how it works.

ANSWER: Radiation damage tends to kill bacteria and viruses that exist on medical supplies or food. During the procedure, the medical supply or food will be sterilized. However, the medical supplies or food will not become radioactive as a result of the sterilization procedure.

EXAMPLE PROBLEM 6. (Section 31-5) At one time it was possible to purchase a watch that contained the radium isotope $^{226}_{88}\text{Ra}$. This isotope emits 4.76 MeV alpha particles that struck a special material in the paint used for the dial of the watch. This material gave off visible light when struck by the alpha particles. Assume that the dial contained 5.00 microcuries of the isotope and that a person wearing the watch had a mass of 70.0 kg and wore the watch 24 hours per day. Determine the (*a*) number of decays per second and (*b*) yearly dosage in rem to the wearer if 10% of the disintegrations interact with the person's body and deposit all of their energy. Assume an RBE of 10 for the alpha particles.

Part a. Step 1.

Determine the number of decays per second.	$1.0\text{ Ci} = 3.7\times10^{10}$ decays/s; therefore, $(5.00\times10^{-6}\text{ Ci})\left(\dfrac{3.7\times10^{10}\text{ decays/s}}{1\text{ Ci}}\right) = 1.85\times10^{5}\text{ decays/s}$

Part b. Step 1.

Determine the amount of energy absorbed by the person's body each second in MeV. Express this energy in joules.	Only 10% (i.e., 0.10) of the disintegrations deposit their energy in the person's body. $(0.10)(1.85\times10^{5}\text{ decays/s}) = (1.85\times10^{4}\text{ decays/s})$ $(1.85\times10^{4}\text{ decays/s})(4.76\text{ MeV/decay}) = 8.81\times10^{4}\text{ MeV/s}$ $(8.81\times10^{4}\text{ MeV/s})(1.6\times10^{-13}\text{ J/MeV}) = 1.41\times10^{-8}\text{ J/s}$

Part b. Step 2.	
Determine the amount of energy absorbed per kg of mass by the person's body.	$\frac{1.41 \times 10^{-8}\ \text{J/s}}{70.0\ \text{kg}} = 2.01 \times 10^{-10}\ \text{J/kg} \cdot \text{s}$
Part b. Step 3.	
Convert the answer in Step 2 to rad/s.	Note: $1\ \text{rad} = 10^{-2}\ \text{J/kg}$. $(2.01 \times 10^{-10}\ \text{J/kg} \cdot \text{s})\left(\frac{1\,\text{rad}}{10^{-2}\ \text{J/kg}}\right) = 2.01 \times 10^{-8}\ \text{rad/s}$
Part b. Step 4.	
Determine the number of rads absorbed by the body in 1 year.	$1\ \text{year} = 3.156 \times 10^{7}\ \text{seconds}$ $(2.01 \times 10^{-8}\ \text{rad/s})(3.156 \times 10^{7}\ \text{s/year}) = 0.634\ \text{rad/year}$
Part b. Step 5.	
Determine the dosage in rem/year.	Dosage in rem = rad × RBE $= (0.634\ \text{rad/year})(10)$ Dosage in rem = 6.34 rem/year

A number of ways have been developed to measure radiation dose. A film badge is worn by people who work around sources of radiation. When developed, the darkness of the film is a measure of the radiation dose received. The Geiger counter, scintillation counter, and semiconductor detector all detect the presence and level of activity of radiation.

PROBLEM SOLVING SKILLS

For problems that involve determination of a missing particle in a nuclear reaction:

1. Apply the law of conservation of charge to determine the atomic number (Z) of the nuclide.
2. Apply the law of conservation of nucleon number to determine the mass number of the nuclide.
3. Use the Periodic Table to identify the unknown nuclide.

For problems that involve the calculation of the Q-value of a nuclear reaction:

1. Determine the total mass of the reactants and the total mass of the products.
2. Determine the mass difference between the sum of the reactants and the sum of the products.

3. If the mass difference is expressed in kg, then use $E = (\Delta m)c^2$ to determine the Q-value in joules. The Q-value may be expressed in MeV by using the conversion factor $1\ \text{MeV} = 1.6 \times 10^{-13}$ J.
4. If the mass difference is expressed in atomic mass units (u), then the Q-value may be expressed in MeV by multiplying the mass difference by 931.5 MeV/u.
5. If the Q-value is positive, then the reaction is exothermic or exoergic and can occur spontaneously. If the Q-value is negative, then the reaction is endothermic or endoergic and energy must be added for the reaction to occur.

If the problem involves dosimetry:

1. Determine the number of decays that occur per second.
2. Determine the amount of energy absorbed per kg of mass per second.
3. Convert the energy absorbed per kg of mass per second to rads per second.
4. If required, determine the number of rads absorbed per year.
5. If the RBE value is given, then the yearly dosage in rem may be determined.

SOLUTIONS TO SELECTED TEXTBOOK PROBLEMS

TEXTBOOK PROBLEM 3. (Section 31-1) Is the reaction $n + {}^{238}_{92}\text{U} \rightarrow {}^{239}_{92}\text{U} + \gamma$ possible with slow neutrons? Explain.

Part a. Step 1.

Determine the total mass of the reactants and the total mass of the products.

Reactants		Products	
${}^{238}_{92}\text{U}$	238.050788 u	${}^{239}_{92}\text{U}$	239.054294 u
${}^{1}_{0}\text{n}$	1.008665 u		
	239.059453 u		239.054294 u

Part a. Step 2.

Determine the mass difference between the products and the reactants.

$$\text{Mass difference} = 239.059453\ \text{u} - 239.054294\ \text{u}$$

$$\text{Mass difference} = 0.005159\ \text{u}$$

The products have less mass than the reactants; therefore, the reaction is exothermic.

Part a. Step 3.

Determine the value of Q in joules and MeV. Note: If $Q > 0$, then the reaction is exothermic and no threshold energy is required.

$$Q = (M_a + M_X - M_b - M_Y)c^2$$

$$Q = (\Delta m)c^2$$

$$\Delta m = (0.005159 \text{ u})(1.66 \times 10^{-27} \text{ kg/u}) = 8.56394 \times 10^{-30} \text{ kg}$$

$$Q = (8.56394 \times 10^{-30} \text{ kg})(3.0 \times 10^{8} \text{ m/s})^2 = 7.707546 \times 10^{-13} \text{ J}$$

$$Q = (7.707546 \times 10^{-13} \text{ J})(1 \text{ MeV}/1.6 \times 10^{-13} \text{ J}) = 4.817 \text{ MeV}$$

Alternate solution:

$$Q = (0.005159 \text{ u})(931.5 \text{ MeV/u}) = 4.806 \text{ MeV}$$

Since $Q > 0,$ the reaction is possible.

TEXTBOOK PROBLEM 16. (Section 31-2) What is the energy released in the fission reaction of Eq. 31-4? (The masses of $^{141}_{56}\text{Ba}$ and $^{92}_{36}\text{Kr}$ are 140.914411 u and 91.926156 u, respectively).
$^{235}_{92}\text{U} + ^{1}_{0}\text{n} \rightarrow ^{141}_{56}\text{Ba} + ^{92}_{36}\text{Kr} + 3^{1}_{0}\text{n}$

Part a. Step 1.

Determine the total mass of the reactants and the total mass of the products.

Reactants		Products	
$^{235}_{92}\text{U}$	235.043930 u	$^{141}_{56}\text{Bal}$	40.914411 u
$^{1}_{0}\text{n}$	1.008665 u	$^{92}_{36}\text{Kr}$	91.926156 u
		3^{1}_{0}n	3.025995 u
	236.052595 u		235.866562 u

Part a. Step 2.

Determine the mass difference.

Mass difference $= 236.052595 \text{ u} - 235.866562 \text{ u}$

Mass difference $= 0.186033 \text{ u}$

Part a. Step 3.

Determine the energy released in the reaction.

The reactants have a greater mass than the products; therefore, the reaction occurs spontaneously:

$$(0.186033\ \text{u})(931.5\ \text{MeV/u}) = 173.3\ \text{MeV}$$

TEXTBOOK PROBLEM 18. (Section 31-2) How many fissions take place per second in a 240-MW reactor? Assume 200 MeV is released per fission.

Part a. Step 1.

Convert 240 MW to joules per second.

$$(240\ \text{MW})\left(\frac{1.0\times 10^{6}\ \text{W}}{1\ \text{MW}}\right)\left(\frac{1\ \text{J/s}}{1\ \text{W}}\right) = 2.4\times 10^{8}\ \text{J/s}$$

Part a. Step 2.

Determine the number of fissions per second.

Let n = number of fissions per second; then

power = n(energy per fission)

$$2.4\times 10^{8}\ \text{J/s} = n(200\ \text{MeV})\left(\frac{1.6\times 10^{-13}\ \text{J}}{1\ \text{MeV}}\right)$$

$$n = \frac{2.4\times 10^{8}\ \text{J/s}}{3.2\times 10^{-11}\ \text{J}} = 7.5\times 10^{18}\ \text{fissions each second}$$

TEXTBOOK PROBLEM 27. (Section 31-3) What is the average kinetic energy of protons at the center of a star where the temperature is 2×10^{7} K? [*Hint*: See Eq. 13-8.]

Part a. Step 1.

Determine the average kinetic energy of the protons in joules.

$$\text{KE} = \tfrac{3}{2}kT = \tfrac{3}{2}(1.38\times 10^{-23}\ \text{J/K})(2\times 10^{7}\ \text{K})$$

$$\text{KE} = 4.1\times 10^{-16}\ \text{J}$$

Part a. Step 2.

Express this energy in electron volts (eV).

$$\text{KE} = (4.1 \times 10^{-16}\ \text{J})\left(\frac{1\ \text{eV}}{1.6 \times 10^{-19}\ \text{J}}\right)$$

$$\text{KE} \approx 2600\ \text{eV}$$

TEXTBOOK PROBLEM 37. (Section 31-3) How much energy (J) is contained in 1.00 kg of water if its natural deuterium is used in the fusion reaction of Eq. 31-8a? Compare to the energy obtained from the burning of 1.0 kg of gasoline, about 5×10^7 J.

${}^2_1\text{H} + {}^2_1\text{H} \rightarrow {}^3_1\text{H} + {}^1_1\text{H}$ (4.03 MeV) (Eq. 31-8a)

Part a. Step 1.

Determine the number of moles of water containing deuterium in 1000 grams of water.

From Appendix B, relative abundance of deuterium is 0.0115% or 1.15×10^{-4}. Also, 1 mole of H_2O containing deuterium has a MW of 18 grams:

$$(1.15 \times 10^{-4})(1000\ \text{g})\left(\frac{1\ \text{mol}}{18\ \text{g}}\right) = 6.39 \times 10^{-3}\ \text{mol}$$

Part a. Step 2.

Determine the number of deuterium nuclei in 1000 g of water.

The formula for water is H_2O. Therefore, there are two deuterium nuclei in each molecule:

$$(6.39 \times 10^{-3}\ \text{mol})\left(\frac{6.02 \times 10^{23}\ \text{molecules}}{1\ \text{mol}}\right)\left(\frac{2\ \text{nuclei}}{1\ \text{molecule}}\right)$$

$$= 7.69 \times 10^{21}\ \text{nuclei}$$

Part a. Step 3.

Determine the energy contained in the 1.0 kg of water due to the deuterium.

As shown in Eq. 31-8, two deuterium nuclei are involved in each reaction:

$$(7.69 \times 10^{21}\ \text{nuclei})\left(\frac{4.03\ \text{MeV}}{1\ \text{reaction}}\right)\left(\frac{1\ \text{reaction}}{2\ \text{nuclei}}\right)$$

$$= 1.55 \times 10^{22}\ \text{MeV}$$

$$(1.55 \times 10^{22}\ \text{MeV})\left(\frac{1.6 \times 10^{-13}\ \text{J}}{1\ \text{MeV}}\right) = 2.48 \times 10^{9}\ \text{J}$$

Part a. Step 4.

Compare the energy found in Step 3 to the energy found in 1.0 kg of gasoline.

$$\left(\frac{2.48\times 10^9\ \text{J deuterium in water}}{5.0\times 10^7\ \text{J gasoline}}\right)\approx 50$$

There is approximately 50 times more energy contained in 1.0 kg of water than in 1.0 kg of gasoline.

TEXTBOOK PROBLEM 42. (Section 31-5) How much energy is deposited in the body of a 65-kg adult exposed to a 2.5-Gy dose?

Part a. Step 1.

Convert Gy to J/kg.

From Section 31-5, 1 Gy is that amount of radiation that deposits energy at a rate of 1 J/kg in any absorbing material:

Therefore, $2.5\ \text{Gy} = 2.5\ \text{J/kg}$.

Part a. Step 2.

Determine the energy in joules absorbed by a 65-kg person.

$$E = (2.5\ \text{J/kg})(65\ \text{kg})$$

$$E \approx 160\ \text{J}$$

TEXTBOOK PROBLEM 47. (Section 31-5) What is the mass of a 2.50-μCi ${}^{14}_{6}\text{C}$ source?

Part a. Step 1.

Determine the decay constant of C-14 in decays per second.

Note: There are 3.156×10^7 seconds in 1 year.

$$\lambda = \frac{0.693}{T_{\frac{1}{2}}},\ \text{where}\ T_{2} = 5730\ \text{years}$$

$$= \frac{0.693}{5730\ \text{years}} = 1.21\times 10^{-4}\ \text{decays/year}$$

$$\lambda = (1.21\times 10^{-4}/\text{year})\left(\frac{1\ \text{year}}{3.156\times 10^7\ \text{s}}\right) = 3.83\times 10^{-12}\ \text{decays/s}$$

Part a. Step 2.

Determine the number (N) of C-14 nuclei present in the sample.

activity = λN

$$(2.5\times10^{-6}\ \text{Ci})\left(\frac{3.7\times10^{10}\text{decays/s}}{1\ \text{Ci}}\right)$$
$$= (3.83\times10^{-12}\ \text{decays/s})N$$

$$N = 2.42\times10^{16}\ \text{nuclei}$$

Part a. Step 3.

Determine the number of moles of ${}^{14}_{6}\text{C}$ present.

$$(2.42\times10^{6}\ \text{nuclei})\left(\frac{1\ \text{mol}}{6.02\times10^{23}\ \text{nuclei}}\right) = 4.0\times10^{-8}\ \text{mol}$$

Part a. Step 4.

Determine the mass of the sample.

$$m = (4.0\times10^{-8}\ \text{mol})\left(\frac{14\ \text{g}}{1\ \text{mol}}\right) = 5.6\times10^{-7}\ \text{g}$$

$$m = (5.6\times10^{-7}\ \text{g})\left(\frac{1\ \text{kg}}{1000\ \text{g}}\right) = 5.6\times10^{-10}\ \text{kg}$$

TEXTBOOK PROBLEM 57. (Section 31-1) One means of enriching uranium is by diffusion of the gas UF_6. Calculate the ratio of the speeds of molecules of this gas containing ${}^{235}_{92}\text{U}$ and ${}^{238}_{92}\text{U}$, on which this process depends.

Part a. Step 1.

Determine the molecular weight of each molecule.

For ${}^{235}_{92}\text{U}$ in UF_6: $235\ \text{u} + 6(19\ \text{u}) = 349\ \text{u}$

For ${}^{238}_{92}\text{U}$ in UF_6: $238\ \text{u} + 6(19\text{u}) = 352\ \text{u}$

Part a. Step 2.

Derive a formula for the rms speed of each molecule as a function of temperature.

$$\text{KE} = \tfrac{1}{2}mv^2 \quad \text{but} \quad \text{KE} = \tfrac{3}{2}kT$$

$$\tfrac{1}{2}mv^2 = \tfrac{3}{2}kT$$

$$v^2 = \frac{3kT}{m}$$

$$v_{\text{rms}} = \sqrt{\frac{3kT}{m}}$$

Part a. Step 3.

Determine the ratio of the rms speeds of the molecules.

$$\frac{v_{235}}{v_{238}} = \sqrt{\frac{3kT/m_{235}}{3kT/m_{238}}}$$

$$\frac{v_{235}}{v_{238}} = \sqrt{\frac{m_{238}}{m_{235}}} = \sqrt{\frac{352\text{ u}}{349\text{ u}}} = 1.0043$$

32

ELEMENTARY PARTICLES

OBJECTIVES

After studying the material of this chapter, the student should be able to:

- describe how the cyclotron, synchrotron, and linear accelerator can be used to accelerate charged particles. Explain the advantages to be gained by using one device over the others.
- describe each of the four forces found in nature. List the fundamental particle associated with each force.
- distinguish between a particle and its antiparticle.
- classify the fundamental particles and the composite particles.
- write out the law of conservation of lepton number and the law of conservation of baryon number.
- list the quantities that must be conserved in order for a particular reaction or decay process to occur. Apply the conservation laws to determine if a particular reaction occurs.
- distinguish between a particle and a resonance.
- explain what is meant by strangeness and conservation of strangeness.
- give the names associated with the six types of quarks and describe properties associated with each quark.
- explain the concept of quark color.
- describe what is meant by the grand unified theory.

KEY TERMS AND PHRASES

cyclotron accelerates protons or electrons through a potential difference into a magnetic field directed perpendicular to their path.

cyclotron frequency is the number of revolutions per second made by the charged particle.

synchrotron was developed to accelerate relativistic particles, increasing the magnetic field B to keep f constant. In this type of accelerator, the radius of the particle's path is held constant while the magnetic field increases.

linear accelerator, or linac, accelerates a charged particle in steps along a straight line.

colliding beam accelerator causes two beams of charged particles moving in opposite directions to collide head-on. The kinetic energy of the colliding particles is available for causing reactions or creating new particles.

graviton is postulated to be responsible for the gravitational force.

positron is the antiparticle of the electron. The positron has the same mass as an electron but has the opposite charge. When the positron encounters an electron, the two annihilate each other and a gamma ray is produced.

antiparticle of the proton is the antiproton, and the antiparticle of the neutron is the antineutron. A few particles, for example, the photon and the neutral pion (π^0), have no antiparticle.

leptons (light particles) include the electron, muon, and two types of neutrino: the electron neutrino (v_e) and muon neutrino (v_μ) as well as their antiparticles.

hadrons are composed of two subgroups, the **baryons** (heavy particles) and the **mesons** (intermediate particles). The baryons include the proton and all heavier particles through the omega particle plus their antiparticles. The mesons include the pion, kaon, and eta particles plus their antiparticles.

conservation laws for reaction and decay processes to occur include energy (including mass), linear momentum, charge, angular momentum (including spin), baryon number, electron lepton number, and muon lepton number.

resonances are extremely short-lived particles that decay via the strong interaction. They do not travel far enough in a bubble chamber or a spark chamber to be detected, and their existence is inferred because of decay products that can be detected.

strangeness and the principle of **conservation of strangeness** provide an explanation for the observation of certain reactions and also why other reactions do not occur. They also provide an explanation for the "long" lifetimes, 10^{-10} s to 10^{-8} s, of certain particles that interact via the strong interaction.

quarks are composed of six particles that are named up (u), down (d), strange (s), charm (c), top (t), and bottom (b) and carry a fractional charge, either $+\frac{2}{3}$ or $-\frac{1}{3}$ the charge on the electron. The quarks have antiparticles called antiquarks.

color and **flavor** are properties associated with quarks. Each flavor of quark can have one of three colors—red, green, or blue—while the antiquarks are colored antired, antigreen, or antiblue.

gluons are particles that transmit the color force, or strong force. According to theory, there are eight gluons. They are massless and have color charge.

electroweak theory proposes that the weak force and the electromagnetic force are two different manifestations of a single, more fundamental, electroweak interaction.

grand unified theory (GUT) suggests that at very short distances $(10^{-31}$ m) and very high energy, the electromagnetic, weak, and strong forces are different aspects of a single underlying force. In this theory the fundamental difference between quarks and leptons disappears. Attempts are presently being made to incorporate the four forces found in nature—gravity, electromagnetic, weak, and strong forces—into a single theory.

SUMMARY OF MATHEMATICAL FORMULAS

cyclotron frequency	$f=\frac{1}{T}$ $f=\frac{qB}{2\pi m}$	The cyclotron frequency (f) is the number of revolutions per second made by the charged particle.

Type of force	Relative strength	Field particle
strong nuclear	1	mesons/gluons, very short distance ($\approx 10^{-15}$ m)
electromagnetic	10^{-2}	photon
weak nuclear	10^{-9}	$W^{\pm}$ and Z^{0}, extremely short range ($\approx 10^{-15}$ m)
gravitational	10^{-38}	graviton

CONCEPT SUMMARY

Particle Accelerators

As discussed in Chapter 25, the sharpness or resolution of the details of an image is limited by the wavelength of the incident radiation. The de Broglie wavelength of particles such as electrons or protons is given by $\lambda = \frac{h}{mv}$. Particles that have greater momentum have a shorter wavelength. There are a number of different types of particle accelerators available to produce the high-energy, short-wavelength particles needed to investigate nuclear structure.

The **cyclotron** accelerates protons or electrons through a potential difference into a magnetic field directed perpendicular to their path. The charged particles move within two D-shaped cavities, called **dees**, and each time they move into the space between the dees a potential difference is applied that increases their speed. As their speed increases, the radius of the circle in which they travel increases until they reach the outer edge of the dee and either strike a target within the cyclotron or leave the cyclotron and strike an external target.

$$f = \frac{1}{T} \text{ and } f = \frac{qB}{2\pi m}$$

where f is the cyclotron frequency, which is the number of revolutions per second made by the charged particle. T is the period in seconds of the particle's motion. q is the charge in coulombs on the particle. B is the strength in teslas of the magnetic field that is directed perpendicular to the particle's path, and m is the mass of the particle in kilograms.

The **cyclotron frequency** determines the frequency of the applied voltage needed to accelerate the charged particle. The frequency does not depend on the radius of the circle in which the particle is traveling, but it does depend on the particle's momentum.

The **synchrotron** was developed to accelerate relativistic particles, increasing the magnetic field B to keep f constant. In this type of accelerator, the radius of the particle's path is held constant while the magnetic field increases. The radius of a synchrotron may exceed 1.0 km and accelerate protons to energies exceeding 500 GeV.

In both the cyclotron and the synchrotron, the charged particle travels in a circle and undergoes centripetal acceleration. The accelerated charges radiate electromagnetic energy, and considerable energy is lost through radiation. This effect is called **synchrotron radiation**.

In a **linear accelerator**, or **linac**, a charged particle is accelerated in steps along a straight line. Linear accelerators have been constructed that have accelerated electrons to over 50 GeV. No magnetic fields are used in a linear accelerator.

In a **colliding beam accelerator**, two beams of charged particles moving in opposite directions collide head-on. If the particles have the same mass and energy, then the momentum before and after is zero. If the kinetic energy after impact is zero, then all of the initial kinetic energy of the colliding particles is available for causing reactions or creating new particles.

EXAMPLE PROBLEM 1. (Section 32-1) Determine the kinetic energy in GeV of a proton whose de Broglie wavelength is 1.50 fm.

Part a. Step 1.

Use the de Broglie formula to determine the energy. Note: The energy of the proton is on the order of GeVs (so convert from joules). Assume that the proton's velocity is approximately equal to the speed of light.

$E = h\frac{c}{\lambda}$, where $\lambda = 1.50 \text{ fm} = 1.50 \times 10^{-15}$ m

$$E = \frac{(6.63 \times 10^{-34} \text{ J}\cdot\text{s})(3 \times 10^{8} \text{ m/s})}{1.5 \times 10^{-15} \text{ m}} = 1.36 \times 10^{-10} \text{ J}$$

$$E = (1.36 \times 10^{-10} \text{ J})\left(\frac{1.00 \text{ eV}}{1.6 \times 10^{-19} \text{ J}}\right) = 8.50 \times 10^{8} \text{ eV}$$

$$= (8.50 \times 10^{8} \text{ eV})\left(\frac{1.00 \text{ GeV}}{1.0 \times 10^{9} \text{ eV}}\right)$$

$$E = 0.85 \text{ GeV}$$

EXAMPLE PROBLEM 2. (Section 32-2) A cyclotron is used to accelerate protons to an energy of 1.0 MeV. The strength of the magnetic field is 0.85 T. Determine the (*a*) radius of the proton's orbit when the proton's energy reaches 1.0 MeV, (*b*) cyclotron frequency, and (*c*) period of the proton's motion.

Part a. Step 1.

Determine the proton's velocity when its energy reaches 1.0 MeV.

$\text{KE} = \frac{1}{2}mv^2$

$(1.0 \text{ MeV})(1.6 \times 10^{-13} \text{ J/MeV}) = \frac{1}{2}(1.67 \times 10^{-27} \text{ kg})v^2$

$v^2 = 1.9 \times 10^{14} \text{ m}^2/\text{s}^2$, and solving for v gives

$v = 1.4 \times 10^{7}$ m/s

Part a. Step 2.

Solve for the radius of the proton's orbit.

$F = qvB \sin\theta$, where $\theta = 90°$ and $\sin 90° = 1$. The motion is circular; therefore, $F = mv^2/r$. Then

$qvB = mv^2/r$, and solving for r gives

$$r = \frac{mv}{qB} = \frac{(1.67 \times 10^{-27} \text{kg})(1.4 \times 10^{7} \text{m/s})}{(1.6 \times 10^{-19} \text{C})(0.85 \text{T})}$$

$r = 0.17$ m, or 17 cm

Part b. Step 1.

Determine the cyclotron frequency.

For a particle traveling in a circle at constant speed (v),

$$v = \frac{2\pi r}{T} = 2\pi r f$$

From Part a, Step 2, $r = \frac{mv}{qB}$, therefore,

$$r = \frac{m 2\pi r f}{qB}; \text{ solving for } f \text{ gives}$$

$$f = \frac{qB}{2\pi m} = \frac{(1.6\times 10^{-19}\ \text{C})(0.85\ \text{T})}{2\pi(1.67\times 10^{-27}\ \text{kg})}$$

$$f = 1.3\times 10^{7}\ \text{Hz}$$

Part c. Step 1.

Determine the period of the proton's motion.

The period is inversely related to the frequency; therefore,

$$T = \frac{1}{f} = \frac{1}{1.3\times 10^{7}\ \text{Hz}}$$

$$T = 7.7\times 10^{-8}\ \text{s}$$

EXAMPLE PROBLEM 3. (Section 32-3) A head-on collision occurs between two protons having equal kinetic energy. Determine the minimum kinetic energy of each proton in order to produce the following reaction: $\text{p}+\text{p}\rightarrow \text{n}+\text{p}+\pi^{+}$

Part a. Step 1.

Determine the total mass of the reactants and the total mass of the products. Note: In Table 32-2 of the text, the mass is expressed in MeV/c^2.

Reactants	Products
p 938.3 MeV/c^2	n 939.6 MeV/c^2
p 938.3 MeV/c^2	p 938.3 MeV/c^2
	π^{+} 139.6 MeV/c^2
1876.6 MeV/c^2	2017.5 MeV/c^2
total mass of reactants (in MeV/c^2)	total mass of products (in MeV/c^2)

Part a. Step 2.

Determine the mass difference (expressed in MeV/c^2) between the products and the reactants.	Mass difference $= 2017.5\ \text{MeV}/c^2 - 1876.6\ \text{MeV}/c^2$ Mass difference $= 140.9\ \text{MeV}/c^2$

Part a. Step 3.

Determine the kinetic energy of each of the colliding protons.	The mass of the products is greater than the mass of the reactants; therefore, the kinetic energy of the colliding protons must provide the energy necessary for the reaction to occur. Since the protons have equal kinetic energy, each proton must have $[140.9\ \text{MeV}/c^2](c^2/2) = 70.5\ \text{MeV}$ of kinetic energy

The Four Forces in Nature

Because of the wave–particle duality, it is possible to suggest that the electromagnetic force between charged particles is the result of an electromagnetic field produced by one particle that affects the second particle (wave theory) or by an exchange of photons (γ particles) between the charged particles (particle theory).

In 1935, Japanese physicist Hideki Yukawa postulated the existence of a particle that produces the strong nuclear force that holds the atomic nucleus together. This particle, now known as the pi meson or pion and represented by the symbol π, was discovered in 1947. The particle can be charged positively (π^+) or negatively (π^-) or be uncharged (π^0). The mass of the π^+ or π^- particle is 140 MeV/c^2, 273 times the mass of the electron, while the mass of the π^0 particle is 135 MeV/c^2, which is 264 times the mass of the electron.

The weak nuclear force is used to account for beta decay (β). The weak nuclear force is presumed to be due to particles known as W^+, W^-, and Z^0. In 1983, the discovery of the W and Z particles was announced by a group of scientists led by Carlo Rubbia. The group used the high-energy accelerator at CERN.

The gravitational force is believed to be due to a particle known as the **graviton**. The graviton has not yet been observed. The electromagnetic force and the gravitational force are known as long-range forces, decreasing as the square of the distance between interacting particles. The strong nuclear force and the weak nuclear force are very short range forces, limited to distances of approximately 10^{-15} m. This distance is the approximate size of the atomic nucleus. The following table lists each type of force, the relative strength of the particular force as compared to the strong nuclear force, and the name of the particle credited with producing the force.

Type	Relative strength	Field particle
strong nuclear	1	mesons/gluons
electromagnetic	10^{-2}	photon
weak nuclear	10^{-9}	$W^{\pm}$ and Z^0 particle
gravitational	10^{-38}	graviton

TEXTBOOK QUESTION 7. (Section 32-10) Which of the four interactions (strong, electromagnetic, weak, gravitational) does an electron take part in? A neutrino? A proton?

ANSWER: The electron and proton both possess electric charge and therefore interact with other charged particles via the electromagnetic force and photon–photon exchange. The neutrino does not carry electric charge and therefore does not interact via the electromagnetic force. All three particles interact via the gravitational force and the weak nuclear force. The proton is a baryon and interacts via the strong nuclear force as well as the weak nuclear force. The electron and neutrino are leptons and do not take part in interactions via the strong force.

Particles and Antiparticles

The first **antiparticle**, the **positron**, was discovered in 1932. The positron has the same mass as an electron but has the opposite charge. When the positron encounters an electron, the two annihilate one another, and a gamma ray is produced. The energy of the gamma ray is equal to the mass equivalent of the electron and positron plus any kinetic energy the two possessed at the time of interaction.

It is now known that most particles have antiparticles, for example, proton (p)–antiproton $(\overline{\text{p}})$ and neutron (n)–antineutron $(\overline{\text{n}})$. A few particles, for example, the photon and the neutral pion (π^0), have no antiparticle.

TEXTBOOK QUESTION 3. (Section 32–4) What would an "antiatom," made up of the antiparticles to the constituents of normal atoms, consist of? What might happen if *antimatter*, made of such antiatoms, came in contact with our normal world of matter?

ANSWER: The antiatom would contain the antiparticles to those found in the normal atom. A normal atom contains electrons, protons, and neutrons. Therefore, the antiatom would contain positrons, antiprotons, and antineutrons. The conservation laws apply to antiatoms as well as to atoms. It is known that antimatter–matter interactions, such as electron–positron interactions, result in the annihilation of the particles involved with the production of their energy equivalent.

Particle Classification

Table 32-2 in the textbook lists a number of particles according to their family group and the way they interact. Photons interact only through the electromagnetic force, while **leptons** (light particles) interact only through the weak nuclear force. The leptons include the electron, **muon**, and two types of **neutrino**: the electron neutrino (v_e) and muon neutrino (v_μ) as well as their antiparticles.

The **hadrons** interact through the strong nuclear force. The hadrons are composed of two subgroups, the **baryons** (heavy particles) and the **mesons** (intermediate particles). The baryons include the proton and all heavier particles through the omega particle plus their antiparticles. The mesons include the pion, kaon, and eta particles plus their antiparticles.

Particle Interactions and Conservation Laws

In order for reaction and decay processes to occur, seven quantities must be conserved: energy (including mass), linear momentum, charge, angular momentum (including spin), baryon number, electron lepton number, and muon lepton number.

Energy is conserved if the sum of the rest energies and total kinetic energy of the reactants equals the sum of the rest energies and total kinetic energy of the products. Linear momentum must be conserved. In the case of a decay of a nucleus at rest, linear momentum can be satisfied if the decay products travel in different directions such that the total final momentum equals zero.

The law of conservation of charge is satisfied if the sum of the charges on the reactants equals the sum of the charges on the products.

Angular momentum is satisfied if the sum of the spin angular momentum quantum numbers of the reactants equals that of the products. The spin angular momentum of each particle is a multiple of $\dfrac{h}{2\pi}$ or $\hbar$. For example, the spin angular momentum of an electron equals $\pm\frac{1}{2}\hbar$, where the + or − is used since the spin can be up (+) or down (−). The spin angular momentum quantum number for each particle listed in Table 32-2 is as follows:

Category	Particle name	Spin quantum number
leptons	electron neutrino	$\frac{1}{2}$
	muon neutrino	$\frac{1}{2}$
	electron	$\frac{1}{2}$
	muon	$\frac{1}{2}$
photons	photon	1
hadrons		
mesons	pion	0
	kaon	0
	eta	0
baryons	proton	$\frac{1}{2}$
	neutron	$\frac{1}{2}$
	lambda	$\frac{1}{2}$
	sigma	$\frac{1}{2}$
	xi	$\frac{1}{2}$
	omega	$\frac{3}{2}$

The law of conservation of baryon number states that in every interaction the total baryon number remains unchanged. The baryon number for a baryon is $B = +1$ and for an antibaryon $B = -1$. All non-baryons have a baryon number $B = 0$.

The law of conservation of lepton number must be applied separately to each lepton family. In the electron lepton family the electron and electron neutrino (v_e) are $L_e = +1$, while the positron and electron antineutrino ($\overline{v}_e$) are assigned $L_e = -1$. The electron lepton number must be conserved in order for a reaction to occur.

The law of conservation of muon lepton number states that the muon lepton number (L_μ) must be conserved for a process to occur. The μ^- and v_μ particles have numbers $L_\mu = +1$, while for μ^+ and μ^- the number is $L_\mu = -1$. All other particles have a lepton number equal to zero.

TEXTBOOK QUESTION 4. (Section 32-2) What particle in a decay signals the electromagnetic interaction?

ANSWER: The appearance of a photon (γ) indicates that an electromagnetic interaction has occurred.

EXAMPLE PROBLEM 4. (Sections 32-5 and 32-6) Use the conservation laws to determine whether the following reaction may occur. As part of your answer, demonstrate which conservation laws are satisfied and which are violated. Assume that the nucleus that decays is at rest. $\kappa^- \rightarrow \mu^- + \overline{v}_\mu$

Part a. Step 1.

Apply the law of conservation of charge.		$\kappa^- \rightarrow \mu^- + \overline{v}_\mu$
	charge	$-1 \rightarrow -1 + 0$ (allowed)

Part a. Step 2.

Apply the law of conservation of angular momentum.	spin	$0 \rightarrow \pm\frac{1}{2} + \pm\frac{1}{2}$ (allowed)

Part a. Step 3.

Apply the law of conservation of baryon number.	baryon number	$0 \rightarrow 0 + 0$ (not applicable)

Part a. Step 4.

Apply the law of conservation of lepton number.	muon lepton number	$0 \rightarrow +1 + (-1)$ (allowed)

Part a. Step 5.

Apply the laws of conservation of energy and conservation of momentum.

energy 493.8 MeV $\rightarrow$ 105.7 MeV + 0 (allowed)

The missing energy can be accounted for in the kinetic energy of the product particles.

momentum (allowed)

The product particles could be moving such that they have equal but opposite momenta.

All of the conservation laws are satisfied; therefore, the reaction is allowed.

Resonances

In addition to the particles listed in Table 32-2, there are a great many particles that are short-lived and decay in 10^{-23} seconds or less. Such super short-lived particles are known as **resonances** and decay via the strong interaction. They do not travel far enough in a bubble chamber or a spark chamber to be detected, and their existence is inferred because of decay products that can be detected.

EXAMPLE PROBLEM 5. (Section 32-7) (*a*) The estimated lifetime of a certain resonance is 5.0×10^{-23} s. Estimate the width of the resonance in joules and MeV. (*b*) Express the width of the resonance in terms of the mass of an electron.

Part a. Step 1.

Use the uncertainty principle to determine the width of the resonance in joules.

If the particle's lifetime is uncertain by an amount Δt, then its rest energy will be uncertain by an amount given by

$$\Delta E\Delta t \approx \frac{h}{2\pi}$$

$$\Delta E(5.0\times10^{-23}\text{ s}) \approx \frac{6.63\times10^{-34}\text{ J}\cdot\text{s}}{2\pi}$$

$$\Delta E \approx 2.1\times10^{-12}\text{ J}$$

Part a. Step 2.

Express the answer in MeV.

$$\Delta E \approx (2.1\times10^{-12}\text{ J})\left(\frac{1\text{ MeV}}{1.6\times10^{-13}\text{ J}}\right)$$

$$\Delta E \approx 13\text{ MeV}$$

Part b. Step 1.

Express the width of the resonance in terms of the mass of an electron.

The rest energy of an electron is 0.51 MeV. Therefore, the width of this particular resonance is approximately

$$\frac{13\text{ MeV}}{0.51\text{ MeV}} \approx 26\text{ times the mass of an electron}$$

Strange Particles

The production and decay of certain particles led to the introduction of a new quantum number called **strangeness** (*S*). Particles are assigned a number called the strangeness number (see Table 32-2). The use of strangeness and the principle of **conservation of strangeness** provide an explanation for the observation of certain reactions and also why other reactions do not occur. It also provides an explanation for the relatively long lifetimes, 10^{-10} s to 10^{-8} s, of certain particles that interact via the strong interaction. Strangeness is conserved in the strong interaction but not in the weak interaction. Thus conservation of strangeness is an example of a **partially conserved** quantity.

Quarks

The six leptons $(e^-, \mu^-, \nu_e, \nu_\mu, \tau^-, \nu_\tau)$ seem to be truly elementary particles. There is no evidence that they have internal structure; also, they have no measurable size and do not decay.

Experiments indicate that hadrons, mesons, and baryons do have internal structure and a definite size (10^{-15} m in diameter). It was proposed in 1963 that the hadrons are combinations of three constituent particles known as **quarks**. The quarks were given the names **up** (u), **down** (d), and **strange** (s) and have a fractional charge, either $+\frac{2}{3}$ or $-\frac{1}{3}$ of the charge on the electron. The quarks have antiparticles called **antiquarks**, and the properties of both quarks and antiquarks are listed in Table 32-3. All of the hadrons known in 1963 could be constructed from a combination of quarks and antiquarks.

The up and down quarks were named because of their spin directions, while the strange quark was named because it is associated with the concept of strangeness. In 1964, the existence of a fourth quark, called **charm** (c) was postulated. The c quark has a charm number $c = +1$ while the $\overline{c}$ antiquark is $c = -1$. Theory today includes two more quarks, called **top** (t) and **bottom** (b).

Quantum Electrodynamics

An extension of the quark theory suggests that quarks have a property called **color** with the distinction between the different quarks called **flavor**. Each of the flavors can have one of three colors—red, green, or blue—while the antiquarks are colored antired, antigreen, or antiblue.

The concept of color is used to explain the force that binds the quarks together in a hadron. Each quark carries a color charge, and the strong force between the quarks is called the **color force**. This theory of a strong force is called **quantum chromodynamics**, or **QCD**, in order to indicate that the theory refers to the force between color charges.

The particles that transmit the strong force are called **gluons**. According to the theory there are eight gluons, which are massless and have color charge.

Grand Unified Theories

In the 1960s, a theory called the **electroweak theory** was proposed in which the weak force and the electromagnetic force are viewed as two different manifestations of a single, more fundamental, electroweak interaction. Attempts have been made to incorporate the electroweak force and the strong (color) force into a theory known as the **grand unified theory (GUT)**. One such theory suggests that at very short distances (10^{-31} m) and very high energy, the electromagnetic, weak, and strong forces are different aspects of a single underlying force. In this theory, the fundamental difference between quarks and leptons disappears.

Attempts are presently being made to incorporate the four forces found in nature—gravity, electromagnetic, weak, and strong forces—into a single theory.

TEXTBOOK QUESTION 15. (Section 32-10) Suppose there were a kind of "neutrinolet" that was massless, had no color charge or electrical charge, and did not feel the weak force. Could you say that this particle even exists?

ANSWER: Laymen, as well as scientists, depend on their five senses and the instruments that extend the senses to detect the existence of objects. As described in the question, the "neutrinolet" has no mass, has no color charge or electrical charge, and does not feel the weak force. Therefore, the particle does not interact via any one of the four forces previously discussed. At the present time, there would be no way to detect the existence of this particle. Even if such a particle exists in theory, a method would have to be found to either directly or indirectly detect it.

PROBLEM SOLVING SKILLS

For problems involving high-energy projectiles:

1. Express the kinetic energy of particles in the 1.0-GeV range in joules. The speed of such particles is approximately equal to the speed of light.
2. The de Broglie wavelength of such a particle (or the energy if the wavelength is given) can be found by using $\lambda = \dfrac{h}{mv} \approx \dfrac{hc}{mc^2}$, where $mc^2 = E$, the particle's energy.

For problems involving the cyclotron:

1. Determine the particle's velocity. If the velocity is above $0.1c$, then equations used for special relativity must be applied.
2. Derive an equation for the radius of the orbit, the cyclotron frequency, and the period of the motion.
3. Use the equations derived in Step 2 to solve the problem.

For problems involving particle interactions:

1. Apply each of the conservation laws.
2. Each of the conservation laws must hold for the interaction to be allowed.

For problems involving the width of a resonance:

1. Use the uncertainty principle, $(\Delta E)(\Delta t) = \dfrac{h}{2\pi}$, to determine the width of the resonance in joules.
2. Express the width of the resonance in MeV.
3. Express the width of the resonance in terms of the mass of an electron.

SOLUTIONS TO SELECTED TEXTBOOK PROBLEMS

TEXTBOOK PROBLEM 2. (Section 32-2) Calculate the wavelength of 28-GeV electrons.

Part a. Step 1.

Determine the rest energy of an electron in joules.

$$E_{\text{rest}} = m_e c^2$$
$$= (9.1 \times 10^{-31}\ \text{kg})(3 \times 10^8\ \text{m/s})^2$$
$$E_{\text{rest}} = 8.19 \times 10^{-14}\ \text{J}$$

Part a. Step 2.

Determine the kinetic energy of an electron in joules.

$$\text{KE} = (28\ \text{GeV})\left(\frac{10^9\ \text{eV}}{1\ \text{GeV}}\right)\left(\frac{1.6 \times 10^{-19}\ \text{J}}{1\ \text{eV}}\right)$$
$$\text{KE} = 4.5 \times 10^{-9}\ \text{J}$$

Part a. Step 3.

Determine the total energy of an electron in joules.

$$E_{\text{total}} = \text{KE} + E_{\text{rest}}$$
$$= 4.5 \times 10^{-9}\ \text{J} + 8.19 \times 10^{-14}\ \text{J}$$
$$E_{\text{total}} \approx 4.5 \times 10^{-9}\ \text{J}$$

Part a. Step 4.

Use the de Broglie formula to determine the wavelength.

At 28 GeV, the speed of an electron $\approx c$ (the speed of light):

$$\lambda = \frac{h}{mv} \approx \frac{h}{mc} \approx \frac{hc}{mc^2} \quad \text{but } E = mc^2$$
$$\lambda \approx \frac{hc}{E}$$
$$\lambda \approx \frac{(6.63 \times 10^{-34}\ \text{J}\cdot\text{s})(3 \times 10^8\ \text{m/s})}{4.5 \times 10^{-9}\ \text{J}}$$
$$\lambda \approx 4.5 \times 10^{-17}\ \text{m}$$

TEXTBOOK PROBLEM 5. (Section 32-1 and 32-2) What strength of magnetic field is used in a cyclotron in which protons make 3.1×10^{7} revolutions per second?

Part a. Step 1.

Derive a formula for the cyclotron frequency.

The direction of motion of the protons is perpendicular to the magnetic field. As a result, the protons travel in a circular orbit. The force causing the circular motion is $F = qvB \sin\theta$,

where $\theta = 90°$ and $\sin 90° = 1$

$F = ma,\quad$ where $a = \dfrac{v^2}{r}$, then

$$qvB = m\frac{v^2}{r}$$

$$qB = m\frac{v}{r}, \qquad \text{where } v = \frac{2\pi r}{T} = 2\pi rf$$

$$qB = m\frac{2\pi rf}{r}$$

$$f = \frac{qB}{2\pi m}$$

Part a. Step 2.

Rearrange the formula derived in Step 1 and solve for the magnetic field strength.

$$B = \frac{m2\pi f}{q}$$

$$= \frac{2\pi(1.67\times10^{-27}\text{ kg})(3.1\times10^{7}\text{ rev/s})}{1.6\times10^{-19}\text{ C}}$$

$$B = 2.03\text{ T}$$

TEXTBOOK PROBLEM 8. (Section 32-1) What is the wavelength (= minimum resolvable size) of 7.0-TeV protons at the LHC?

Part a. Step 1.

Determine the rest energy of a proton in joules.

$$E_{\text{rest}} = m_{\text{p}}c^2$$

$$= (1.67\times10^{-27}\text{ kg})(3\times10^{8}\text{ m/s})^2$$

$$E_{\text{rest}} = 1.5\times10^{-10}\text{ J}$$

Part a. Step 2.

Determine the kinetic energy of an electron in joules.

$$\text{KE} = (7.0\ \text{TeV})\left(\frac{10^{12}\ \text{eV}}{1\ \text{TeV}}\right)\left(\frac{1.6\times10^{-19}\ \text{J}}{1\ \text{eV}}\right)$$

$$\text{KE} = 1.1\times10^{-6}\ \text{J}$$

Part a. Step 3.

Determine the total energy of an electron in joules.

$$E_{\text{total}} = \text{KE} + E_{\text{rest}}$$

$$= 1.1\times10^{-6}\ \text{J} + 1.5\times10^{-10}\ \text{J}$$

$$E_{\text{total}} \approx 1.1\times10^{-6}\ \text{J}$$

Part a. Step 4.

Use the de Broglie formula to determine the wavelength.

At 7.0 TeV, the speed of an electron $\approx c$ (the speed of light):

$$\lambda = \frac{h}{mv} \approx \frac{h}{mc} \approx \frac{hc}{mc^2} \quad \text{but} \quad E = mc^2$$

$$\lambda \approx \frac{hc}{E} \approx \frac{(6.63\times10^{-34}\ \text{J}\cdot\text{s})(3\times10^{8}\ \text{m/s})}{1.1\times10^{-6}\ \text{J}}$$

$$\lambda \approx 1.8\times10^{-19}\ \text{m}$$

TEXTBOOK PROBLEM 21. (Section 32-2 to 32-6) What would be the wavelengths of the two photons produced when an electron and a positron, each with 420 keV of kinetic energy, annihilate in a head-on collision?

Part a. Step 1.

Determine the total KE of the electron–positron in joules.

$$2(420\ \text{keV})\left(\frac{1000\ \text{eV}}{1\ \text{KeV}}\right)\left(\frac{1.6\times10^{-19}\ \text{J}}{1\ \text{eV}}\right) = 1.34\times10^{-13}\ \text{J}$$

Part a. Step 2.

Determine the rest energy of the electron–positron in joules.

$$E_{\text{rest}} = (m_{e-}c^2)_{\text{electron}} + (m_{e+}c^2)_{\text{positron}}$$

$$= 2(9.1\times10^{-31}\ \text{kg})(3\times10^{8}\ \text{m/s})^2$$

$$E_{\text{rest}} = 1.64\times10^{-13}\ \text{J}$$

Part a. Step 3.

Determine the total energy of the electron–positron in joules.

$E_{\text{total}} = \text{KE} + E_{\text{rest}}$

$= 1.34 \times 10^{-13}\ \text{J} + 1.64 \times 10^{-13}\ \text{J}$

$E_{\text{total}} = 2.98 \times 10^{-13}\ \text{J}$

Part a. Step 4.

Use the de Broglie formula to determine the wavelength.

The speed of a photon $= c$ (the speed of light):

$$\lambda = \frac{h}{mv} \approx \frac{h}{mc} \approx \frac{hc}{mc^2} \quad \text{but} \quad E = mc^2$$

$\lambda \approx \dfrac{hc}{E}$ but each photon shares $\frac{1}{2}$ the total energy

$$\lambda \approx \frac{(6.63 \times 10^{-34}\ \text{J}\cdot\text{s})(3 \times 10^8\ \text{m/s})}{1.5 \times 10^{-13}\ \text{J}}$$

$\lambda = 1.33 \times 10^{-12}\ \text{m}$

TEXTBOOK PROBLEM 31. (Section 32-7) What is the energy width (or uncertainty) of (a) η^0, and (b) ρ^+? See Table 32-2.

Part a. Step 1.

Use Table 32-2 to determine the mean lifetime of η^0.

From Table 32-2, $\Delta t = 5.1 \times 10^{-19}$ s.

Part a. Step 2.

Use the uncertainty principle to estimate the energy.

$$\Delta E \Delta t \approx \frac{h}{2\pi}$$

$$\Delta E(5.1 \times 10^{-19}\ \text{s}) \approx \frac{6.63 \times 10^{-34}\ \text{J}\cdot\text{s}}{2\pi}$$

$$\Delta E \approx (2.1 \times 10^{-16}\text{J})\left(\frac{1\ \text{KeV}}{1.6 \times 10^{-16}\ \text{J}}\right) \approx 1.3\ \text{KeV}$$

Part b. Step 1.

Use Table 32-2 to determine the mean lifetime of ρ^+.

From Table 32-2, $\Delta t \approx 4.4 \times 10^{-24}$ s.

Part b. Step 2.

Use the uncertainty principle to estimate the energy.

$$\Delta E \Delta t \approx \frac{h}{2\pi}$$

$$\Delta E(4.4 \times 10^{-24}\ \text{s}) \approx \frac{6.63 \times 10^{-34}\ \text{J}\cdot\text{s}}{2\pi}$$

$$\Delta E \approx (2.4 \times 10^{-11}\ \text{J})\left(\frac{1\ \text{MeV}}{1.6 \times 10^{-13}\ \text{J}}\right) \approx 150\ \text{MeV}$$

TEXTBOOK PROBLEM 32a. (Section 32-7 and 32-11) Which of the following decays are possible? For those that are forbidden, explain which laws are violated.
(*a*) $\Xi^0 \rightarrow \Sigma^+ + \pi^-$

Part a. Step 1.

Apply the law of conservation of charge. Note: Use Table 32-2.

$$\Xi^0 \rightarrow \Sigma^+ + \pi^-$$

$$0 \rightarrow +1 + -1 \ \text{(allowed)}$$

The left side has a net charge of zero, while the right side has a net charge of zero. The law of conservation of charge is not violated.

Part a. Step 2.

Apply the law of conservation of angular momentum.

$$\pm\tfrac{1}{2} \rightarrow \pm\tfrac{1}{2} + 0 \ \text{(allowed)}$$

By inspection, it is possible to combine the possible values so that the left side will equal the right side of the equation. The law of conservation of angular momentum is not violated.

Part a. Step 3.

Apply the law of conservation of baryon number.

$$1 \rightarrow 1 + 0 \ \text{(allowed)}$$

The left side has a net result of $+1$, while the right side has a net value of $+1$. The law of conservation of baryon number is not violated.

Part a. Step 4.

Apply the law of conservation of lepton number.

No leptons are involved in this process. The law is not applicable.

Part a. Step 5.

Apply the law of conservation of energy.

$1314.9 \text{ MeV} < 1189.4 \text{ MeV} + 139.6 \text{ MeV}$

The law of conservation of energy is not conserved.

For a reaction to be allowed, every law must be satisfied. Therefore, this reaction is not allowed.

TEXTBOOK PROBLEM 43. (Sections 32-3 to 32-6) (*a*) How much energy is released when an electron and a positron annihilate each other? (*b*) How much energy is released when a proton and an antiproton annihilate each other? (All particles have KE $\approx$ 0.)

Part a. Step 1.

Determine the rest energy of the electron plus positron in joules.

$$E_{\text{rest}} = (m_{\text{e-}}c^2)_{\text{electron}} + (m_{\text{e+}}c^2)_{\text{positron}}$$

$$= 2(9.1 \times 10^{-31} \text{ kg})(3 \times 10^8 \text{ m/s})^2$$

$$E_{\text{rest}} = 1.64 \times 10^{-13} \text{ J}$$

Part a. Step 2.

Determine the total energy of the electron–positron pair in joules.

The statement of the problem implies that the particles are initially at rest; therefore, KE = 0:

$$E_{\text{total}} = \text{KE} + E_{\text{rest}}$$

$$= 0 + 1.64 \times 10^{-13} \text{ J}$$

$$E_{\text{total}} = 1.64 \times 10^{-13} \text{ J}$$

Part a. Step 3.

Express the energy released in MeV.

$$(1.64 \times 10^{-13} \text{ J})\left(\frac{1 \text{ MeV}}{1.6 \times 10^{-13} \text{ J}}\right) = 1.02 \text{ MeV}$$

Part b. Step 1.

Determine the rest energy of the proton plus antiproton in joules.

$$E_{\text{rest}} = (m_{\text{p}}c^2)_{\text{protron}} + (m_{\bar{\text{p}}}c^2)_{\text{antiproton}}$$

$$= 2(1.67 \times 10^{-27} \text{ kg})(3 \times 10^8 \text{ m/s})^2$$

$$E_{\text{rest}} = 3.0 \times 10^{-10} \text{ J}$$

Part b. Step 2.

Determine the total energy of the proton–antiproton pair in joules.

$$E_{\text{total}} = \text{KE} + E_{\text{rest}}$$
$$= 0 + 3.0 \times 10^{-10}\ \text{J}$$
$$E_{\text{total}} = 3.0 \times 10^{-10}\ \text{J}$$

Part b. Step 3.

Express the energy released in MeV.

$$(3.0 \times 10^{-10}\ \text{J})\left(\frac{1\ \text{MeV}}{1.6 \times 10^{-13}\ \text{J}}\right) = 1877\ \text{MeV}$$

TEXTBOOK PROBLEM 44. (Sections 32-1 and 32-1) If 2×10^{14} protons moving at $v \approx c$, with $\text{KE} = 4.0\ \text{TeV}$, are stored in the 4.3-km-radius ring of the LHC, (*a*) how much current (amperes) is carried by this beam? (*b*) How fast would a 1500-kg car have to move to carry the same kinetic energy as this beam?

Part a. Step 1.

Determine the period of motion of a proton as it orbits the LHC.

At an energy of 4.0 TeV, the protons are moving at approximately the speed of light:

$$T = \frac{2\pi r}{v} = \frac{2\pi(4300\ \text{m})}{3.0 \times 10^{8}\ \text{m/s}}$$
$$T = 9.0 \times 10^{-5}\ \text{s}$$

Part a. Step 2.

Determine the current in amperes carried by this beam.

$I = \dfrac{Q}{t}$, where $Q = ne$ n = number of protons in the beam

e = charge on a proton

$$Q = (2 \times 10^{14}\ \text{protons})\left(\frac{1.6 \times 10^{-19}\text{C}}{1\ \text{proton}}\right) = 3.2 \times 10^{-5}\ \text{C}$$
$$I = \frac{3.2 \times 10^{-5}\ \text{C}}{9.0 \times 10^{-5}\ \text{s}} = 0.35\ \text{A}$$

Part b. Step 1.

Convert 4.0 TeV to joules.

$$(4.0\ \text{TeV})\left(\frac{1 \times 10^{12}\ \text{eV}}{1\ \text{TeV}}\right)\left(\frac{1.6 \times 10^{-19}\ \text{J}}{1\ \text{eV}}\right) = 6.4 \times 10^{-7}\ \text{J}$$

Part b. Step 2.

Determine the total energy of the beam in joules.	$(2\times10^{14}\text{ protons})\left(\dfrac{6.4\times10^{-7}\text{ J}}{\text{proton}}\right)=1.3\times10^{8}\text{ J}$

Part b. Step 3.

Determine speed of a 1500-kg car with 5.0×10^{7} J of kinetic energy.	$\text{KE}=\frac{1}{2}mv^2$ $1.3\times10^{8}\text{ J}=\frac{1}{2}(1500\text{ kg})v^2$ $v^2=\left(\dfrac{2(1.3\times10^{8}\text{ J})}{1500\text{ kg}}\right)=1.7\times10^{5}\text{ m}^2/\text{s}^2$ $v=410\text{ m/s}$

ASTROPHYSICS AND COSMOLOGY

OBJECTIVES

After studying the material of this chapter, the student should be able to:

- use parallax to determine the distance to nearby stars.
- distinguish between a star, a galaxy, and a supercluster.
- given the intrinsic luminosity and distance from the Earth to a star, calculate the apparent brightness of the star.
- calculate the apparent and absolute magnitude of a star.
- draw a Hertzsprung–Russell diagram. Explain what is meant by the main sequence and locate the current position of our Sun on the main sequence.
- describe the current theory of stellar evolution as applied to our Sun. Describe the possible "death" of a star.
- state Einstein's principle of equivalence.
- explain what is meant by gravity as curvature in space and time.
- determine the Schwarzschild radius of the black hole formed by a star of given mass.
- explain how the Doppler effect indicates that the universe is expanding. Use Hubble's law to determine the velocity of recession of a star a known distance from the Earth.
- cite evidence that indicates that the universe evolved from an initial explosion called the Big Bang.
- use the Standard Cosmological Model to describe the evolution of the universe from 10^{-43} s to the present time.
- distinguish between an open universe and a closed universe. Describe how the critical density can be used to determine whether the universe is open or closed.

KEY TERMS AND PHRASES

astrophysics is a branch of astronomy that applies the techniques and theories of physics to the study of celestial objects.

cosmology is the study of the organization and structure of the universe and its evolution.

parallax is a method for measuring the distance to nearby stars. Parallax refers to the apparent shift of position of a star against the background of more distant stars as the Earth moves about the Sun.

parsec is a measure of stellar distance. The distance to a star that changes its apparent position in the sky through an angle of 1" (1 second of arc) in the course of the year is one parallax second, or parsec; 1 parsec = 3.26 light-years.

redshift refers to the displacement of a star's spectrum toward the red end of the spectrum. Based on the Doppler effect, the redshift indicates that the universe is expanding.

A **galaxy** contains billions of stars. Our own Milky Way Galaxy contains about 10^{11} stars.

intrinsic luminosity, or simply **luminosity**, is to the total power radiated in watts by a star or galaxy.

apparent brightness is the power crossing per unit area at the Earth perpendicular to the path of the light.

Hertzsprung–Russell (H–R) diagram is a graph of absolute magnitude of stars versus temperature, with stars represented by a point on the diagram. Most stars fall along the diagonal band called the **main sequence**.

black hole is the residual mass of a star that is so dense that no matter or light can escape from its gravitational field.

Schwarzschild radius represents the event horizon of a black hole. The event horizon is the surface beyond which no signals can ever reach us.

Hubble's law states that galaxies are moving away from one another at speeds proportional to the distance between them.

Big Bang theory suggests that the expansion of the universe is due to an explosion that probably occurred about 14 billion years ago.

Standard Cosmological Model gives an explanation of how the universe evolved after the Big Bang.

SUMMARY OF MATHEMATICAL FORMULAS

redshift	$\lambda_{\text{obs}} = \lambda_{\text{rest}}\sqrt{\dfrac{1+v/c}{1-v/c}}$	The observed wavelength (λ_{obs}) depends on the wavelength as seen in a reference frame at rest with respect to the source (λ_{rest}), the velocity of recession (v), and the speed of light (c).
apparent brightness	$b = \dfrac{L}{4\pi d^2}$	The apparent brightness (b) is the total power (L) crossing per unit area at the Earth perpendicular to the path of the light; d is the distance from the Earth to the star or galaxy, and the total area of a sphere of radius d is $4\pi d^2$. Note: The total power radiated in watts by a star or galaxy is called the intrinsic luminosity, or simply luminosity (L).
Schwarzschild radius	$R = \dfrac{2GM}{c^2}$	Schwarzschild radius (R) represents the radius of the event horizon of a black hole. M is the object's mass, G is the gravitational constant, and c is the speed of light in a vacuum.

CONCEPT SUMMARY

Astrophysics is a branch of astronomy that applies the techniques and theories of physics to the study of celestial objects. **Cosmology** is the study of the organization and structure of the universe and its evolution.

Stellar Distances

One method for measuring the distance to nearby stars is **parallax**. Parallax refers to the apparent shift of position of a star against the background of more distant stars as the Earth moves about the Sun. The distance to a star that changes its apparent position in the sky through an angle of 1" (1 second of arc) in the course of the year is one parallax second, or **parsec**. 1 parsec = 3.26 light-years, where 1 **light-year** is the distance that light travels through a vacuum in one year (approximately 10^{13} km or 6 trillion miles).

As shown in the diagram, the sighting angle of a star relative to the plane of the Earth's orbit (θ) is measured at different times of the year. The distance from the Earth to the Sun (d) is known, and trigonometry can be used to determine the distance (D) to the star.

Parallax is useful for stars up to 100 light-years (about 30 parsecs) from the Earth. The apparent brightness of more distant stars can give an approximate measurement of distance. Also useful is analysis of the shift of a star's spectrum, the so-called redshift.

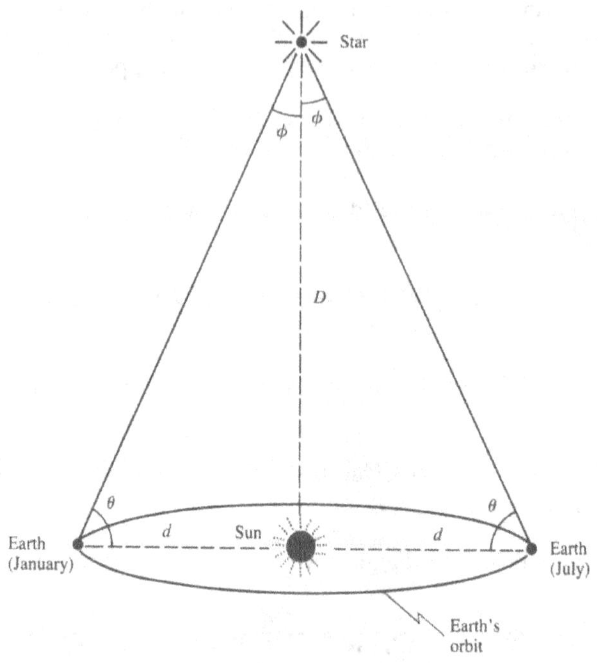

TEXTBOOK QUESTION 1. (Section 33-1) The Milky Way was once thought to be "murky" or "milky" but is now considered to be made up of point sources. Explain.

ANSWER: As seen by an Earth-bound observer, the stars that make up the Milky Way are so numerous and so close together that the human eye is incapable of resolving individual stars. It was not until the introduction of the telescope that astronomers were able to distinguish individual stars.

EXAMPLE PROBLEM 1. (Section 33-1) The star Vega is a star in the constellation Lyra. Vega is 25.0 light-years from the Earth. Determine the (*a*) parallax angle for Vega and (*b*) distance from Earth to Vega in parsecs.

Part a. Step 1.

Determine the parallax angle in radians.	Using the previous diagram, note that tan $\phi = d/D$, where d is the radius of the Earth's orbit about the Sun: $d = 1.5\times10^{8}$ km. D is the distance from the Earth to Vega:

$D = (25.0 \text{ ly})(9.46 \times 10^{12} \text{ km/ly}) = 2.37 \times 10^{14} \text{ km}$

$\tan\phi = (1.5 \times 10^{8} \text{ km})/(2.37 \times 10^{14} \text{ km}) = 6.3 \times 10^{-7}$

The angle is small; therefore, $\tan\phi \approx \phi$ expressed in radians:

$\phi = 6.3 \times 10^{-7}$ radians

Part a. Step 2.

Express the angle in seconds of arc.

$$\phi = (6.3 \times 10^{-7} \text{ rad})\left(\frac{360°}{2\pi \text{ rad}}\right)\left(\frac{3600''}{1°}\right)$$

$\phi = 0.131''$, or 0.131 seconds of arc

Part b. Step 1.

Determine the distance to Vega in parsecs.

1 parsec (pc) $= 1/\phi$, where ϕ is expressed in seconds of arc:

distance in parsecs $= 1/(0.131'') = 7.66$ pc

Alternate solution:

$$(25.0 \text{ ly})\left(\frac{1 \text{ pc}}{3.26 \text{ ly}}\right) = 7.66 \text{ pc}$$

Redshift

Analysis of the spectra of stars and galaxies indicates a shift toward the red end of the spectrum. The Doppler effect suggests that this shift toward the red, or **redshift,** of light indicates that the star or galaxy is receding from the Earth and that the universe is expanding in size.

Based on equations derived from the Doppler effect, the observed wavelength (λ_{obs}) is related to the wavelength as seen in a reference frame at rest with respect to the source (λ_{rest}) by the equation

$$\lambda_{\text{obs}} = \lambda_{\text{rest}}\sqrt{\frac{1+v/c}{1-v/c}}$$

where v = velocity of recession and c = speed of light.

The shift in the wavelength ($\Delta\lambda$) can be determined from the equation $\Delta\lambda = \lambda_{\text{obs}} - \lambda_{\text{rest}}$.

EXAMPLE PROBLEM 2. (Section 33-4) (*a*) Calculate the observed wavelength for the 410-nm line in the Balmer series for hydrogen in the spectrum of a galaxy that is receding at $0.25c$ from the Earth. (*b*) Determine the change in the wavelength.

Part a. Step 1.

Calculate the observed wavelength.

$$\lambda_{\text{obs}} = \lambda_{\text{rest}}\sqrt{\frac{1+v/c}{1-v/c}}$$

$$\lambda_{\text{obs}} = (410\text{ nm})\sqrt{\frac{1+0.25c/c}{1-0.25c/c}} = (410\text{ nm})\sqrt{\frac{1+0.25}{1-0.25}}$$

$$\lambda_{\text{obs}} = (410\text{ nm})\sqrt{\frac{1.25}{0.75}} = (410\text{ nm})(1.29)$$

$$\lambda_{\text{obs}} = 530\text{ nm}$$

Part b. Step 1.

Determine the shift in the wavelength.

$$\Delta\lambda = \lambda_{\text{obs}} - \lambda_{\text{rest}}$$
$$= 530\text{ nm} - 410\text{ nm}$$
$$\Delta\lambda = 120\text{ nm (redshift)}$$

Stars and Galaxies

Our Sun is a single star in a **galaxy** of stars known as the Milky Way Galaxy. The Milky Way Galaxy contains about 10^{11} stars. It has a diameter of 100,000 light-years and a thickness of about 2000 light-years. Our Sun is located about halfway from the center to the edge, about 26,000 light-years from the center. The Sun orbits the galactic center at a speed of 200 km/s relative to the center of the galaxy.

Galaxies tend to be grouped in **galaxy clusters**, which are organized into clusters of clusters called **superclusters**.

Luminosity and Magnitude

The total power radiated in watts by a star or galaxy is called the **intrinsic luminosity**, or simply **luminosity** (L). The **apparent brightness** (b) is the power crossing per unit area at the Earth perpendicular to the path of the light. Ignoring any absorption of the light as it travels through space, the apparent brightness is given by the equation

$$b = \frac{L}{4\pi d^2}$$

where d is the distance from the Earth to the star or galaxy and the total area of a sphere of radius d is $4\pi d^2$.

Hertzsprung–Russell (H–R) Diagram

For most stars, the color is related to its absolute luminosity and therefore to its mass. This relationship is represented on the Hertzsprung–Russell, or H–R, diagram. On the H–R diagram, one axis represents the absolute magnitude while the other represents the temperature, and each star is represented by a point on the diagram.

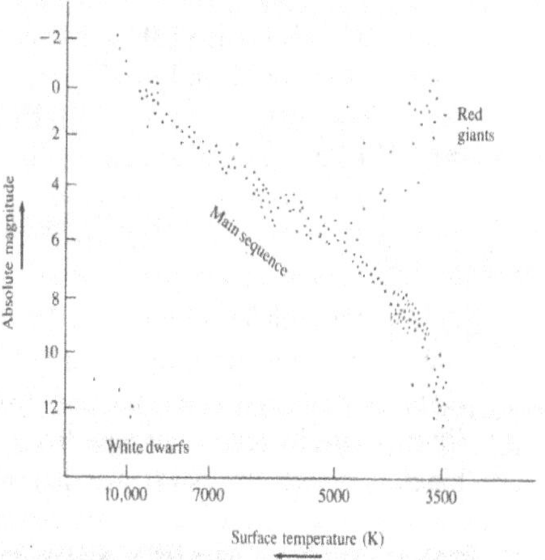

Most stars fall along the diagonal band called the **main sequence**. The coolest stars, reddish in color, are located on the lower right. These stars are the least luminous and therefore must be of low mass. Stars on the upper left, bluish in color, are much more massive and have a higher luminosity. Stars that are not main-sequence type stars include red giants and white dwarfs.

TEXTBOOK QUESTION 4. (Section 33-2) Does the H–R diagram directly reveal anything about the core of a star?

ANSWER: The H–R diagram relates only the surface temperature of a star to the absolute luminosity of the surface of the star. Theories on stellar evolution provide ideas on the core of a star and its place on the main sequence of the H–R diagram during its life cycle. Therefore, knowledge of a star's position on the H–R diagram might give some indirect knowledge of the stellar interior.

Stellar Evolution

Current theories suggest that protostars form from collapsing masses of hydrogen gas that exist in great clouds in space. During collapse, gravitational potential energy is transformed into kinetic energy, causing a heating of the gas. When the temperature reaches 10 million kelvins, nuclear fusion begins. The dominant early reaction that takes place in the core is the proton–proton cycle, in which four protons fuse to form a helium nucleus. After a time, the protostar becomes stable and can be placed on the main sequence of the H–R diagram. A star with the mass of our Sun takes about 30 million years to reach the main sequence and will remain there for about 10 billion years.

When the supply of hydrogen in the core is depleted, the core contracts and the decrease in gravitational potential energy results in higher temperatures. This leads to an increase in nuclear reactions, some of which result in the formation of more massive nuclei such as carbon. The outer part of the star expands and cools, and the star's position on the H–R diagram moves from the main sequence to the **red giant** stage. In about 5 billion years our Sun should become a red giant, and its size could increase beyond the orbit of some if not all of the inner planets.

Stars with residual mass less than 1.4 solar masses will cool and become **white dwarfs**. The white dwarf will continue radiating energy until it eventually becomes a black dwarf, a dark, cold piece of ash.

More massive stars can continue to contract, due to their greater gravitational fields, and will eventually approach nuclear densities. Under sufficient pressure, the electrons combine with protons to form neutrons and the result may be a **neutron star**. Neutron stars are thought to have diameters on the order of tens of kilometers. It is thought that **supernovae** are the result of the energy released in the final contraction of a neutron star.

If the star's residual mass is greater than two or three solar masses, it may contract still further and form a **black hole**, which is so dense that no matter or light can escape from its gravitational field.

TEXTBOOK QUESTION 5. (Section 33-2) Why do some stars end up as while dwarfs, and others as neutron stars or black holes?

ANSWER: The answer depends on a star's residual mass. Stars with residual mass less than 1.4 solar masses will cool and become white dwarfs. The white dwarf will continue radiating energy until it eventually becomes a black dwarf, a dark, cold piece of ash.

A star with a residual mass between 1.4 and 2 to 3 solar masses will eventually became a neutron star. A neutron star may eventually explode as a supernova and become a white dwarf. If a star's residual mass is greater than 2 to 3 solar masses, the gravitational force could become so great that no light would be able to escape from it. In this case it is possible that the star could eventually form a black hole, which is so dense that no matter or light can escape from its gravitational field.

General Relativity: Gravity and the Curvature of Space

Einstein's **principle of equivalence** states that no experiment can be performed in a small region of space that could distinguish between a gravitational field and an equivalent uniform acceleration. As a result, there is an equivalence between **gravitational mass** and **inertial mass**.

Gravity is considered to be curvature in space-time. The curvature is greater near massive bodies. The theory requires the use of non-Euclidian geometries where the shortest distance between two points is not a straight line but a curve called a **geodesic**. In such a space the sum of the angles of a triangle is not equal to 180°. The theory predicts that light rays passing near a massive object will be deflected. Confirmation came in 1919 when starlight passing the edge of the Sun during a solar eclipse was deflected by an amount consistent with the theory. In addition, the universe as a whole may be curved. If there is sufficient mass, then the curvature of the universe is positive and the universe is closed and finite. Otherwise, the curvature is negative and the universe is open and infinite.

A black hole could produce extreme curvature of space-time. A black hole is the residual mass of a star that is so dense that no matter or light can escape from it. The **Schwarzschild radius** represents the **event horizon** of a black hole. This is the surface beyond which no signals can ever reach us. Inside the event horizon, at $r = 0$, there is an infinitely dense singularity. Whether such singularities can form is still unknown. The Schwarzschild radius (R) is

$$R = \frac{2GM}{c^2}$$

where G is the gravitational constant and c is the speed of light in a vacuum.

EXAMPLE PROBLEM 3. (Section 33-3) The second largest planet in our solar system is Saturn. The mass of Saturn is 95.2 times the mass of the Earth. Determine the Schwarzschild radius for Saturn.

Part a. Step 1.

Determine the mass of Saturn in kg.	$M_S = 95.2M_E = (95.2)(5.98 \times 10^{24} \text{ kg}) = 5.69 \times 10^{26} \text{ kg}$

Part a. Step 2.

Determine the Schwarzschild radius for Saturn.

$$R = \frac{2GM}{c^2} = \frac{(2)(6.67\times10^{-11}\ \text{N}\cdot\text{m}^2/\text{kg}^2)(5.69\times10^{26}\ \text{kg})}{(3.0\times10^{8}\ \text{m/s})^2}$$

$$R = 0.843\ \text{m}$$

This compares with a Schwarzschild radius of 2.95 km for the Sun and 8.9 mm for the Earth.

The Expanding Universe

Distant galaxies display a redshift of spectral lines. The Doppler effect is used to interpret the redshift, and the interpretation indicates that the universe is expanding. The galaxies are moving away from one another at speeds (v) proportional to the distance (d) between them. **Hubble's law** describes the relationship as follows:

$$v = H_0 d,$$

where H_0 is the Hubble parameter. The value of the Hubble parameter is approximately $21\ \text{km/s/Mly} \approx 21\ \text{km/s}$ per million light-years of distance. Note: $21\ \text{km/s/Mly} = 67$ kilometers per second per million parsecs of distance.

The expansion of the universe is thought to be due to an explosive origin of the universe called the **Big Bang**, which occurred approximately 14 billion years ago.

There is a class of objects, called **quasars**, that do not seem to conform to Hubble's law. Quasars are quasistellar objects that are as bright as nearby stars but display very large redshifts. The large redshift indicates that they are very far away, and because of their brightness they must be thousands of times brighter than normal galaxies.

TEXTBOOK QUESTION 11. (Section 33-5) If you were located in a galaxy near the boundary of our observable universe, would galaxies in the direction of the Milky Way appear to be approaching you or receding from you? Explain.

ANSWER: You would see all other galaxies, including the Milky Way Galaxy, receding from you. To use an analogy, imagine a deflated balloon covered with dots. As the balloon is inflated, your particular dot would recede from all of the other dots with a velocity directly proportional to the distance between the dots.

EXAMPLE PROBLEM 4. (Section 33-4) (*a*) Use Hubble's law to estimate the recessional velocity of a galaxy 2.5 billion light-years from the Earth. Note: The Hubble parameter $H_0 = 21\ \text{km/s}$ per million light-years of distance. (*b*) Express the answer to Part a as a fraction of the speed of light.

Part a. Step 1.

Determine the recessional velocity of the galaxy.

$$v = H_0 d$$

$$v = \left(\frac{21\ \text{km/s}}{1\times10^{6}\ \text{ly}}\right)(2.5\times10^{9}\ \text{ly}) \approx 53{,}000\ \text{km/s}$$

Part b. Step 1.

Express the answer to Part a as a fraction of the speed of light.

$$(53{,}000\ \text{km/s})\left(\frac{1000\ \text{m}}{1\ \text{km}}\right) = 5.3 \times 10^{7}\ \text{m/s}$$

$$\frac{5.3 \times 10^{7}\ \text{m/s}}{3.0 \times 10^{8}\ \text{m/s}} = 0.177c = (1.77 \times 10^{-2})c$$

The Big Bang and the Cosmic Microwave Background

In 1964, evidence to support the Big Bang theory was discovered. The discovery came in the form of radiation of wavelength 7.35 cm, which is in the microwave region of the spectrum. The intensity of this radiation was found to be independent of the time of day, and it came from all directions in the universe with equal intensity. The intensity corresponds to a blackbody radiation temperature of 2.725 K.

The **Standard Cosmological Model** gives an explanation of how the universe evolved after the Big Bang. According to the model, starting at 10^{-43} s after the Big Bang, there was a series of phase transitions during which previously unified forces of nature condensed out one by one. The **inflationary scenario** assumes that during one of these phase transitions, the universe underwent a brief but rapid exponential expansion. Until about 10^{-35} s there was no distinction between quarks and leptons. Shortly thereafter, the universe entered the **hadron era**. About 10^{-6} s after the Big Bang the majority of hadrons disappeared, introducing the **lepton era**.

By the time the universe was 10 s old, the electrons too had mostly disappeared, and the universe became **radiation-dominated**. As the universe continued to expand, the radiation density decreased faster than the matter density. This eventually resulted in a **matter-dominated** universe. After about 500,000 years the universe was at a temperature of 3000 K and was cool enough for electrons to combine with nuclei and form atoms. After this time, the universe was cool enough for formation of stars and galaxies. In the 14 billion years since the Big Bang, the radiation has cooled to a temperature of 2.725 K, producing the 7.35-cm background radiation discovered in 1964.

TEXTBOOK QUESTION 15. (Section 33-6) Why were atoms, as opposed to bare nuclei, unable to exist until hundreds of thousands of years after the Big Bang?

ANSWER: As the universe expanded, the energy spread out over an increasingly larger volume and the temperature dropped. At temperatures higher than 3000 K the energy of the electrons and nuclei was too high for atoms to exist. It was necessary for the temperature of the universe to drop to approximately 3000 K before atoms could form.

The Future of the Universe

Whether the universe is open or closed depends on the matter density of the universe. If the average density is above a critical value known as the **critical density**, about $10^{-26}\ \text{kg/m}^3$, then gravity will eventually stop the expansion and the universe will collapse back into a **big crunch**. If the average density is less than this value, the universe will continue to expand forever.

PROBLEM SOLVING SKILLS

For problems involving the parallax angle of a star and the stellar distance expressed in parsecs:

1. The parallax angle ϕ can be determined from the equation $\tan\phi = d/D$, where d is the radius of the Earth's orbit about the Sun in km and D is the distance to the star in km. Note: $1\text{ ly} = 9.46 \times 10^{12}$ km.
2. The parallax angle is small; therefore, $\tan\phi \approx \phi$ when ϕ is expressed in radians.
3. The stellar distance in parsecs $= 1/\phi$, where ϕ is expressed in seconds of arc. An alternative method to use is the conversion $1\text{ pc} = 3.26$ ly.

For problems involving the Schwarzschild radius:

1. Determine the mass of the object in kg.
2. The Schwarzschild radius (R) is related to the object's mass by the equation $R = \dfrac{2GM}{c^2}$.

For problems involving Hubble's law:

1. Hubble's law states that the velocity (v) of a star/galaxy relative to the Earth is related to the distance to the star/galaxy by the equation $v = H_0 d$. Express the Hubble parameter in the appropriate units and solve the problem.

For problems involving the redshift of light from a star or galaxy receding from the Earth:

1. Express the velocity (v) of the galaxy as a fraction of the speed of light.
2. The observed wavelength (λ_{obs}) is related to the wavelength as seen in a reference frame at rest with respect to the source (λ_{rest}) by this equation:

$$\lambda_{\text{obs}} = \lambda_{\text{rest}}\sqrt{\frac{1+v/c}{1-v/c}}$$

3. The shift in the wavelength ($\Delta\lambda$) can be determined from the equation $\Delta\lambda = \lambda_{\text{obs}} - \lambda_{\text{rest}}$.

SOLUTIONS TO SELECTED TEXTBOOK PROBLEMS

TEXTBOOK PROBLEM 2. (Sections 33-1 and 33-2) A star exhibits a parallax of 0.27 seconds of arc. How far away is it?

Part a. Step 1.

Determine the stellar distance in parsecs (pc).

The stellar distance (D) in parsecs $= \dfrac{1}{\phi}$, where ϕ is expressed in seconds of arc:

$$D = \frac{1}{\phi} = \frac{1}{0.27\text{ s}} = 3.7\text{ pc}$$

Part a. Step 2.

Express this distance in light-years (ly), kilometers, and miles.

$$(3.7 \text{ pc})\left(\frac{3.26 \text{ ly}}{1 \text{ pc}}\right) \approx 12 \text{ ly}$$

$$(12 \text{ ly})\left(\frac{10^{13} \text{ km}}{1 \text{ ly}}\right) \approx 1.2 \times 10^{14} \text{ km}$$

$$(1.2 \times 10^{14} \text{ km})\left(\frac{1 \text{ mile}}{1.609 \text{ km}}\right) \approx 7.5 \times 10^{13} \text{ miles}$$

Note: 7.5×10^{13} miles = 75 trillion miles.

TEXTBOOK PROBLEM 13. (Sections 33-1 and 33-2) A star is 85 pc away. How long does it take for its light to reach us?

Part a. Step 1.

Convert 85 parsecs to light-years (ly).

$$(85 \text{ pc})\left(\frac{3.26 \text{ ly}}{1 \text{ pc}}\right) \approx 280 \text{ ly}$$

One light-year is the distance that light travels in one year.

Therefore, it takes approximately 280 years for the light to reach the Earth.

TEXTBOOK PROBLEM 4. (Sections 32-1 and 32-2) What is the relative brightness of the Sun as seen from Jupiter, as compared to its brightness from Earth? (Jupiter is 5.2 times farther from the Sun than the Earth is.)

Part a. Step 1.

Use Section 14-9 to determine the solar constant at Earth's surface.

The solar refers the Sun's energy per second per square meter that strikes the Earth.

From Section 14-9, this value is approximately 1350 W/m^2.

Part a. Step 2.

Determine the apparent brightness of the Sun as seen on Jupiter.

The apparent brightness (b) at the surface of a planet depends on the energy radiated from the Sun—the intrinsic luminosity (L)—and the distance (d) from the Sun to the planet:

$$b_J = \frac{L}{4\pi d^2} \text{ (Jupiter) and } b_E = \frac{L}{4\pi d^2} \text{ (Earth)}$$

$$\frac{b_J}{b_E} = \frac{\frac{L}{4\pi d_J^2}}{\frac{L}{4\pi d_E^2}} = \frac{d_E^2}{d_J^2} \text{ but } d_J = 5.2 d_E$$

$$\frac{\ell_J}{1350\ \text{W/m}^2} = \left(\frac{d_E}{5.2\ d_E}\right)^2$$

$$b_J = (0.19)^2 (1350\ \text{W/m}^2)$$

$$b_J \approx 50\ \text{W/m}^2$$

Part a. Step 3.

Determine the relative brightness of the Sun as observed on each planet.

$$\frac{b_J}{b_E} \approx \frac{50\ \text{W/m}^2}{1350\ \text{W/m}^2} \approx 0.037$$

The energy per second per square meter reaching Jupiter is only 3.7% as much as that striking the Earth.

TEXTBOOK PROBLEM 9. (Sections 33-1 and 33-2) Calculate the density of a white dwarf whose mass is equal to the Sun's and whose radius is equal to the Earth's. How many times larger than the Earth's density is this?

Part a. Step 1.

Refer to the text to determine the mass of the Sun and the mass and radius of the Earth.

From inside the front cover of the textbook:

$$m_{\text{Sun}} = 1.99 \times 10^{30}\ \text{kg} \qquad m_{\text{Earth}} = 5.98 \times 10^{24}\ \text{kg}$$

$$r_{\text{Earth}} = 6.38 \times 10^3\ \text{km} = 6.38 \times 10^6\ \text{m}$$

Part a. Step 2.

Determine the density of the white dwarf.

$$\rho_{\text{dwarf}} = \frac{m_{\text{dwarf}}}{V_{\text{dwarf}}} = \frac{1.99 \times 10^{30}\ \text{kg}}{\frac{4}{3}\pi(6.38 \times 10^{6}\ \text{m})^{3}} = \frac{1.99 \times 10^{30}\ \text{kg}}{1.08 \times 10^{21}\ \text{kg/m}^{3}}$$

$$\rho_{\text{dwarf}} = 1.83 \times 10^{9}\ \text{kg/m}^{3}$$

Part a. Step 3.

Determine the density of the Earth.

$$\rho_{\text{Earth}} = \frac{m_{\text{Earth}}}{V_{\text{Earth}}} = \frac{5.98 \times 10^{24}\ \text{kg}}{\frac{4}{3}\pi(6.38 \times 10^{6}\ \text{m})^{3}} = \frac{5.98 \times 10^{24}\ \text{kg}}{1.08 \times 10^{21}\ \text{m}^{3}}$$

$$\rho_{\text{Earth}} = 5.54 \times 10^{3}\ \text{kg/m}^{3}$$

Part a. Step 4.

Determine the ratio of the densities.

$$\frac{\rho_{\text{dwarf}}}{\rho_{\text{Earth}}} \approx \frac{1.83 \times 10^{9}\ \text{kg/m}^{3}}{5.54 \times 10^{3}\ \text{kg/m}^{3}}$$

$$\frac{\rho_{\text{dwarf}}}{\rho_{\text{Earth}}} \approx 3.3 \times 10^{5}$$

The density of the white dwarf is approximately 330,000 times greater than the density of the Earth.

TEXTBOOK PROBLEM 16. (Section 33-4) Show that the Schwarzschild radius for Earth is 8.9 mm.

Part a. Step 1.

Determine the Schwarzschild radius for Earth.

$$R = \frac{2GM}{c^{2}}$$

$$= \frac{(2)(6.67 \times 10^{-11}\ \text{N}\cdot\text{m}^{2}/\text{kg}^{2})(5.97 \times 10^{24}\ \text{kg})}{(3.0 \times 10^{8}\ \text{m/s})^{2}}$$

$$R \approx 8.86 \times 10^{-3}\ \text{m} \approx 8.9\ \text{mm}$$

TEXTBOOK PROBLEM 24. (Section 33-4) A galaxy is moving away from Earth. The "blue" hydrogen line at 434 nm emitted from the galaxy is measured on Earth to be 455 nm. (*a*) How fast is the galaxy moving? (*b*) How far is it from Earth based on Hubble's law?

Part a. Step 1.

Use the formula for the Doppler shift to determine the galaxy's speed.

$$\lambda_{\text{obs}} = \lambda_{\text{rest}}\sqrt{\frac{1+v/c}{1-v/c}}$$

$$455\text{ nm} = 434\text{ nm}\sqrt{\frac{1+v/c}{1-v/c}}$$

$$1.05 = \sqrt{\frac{1+v/c}{1-v/c}}$$

$$1.10 = \frac{1+v/c}{1-v/c}$$

$$1.10(1-v/c) = 1+v/c$$

$$1.10 - 1.10v/c = 1+v/c$$

$$0.10 = 2.10v/c$$

$$0.048 = v/c$$

$$v = 0.048c$$

Part a. Step 2.

Use Hubble's law to determine its distance from the Earth.

$v = H_0 d$, where $H_0 \approx 67$ km/s/Mpc ≈ 21 km/s/Mly

$$(0.048)(3\times10^8\text{ m/s})\left(\frac{1\text{ km}}{1000\text{ m}}\right) \approx (21\text{ km/s/Mly})d$$

$d \approx 686$ Mly, or 690 million light-years

TEXTBOOK PROBLEM 48. (Section 33-6) (*a*) What temperature would correspond to 14-TeV collisions at the LHC? (*b*) To what era in cosmological history does this correspond? [*Hint*: See Fig. 33-29.]

Part a. Step 1.

Convert the collision energy to joules.

$$(14\text{ TeV})\left(\frac{1\times10^{12}\text{ eV}}{1\text{ TeV}}\right)\left(\frac{1.6\times10^{-19}\text{ J}}{1\text{ eV}}\right)=2.2\times10^{-6}\text{ J}$$

Part a. Step 2.

Determine the temperature in kelvins.

From Section 33-7 of the textbook,

$$\text{KE}\approx kT$$

$$T\approx \text{KE}/k$$

$$T\approx\frac{2.2\times10^{-6}\text{ J}}{1.38\times10^{-23}\text{ J/K}}$$

$$T\approx1.6\times10^{17}\text{ K}$$

Part a. Step 3.

Use Fig. 33-29 to determine the cosmological era.

Based on Fig. 33-29, this temperature corresponds to the hadron era.

TEXTBOOK PROBLEM 50. (Sections 33-1 and 33-2) Consider the reaction

$$^{16}_{8}\text{O}+{}^{16}_{8}\text{O}\rightarrow{}^{28}_{14}\text{Si}+{}^{4}_{2}\text{He}$$

(*a*) How much energy is released in this reaction (see Appendix B)? (*b*) How much kinetic energy must each oxygen nucleus have (assume equal) in a head-on collision if they are to touch (use Eq. 30-1) so that the strong force can come into play? (*c*) What temperature does thois kinetic energy correspond to?

Part a. Step 1.

Determine the total mass of the reactants and the total mass of the products.

Reactants		Products	
$^{16}_{8}\text{O}$	15.994915 u	$^{28}_{14}\text{Si}$	27.976927 u
$^{16}_{8}\text{O}$	15.994915 u	$^{4}_{2}\text{He}$	4.002603 u
	31.989830 u		31.97953 u

Part a. Step 2.

Determine the mass difference (Δm) between the products and the reactants.	$\Delta m = 31.989830\text{ u} - 31.97953\text{ u}$ $\Delta m = 0.01030\text{ u}$

Part a. Step 3.

Determine the value of Q in joules and MeV.	$Q = (M_a + M_X - M_b - M_Y)c^2$ $Q = (\Delta m)c^2$ $\Delta m = (0.01030\text{ u})(1.66 \times 10^{-27}\text{ kg/u}) = 1.7098 \times 10^{-29}\text{ kg}$ $Q = (1.0798 \times 10^{-29}\text{ kg})(3.0 \times 10^{8}\text{ m/s})^2 = 1.5388 \times 10^{-12}\text{ J}$ $Q = (1.5388 \times 10^{-12}\text{ J})(1\text{ MeV}/1.6 \times 10^{-13}\text{ J}) = 9.62\text{ MeV}$

Part b. Step 1.

Use Eq. 30-1 to determine the radius of the oxygen nucleus.	$r = (1.2 \times 10^{-15}\text{ m})A^{\frac{1}{3}}$ $= (1.2 \times 10^{-15}\text{ m})(16)^{\frac{1}{3}} = (1.2 \times 10^{-15}\text{ m})(2.518)$ $r = 3.02 \times 10^{-15}\text{ m}$

Part b. Step 2.

Determine the total potential energy at closest approach.	The kinetic energy of the nuclei is converted to potential energy at closest approach. Note: At closest approach, the distance between the centers of the oxygen nuclei $= 3.02 \times 10^{-15}\text{ m} + 3.02 \times 10^{-15}\text{ m} = 6.04 \times 10^{-15}\text{ m}$ $\text{PE} = \frac{kq_1q_2}{r}$, where $q_1 = q_2 = 8 \times 1.6 \times 10^{-19}\text{ C} = 1.28 \times 10^{-18}\text{ C}$ $= \frac{(9 \times 10^{9}\text{ N}\cdot\text{m}^2/\text{C}^2)(1.28 \times 10^{-18}\text{ C})}{6.04 \times 10^{-15}\text{ m}}$ $\text{PE} = 2.44 \times 10^{-12}\text{ J}$ $\text{PE} = (2.44 \times 10^{-12}\text{ J})\left(\frac{1\text{ MeV}}{1.6 \times 10^{-13}\text{ J}}\right) = 15.26\text{ MeV}$

Part b. Step 3.

Determine the KE of each nucleus.

$KE_{total} = PE_{total}$

but $KE_{total} = KE_1 + KE_2$

Assuming that the KE of each nucleus is the same, the KE of each nucleus $= \frac{1}{2}(15.26\ \text{MeV}) = 7.63\ \text{MeV}$.

Part c. Step 1.

Determine the temperature of the nuclei in kelvins.

From Section 33-7 of the textbook,

$\text{KE} \approx kT$, where $k = 1.38 \times 10^{-23}$ J/K

$$(7.63\ \text{MeV})\left(\frac{1.6\times10^{-13}\ \text{J}}{1\ \text{MeV}}\right) \approx (1.38 \times 10^{-23}\ \text{J/K})T$$

$$1.22 \times 10^{-12}\ \text{J} \approx (1.38 \times 10^{-23}\ \text{J/K})T$$

$$T \approx \frac{1.22 \times 10^{-12}\ \text{J}}{1.38 \times 10^{-23}\ \text{J/K}}$$

$$T \approx 8.8 \times 10^{10}\ \text{K}$$
